Introduction to Building

PEARSON
Education

We work with leading authors to develop the
strongest educational materials in Architecture,
bringing cutting-edge thinking and best
learning practice to a global market.

Under a range of well-known imprints, including
Prentice Hall, we craft high quality print and
electronic publications which help readers to understand
and apply their content, whether studying or at work.

To find out more about the complete range of our
publishing, please visit us on the World Wide Web at:
www.pearsoneduc.com

MITCHELL'S BUILDING SERIES

Introduction to Building

Third edition

DEREK OSBOURN
Dip Arch [Hons], RIBA, MSIAD

Revised by
ROGER GREENO
BA [Hons], MCIOB, Dip (Bldg), C&G

PEARSON
Prentice
Hall

Harlow, England • London • New York • Boston • San Francisco • Toronto
Sydney • Tokyo • Singapore • Hong Kong • Seoul • Taipei • New Delhi
Cape Town • Madrid • Mexico City • Amsterdam • Munich • Paris • Milan

Pearson Education Limited
Edinburgh Gate
Harlow
Essex CM20 2JE
England

and Associated Companies throughout the world

Visit us on the World Wide Web at:
www.pearsoneduc.com

First published by B.T. Batsford 1985
Reprinted by Longman Scientific and Technical 1993
Second edition published under the Addison Wesley Longman imprint 1997
Third edition published 2002

ISBN 0 582 47303 9

British Library Cataloguing-in-Publication Data
A catalogue record for this book is available from the British Library

Library of Congress Cataloging-in-Publication Data
Osbourn, Derek.
    Introduction to building / Derek Osbourn ; revised by Roger Greeno.—3rd ed.
        p. cm. — (Mitchell's building series)
    ISBN 0–582–47303–9 (pbk.)
        1. Building. I. Greeno, Roger. II. Title. III. Series.

TH145 .O726 2002
690—dc21   2002020777

10 9 8 7 6 5 4 3
07 06 05 04 03

Typeset by 35 in 9.5/11pt Times
Produced by Pearson Education Malaysia Sdn Bhd,
Printed in Malaysia, PJB

Contents

Preface to the third edition

The first edition of this book was published in 1985, before the tragic and untimely death of its original author Derek Osbourn. It is a compliment to Derek that his book has become an established part of the Mitchell's Building Series, and that the Higher Education team at Pearson Education acknowledge the benefits and values that its readers will find in the continuity of his work.

This third edition of *Introduction to Building* continues the established format of earlier editions, providing a blend of comprehensive text with supportive illustrations and individual chapter listings of research papers, Building Regulations and other related texts for further reading. Where necessary, the content has been amended and extended to cover the procedural and mandatory changes required by government guidelines and legislation for development and construction work.

Specific changes include initiatives emanating from the Egan Report, with reference to the new Construction Act, the Construction (Design and Management) Regulations, contractual procedures and background to the Joint Contracts Tribunal (JCT 98) forms of contract, new Building Regulations affecting sound insulation, thermal insulation and access for disabled people, and aspects of planning history with current processes.

This book remains a useful professional reference and an essential course reader for students of all construction-related disciplines, including architecture, construction management, building surveying, quantity surveying, estate and facilities management.

Roger Greeno

Acknowledgements

David Clegg gave invaluable guidance on the CI/SfB classification system; information for Figures 7.1 and 7.2 and Tables 7.1–7.3 was readily given by J.E. Moore. Teaching in the subject of this book, it is inevitable that I have also drawn upon much accumulated data supplied by others as documents or during discussions over many years. I apologise to those not credited and can only offer my sincere thanks.

George Dilks, assisted by Jean Marshall, prepared the final drawings and I owe thanks for the skilful and tolerant way in which sketches were interpreted. I am especially grateful to Nori Howard-Butôt for not only typing and retyping the draft, but also for making many important suggestions regarding content. Lastly, of course, my thanks go to the publishers, and particularly Thelma M. Nye for her enthusiasm, patience and extremely able editorial assistance.

Derek Osbourn
London 1984

I would like to thank Alan Murray and Andy Youings of Emap Glenigan, Bournemouth for their help in providing data for Figure 15.2. I am also grateful to Guy Thompson of Norman + Dawbarn Limited (Architects), Guildford for the illustration in Figure 16.6 of one of their designs. I am particularly appreciative of the assistance provided by my former students, Mark Alford and Steve Adcock of Norman + Dawbarn, who helped to resource information. I would also like to thank Colin McGregor of NBS Services for his help in the preparation of Figure 16.17.

I am especially grateful to the Osbourn estate for allowing me the opportunity to develop Derek's work. It has been a valued experience and a privilege to be associated with his concept.

It would be remiss not to acknowledge the publisher, without whom the book could not exist. My thanks to James Newall for inviting me to participate in the prestigious Mitchell's Building Series, and to Brett Gilbert for his enthusiastic development of my revised manuscript.

Roger Greeno
Guildford 1996 and 2001

The publishers are grateful to the following for permission to reproduce copyright material:

- J.S. Foster for our Figures 4.6, 17.50, 17.52 and 17.53 from *Structure and Fabric Part 1*, and our Figures 4.4 and 13.3 from *Structure and Fabric Part 2*.
- Peter Burberry for our Figures 8.1, 8.9, 8.10 and 17.62 from *Environment and Services*.
- The *Architects' Journal* for our Figure 13.5.
- RIBA Publications for our Figures 14.1, 16.1, 16.2, 16.18 and 16.19, our Table 16.1 and material in Appendix II.
- The Ordnance Survey for our Figure 16.3: © Crown copyright (399582).
- Norman + Dawbarn Limited (Architects) for our Figure 16.6.
- The Building Research Establishment and the Controller of The Stationery Office for our Figures 16.7 and 16.13.
- National Building Specification Limited for our Figure 16.17.
- The Chartered Society of Designers for material in Appendix II.

Further information about the National Building Specification may be obtained from NBS Services, Mansion House Chambers, The Close, Newcastle upon Tyne NE1 3RE (Tel: 0191 232 9594).

SI units

All quantities in this volume are given in SI units, which have been adopted by the United Kingdom for use throughout the construction industry as from 1971.

Traditionally, in this and other countries, systems of measurement have grown up employing many different units not rationally related and indeed often in numerical conflict when measuring the same thing. The use of bushels and pecks for volume measurement has declined in this country but pints and gallons, cubic feet and cubic yards are still both simultaneously in use as systems of volume measurement, and conversions between the two must often be made. The subdivision of the traditional units varies widely: 8 pints equal 1 gallon; 27 cubic feet equal 1 cubic yard; 12 inches equal 1 foot; 16 ounces equal 1 pound; 14 pounds equal 1 stone; 8 stones equal 1 hundredweight. In more sophisticated fields the same problem existed. Energy could be measured in terms of foot pounds, British thermal units, horsepower, kilowatt-hour, etc. Conversion between various units of national systems were necessary and complex, and between national systems even more so. Attempts to rationalise units have been made for several centuries. The following stages are the most significant:

- The establishment of the decimal metric system during the French Revolution.
- The adoption of the centimetre and gram as basic units by the British Association for the Advancement of Science in 1873, which led to the CGS system (centimetre, gram, second).
- After circa 1900, the use of metres, kilograms and seconds as basic units (MKS system).
- The incorporation of electrical units between 1933 and 1950 giving metres, kilograms, seconds and amperes as basic units (MKSA system).

- The establishment in 1954 of a rationalised and coherent system of units based on MKSA but also including temperature and light. This was given the title *Système International d'Unites* which is abbreviated to SI units.

The international discussions which have led to the development of the SI system take place under the auspices of the Conference Général des Poids et Mesures (CGPM) which meets in Paris. Eleven meetings have been held since its constitution in 1875.

The United Kingdom has formally adopted the SI system and it will become, as in some 25 countries, the only legal system of measurement. Several European countries, while adopting the SI system, will also retain the old metric system as a legal alternative. The USA has not adopted the SI system.

The SI system is based on the six basic units given in Table 1.

Table 1   The six basic units of the SI system

Quantity	Unit	Symbol
length	metre	m
mass	kilogram	kg
time	second	s
electricity	ampere	A
temperature	kelvin	K
luminous intensity	candela	cd

Kelvins apply to absolute temperature; the Celsius scale is customarily used in many contexts. This book uses degrees kelvin for temperature intervals – the difference between two temperatures. The degree kelvin and the degree Celsius have the same temperature interval; it is only their datums which differ, $0°C =$

273.15K. So K and °C are interchangeable when considering temperature intervals.

Besides the basic units, there are two others, given in Table 2. Degrees °, minutes ′ and seconds ″ will also be used as part of the system.

Table 2   The two supplementary units of the SI system

Quantity	Unit	Symbol
plane angle	radian	rad
solid angle	steradian	sr

From these basic and supplementary units the remainder of the units necessary for measurement are derived, e.g.

- Area is derived from length and breadth, measured in m^2.
- Volume is derived from length, breadth and height, measured in m^3.
- Velocity is derived from length and time, and is measured in m/s.

Some derived units have special symbols and some examples are given in Table 3.

Table 3   Some derived units and their symbols

Quantity	Unit	Symbol	Basic units involved
frequency	hertz	Hz	1 Hz = 1 c/s (1 cycle per second)
force, energy work, quantity	newton	N	$1 \text{ N} = 1 \text{ kg m/s}^2$
of heat	joule	J	1 J = 1 N m
power	watt	W	1 W = 1 J/s
luminous flux	lumen	1m	1 lm = 1 cd sr
illumination	lux	lx	$1 \text{ lx} = 1 \text{ lm/m}^2$

Multiples and submultiples of SI are all formed in the same way and all are decimally related to the basic units. It is recommended that the factor 1000 should be consistently employed as the change point from unit to multiple or from one multiple to another. Table 4 gives the names and symbols of the multiples. When using multiples the description or the symbol is combined with the basic SI unit, e.g. kilojoule kJ.

Table 4   The SI system of multiples and submultiples

Factor		Prefix	
		Name	Symbol
one million million (billion)	10^{12}	tera	T
one thousand million	10^{9}	giga	G
one million	10^{6}	mega	M
one thousand	10^{3}	kilo	k
one thousandth	10^{-3}	milli	m
one millionth	10^{-6}	micro	μ
one thousand millionth	10^{-9}	nano	n
one million millionth	10^{-12}	pico	p

The kilogram departs from the general SI rule with respect to multiples, being already 1000 g. Where more than three significant figures are used, it has been United Kingdom practice to group the digits into three and separate the groups with commas.

This could lead to confusion with calculation from other countries where the comma is used as a decimal point. It is recommended, therefore, that groups of three digits should be separated by thin spaces, not commas. In the United Kingdom the decimal point can still, however, be represented by a point either on or above the baseline.

Introduction

This volume is intended for those who are commencing a serious study of the various mental and physical processes which are involved during the creation of a building. It is therefore primarily for students of building, architecture and interior design. It provides a basic introduction to building; each section is designed to stimulate interest and encourage further reading from the other more advanced volumes of Mitchell's Building Series. Further reading is provided at the end of each chapter, which includes relevant BRE Digests and Information Papers. These can be obtained from the Building Research Establishment's publisher, Construction Research Communications Limited, 151 Rosebery Avenue, London EC1R 4GB.

The factors involved in the creation of a building are complicated, numerous and varied. To understand them in detail it would first be necessary to clearly identify each, and after placing in a sequential order of dependence, provide a thorough analysis of the role they play.

This would be an onerous task, since the factors are closely interdependent and generate subfactors which make a structured sequential order somewhat arbitrary. However, as an *introduction* to the subject, the various factors can be combined and simplified, an order established which relates to purpose and elementary knowledge, and a brief description given to clarify some of the considerations.

The main divisions which have been chosen for this book are as follows:

- **Part A** is an analysis of a building in terms of what it is expected to do: its function and performance.
- **Part B** is an analysis of a building in terms of the processes required, the Building Team which implements them, and the methods used for communicating information.
- **Part C** is an analysis of a building in terms of typical construction methods.

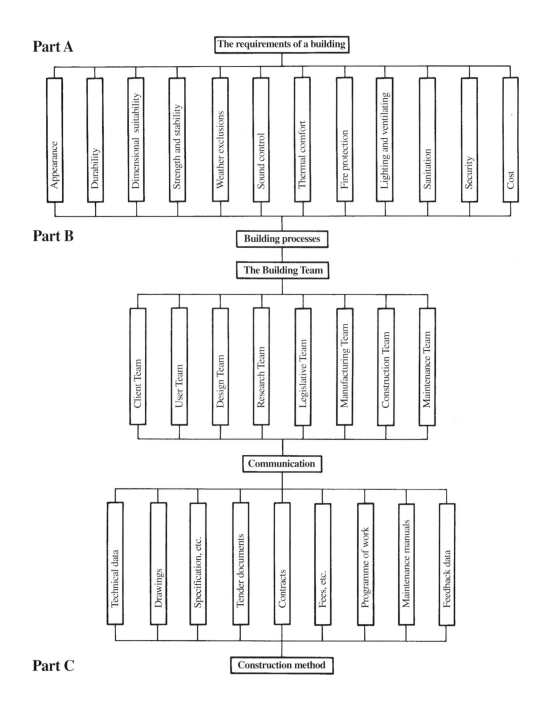

Part A

The requirements of a building

Appearance · Durability · Dimensional suitability · Strength and stability · Weather exclusions · Sound control · Thermal comfort · Fire protection · Lighting and ventilating · Sanitation · Security · Cost

Part B

Building processes

The Building Team

Client Team · User Team · Design Team · Research Team · Legislative Team · Manufacturing Team · Construction Team · Maintenance Team

Communication

Technical data · Drawings · Specification, etc. · Tender documents · Contracts · Fees, etc. · Programme of work · Maintenance manuals · Feedback data

Part C

Construction method

Part A

An analysis of a building in terms of what it is expected to do: its function and performance

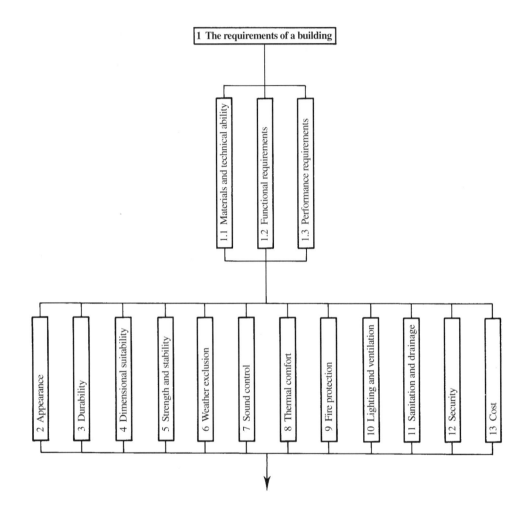

1 The requirements of a building

1.1 Materials and technical ability

1.2 Functional requirements

1.3 Performance requirements

2 Appearance

3 Durability

4 Dimensional suitability

5 Strength and stability

6 Weather exclusion

7 Sound control

8 Thermal comfort

9 Fire protection

10 Lighting and ventilation

11 Sanitation and drainage

12 Security

13 Cost

1 The requirements of a building

CI/SfB (E)/(Y)

1.1 Materials and technical ability

When a building is constructed two main physical resources are involved. These are, *materials* necessary to form the various parts, and *technical ability* to assemble the parts into an enclosure (Fig. 1.1). Initially, the materials employed were those which could most easily be obtained from the accessible areas of the surface of the earth. The technical ability was mostly simple, having evolved from the convenient methods of economically working the rudimentary characteristics of these available materials. The gradual widening in means of communication and corresponding developments in attitudes led to an increased range of these resources becoming available.

The current uses of particular construction methods no longer need to rely on locally available materials or traditional technical ability. Continued investigation has resulted in the enormous range of materials now becoming available which may be used singly, in combination with one another, or even to form new materials. Technological developments are interrelated with this range and use of materials, and enable virtually anything to be constructed.

Nevertheless, there are certain considerations which have always exerted some control on the indiscriminate use of resources. These controls remain, and now that the range of resources is wider, and attitudes towards the function of a building are more complicated, the selection of appropriate construction method becomes much more difficult. For this reason, it is first necessary to understand precisely what is required of a building before selecting an appropriate method of construction.

1.2 Functional requirements

Elaborate shelters have been, and still are, made by most species of insects, reptiles and animals capable

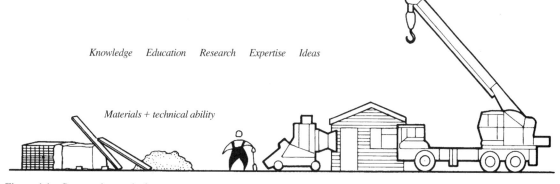

Knowledge   Education   Research   Expertise   Ideas

Materials + technical ability

Figure 1.1   Construction method.

of using the readily available materials (earth, stones, branches and leaves, etc.) with the aid of the inherent manipulative skills (technology) of their arms, legs, wings, claws, beaks and jaws. Early humans also required shelter which provided security for them, their possessions and activities. However, they developed their inherent manipulative skills by inventing tools which led to less indigenous construction methods and also ways of changing the natural state of materials so that they could be used to greater advantage. Each innovation devised usually resulted in shelters, which, although initially providing a good standard of comfort and convenience, eventually became substandard accommodation as requirements became more elaborate. But, regardless of technical developments, the provision of a physically comfortable shelter was not the only or even the principal reason for building. From early times, a building was also required to give an established place of social and religious *identity*: it must indicate culture, status and mood, whilst creating the humanised space in which to learn, experience and carry out normal daily functions in comfort.

Sir Henry Wooten, a fifteenth-century humanist who adapted the writings of Vitruvius for his book, *The Elements of Architecture* (1624), wrote that a good building must satisfy three conditions:

- Commodity: comfortable environment conditions
- Firmness: stability and safety
- Delight: aesthetic and psychological appeal

These *functional requirements* are implicit in the provision of a shelter which is also a building fit for human habitation. *A well-constructed building reflects contemporary attitudes towards environmental control, structural concepts and aesthetic excellence.* And the materials and technical ability used throughout history have normally provided the means of achieving these particular ends.

1.3 Performance requirements

A modern building is expected to be a life-support machine (Fig. 1.2). It is required to provide the facilities necessary for human metabolism such as clean air and water, the removal of waste produce, optimum thermal and humidity control, privacy, security and visual/acoustic comfort. It is generally required to be a source of (perhaps self-generating) energy for appliances, and provide means for communication with television, telephones and postal services. In addition, a building must be safe from collapse, fire, storm and vermin; resistant to the physical forces of snow, rain, wind and earthquakes, etc.; be capable of adaptation to various functions,

external landscaping or internal furniture arrangements. It must also be easily, economically, quickly and well constructed; and allow easy maintenance, alterations and extension as well as having a sustainable form of construction which can be adapted to changing trends and legislative requirements. All this must be accomplished in the context of providing a building which has character and aesthetic appeal.

Criteria of this nature form today's interpretation of the basic functional requirements for a building quoted earlier. In order for them to be conveniently considered, it is necessary to divide a building into the various related duties to be fulfilled and establish the precise *performance requirements* for each. When these duties are incorporated into a building where the *functional requirements* have been clearly defined, the selection of a suitable construction method (materials and technology) can be achieved by using the criteria given by the performance requirements under the following headings:

- Appearance
- Durability
- Dimensional suitability
- Strength and stability
- Weather exclusion
- Sound control
- Thermal comfort
- Fire protection
- Lighting and ventilation
- Sanitation and drainage
- Security
- Cost

Performance requirements cannot be placed in order of importance because any one of them may be more critical than another for a particular element of a building. Priority is normally dictated by the precise function and location of a specific building.

The use of these interrelated performance requirements in establishing a building design was instigated many years ago by the then Building Research Station (now Building Research Establishment, see page 137). Although firm principles have now been established, critical factors arise which result in fundamental changes in attitudes towards construction methods. In recent years, one such influence concerns the availability of energy resources for the production of building materials, and for the heating and lighting of buildings.

During the formation of the earth some five million years ago, only a relatively small amount of hydrocarbon atoms were incorporated and these now form our fossil fuels of gas, oil and coal. These fuels have been continuously used in one form or another during the

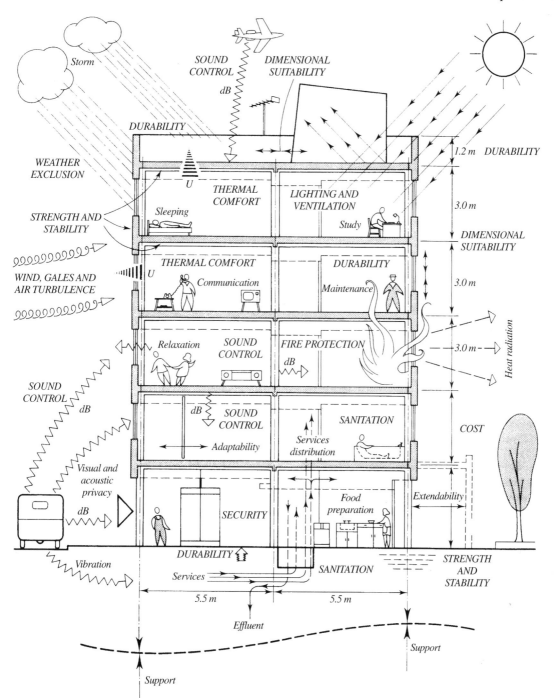

Figure 1.2   Performance requirements for a building.

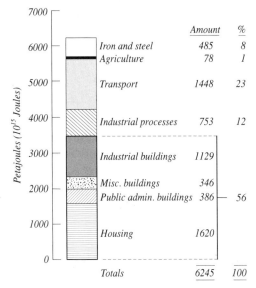

	Amount	%
Iron and steel	485	8
Agriculture	78	1
Transport	1448	23
Industrial processes	753	12
Industrial buildings	1129	
Misc. buildings	346	
Public admin. buildings	386	56
Housing	1620	
Totals	6245	100

Figure 1.3 Energy consumed by the building industry.

development of humankind and the rate of consumption has increased rapidly over the last 40 years. As a result of this situation, even if the world population were stabilised, the requirements for fuel were static, and the poorer countries remained undeveloped, there would be only another 40 years of gas supply, 25–30 years of oil, and 200–300 years of coal obtainable through easy access. Of the net energy consumed, more than 50 per cent is subject to decisions by those involved in the design and construction of buildings (Fig. 1.3).

The current need to consider energy conservation seriously is resulting in new approaches towards the construction and maintenance of buildings. Many prejudices and preconceptions must be shed, and buildings of different appearance from those traditionally accepted

must result. This premise will become more pertinent as the commonly used building materials become scarcer, either through gradual reduction in world availability or (more likely) through difficulties in obtaining them from the more remote regions of the earth. Table 1.1 illustrates the rate of annual consumption for certain metals compared with known reserves, and the total natural occurrences beneath the oceans as well as dry land. Although the future of these metals seems guaranteed for many years to come, it will become increasingly difficult and costly to establish the technology to mine in the more inaccessible regions. It is also highly probable that any future winning of these hitherto inaccessible minerals will produce dramatic ecological changes in the earth's surface and atmosphere. Although aggregates and cements for concretes, clays for bricks and timbers are similarly likely to remain in worldwide abundance for many years, their retrieval for conversion into building materials is already becoming limited by environmental conservation issues. On a positive note as far as timber is concerned, worldwide attempts are being made to reduce indiscriminate deforestation, particularly the equatorial rain forests. Efforts are also being made to replenish the already depleted timber stocks in Europe through replanting, and in future timber may well become the critical building material.

Nevertheless, the development of substitutes for the traditional building materials is likely to increase and these will cause people to reassess construction methods. These changes will be paralleled by changes in the performance requirements for the whole building. Research and development is continually taking place in these areas, and it is the duty of those involved in the design and erection of buildings to be aware of current trends. The achievement of this knowledge cannot be solely through any published building law or contemporary code of good practice, as by the time they have been established other discoveries and experiences may already have taken place.

Table 1.1 Mineral content of oceans and the earth's crust related to present consumption (million tonnes)

Metal	Known reserves	Current annual consumption	Total natural occurrence in oceans, including the seabed	Total natural occurrence in the first mile of the earth's crust (under dry land)
Aluminium	4 250	12	15 500	138 244 000 000
Copper	364	8	4 650	119 000 000
Iron	109 000	700	12 400	84 630 000 000
Lead	94	4	465	27 400 000
Nickel	90	0.5	3 100	135 960 000
Tin	4.4	0.25	4 650	67 760 000
Zinc	306	4.5	15 500	224 000 000

Further specific reading

Mitchell's Building Series

Environment and Services	Chapter 1 Environmental factors
Internal Components	Chapter 1 Component design
External Components	Section 2.8 Component design: accuracy and tolerances
Materials	Chapter 1 General properties
Materials Technology	Chapter 1 Developing an attitude towards materials
Structure and Fabric Part 1	Chapter 1 The nature of buildings and building

Building Research Establishment

Principles of Modern Building Vols 1 and 2, The Stationery Office

2 Appearance

CI/SfB (G)

The appearance of a building is initially determined by the activities to be accommodated, as these strongly influence the scale and proportion of the overall volumetric composition (Fig. 2.1). The shape of the individual spaces forming the collective volume are defined by 'boundaries' which become the walls, floors, roofs, etc. These are ultimately required to conform with precise *aesthetic and technical* criteria, and the materials employed for these purposes are as numerous and varied as the methods which can be adopted for their use. However, the underlying principle remains that both aesthetic and technical criteria are affected by the composition, form, shape, texture, colour and position of the materials employed. To this must be added the skill with which they are placed in a building and the cost, since these factors often provide a deciding role.

2.1 Aesthetic aims or fashions

In a well-designed building, appearance is the reflection of a balance between *aesthetic aims or fashions* and the *construction method* derived from the desire for optimum environmental control, structural stability and logical techniques of instigation. Building designers of an earlier and simpler time than today had available only a comparatively limited choice of technical resources which, through the influences of sociocultural attitudes, led to particular 'styles' in the appearance of buildings. In contrast, the current range of resources, more complicated performance requirements and widening of cultural influences have created multifarious 'styles'. Some of these evolve from an overwhelming bias towards 'high technology' and the absolute economies of industrial forms. However, if technology is allowed to dominate the aesthetics of a building without

compromise, there is a loss of human understanding, scale and proportion and, perhaps, colour, form and texture. By way of comparison, some 'styles' endeavour to imitate the psychological appeal associated with the less complicated requirements of the past. Sometimes this may be done on the premise that a building of today, with its highly technological requirements, is beyond our aesthetic comprehension. The technological advantages may be accepted, but they are combined with efforts which deny their visual influences as well as reduce their performance efficiency.

A building design will invariably be unsuccessful if it relies purely on technology for aesthetic appeal, or purely on outmoded conventions of a past era. A skilful building designer, whatever his or her aesthetic leanings, resolves conflicts through detailed understanding and sympathetic consideration of *all* the performance requirements. None must be ignored or denigrated.

2.2 Relationship to other performance requirements

Although appearance is only an aspect of the total aesthetic quality, it is generally *the* one on which most first impressions of a building are formed. The majority of users have definitive ideas about what a building should look like and, apart from rules concerning 'correctness', their satisfaction or otherwise may be entirely subjective. In this respect, therefore, the requirement for 'appearance' to some extent contrasts with other performance requirements, although from a designer's viewpoint it is the factor which unites them all.

For this reason, reference is made in most of the chapters of the book to the effects on appearance arising from decisions about other performance requirements.

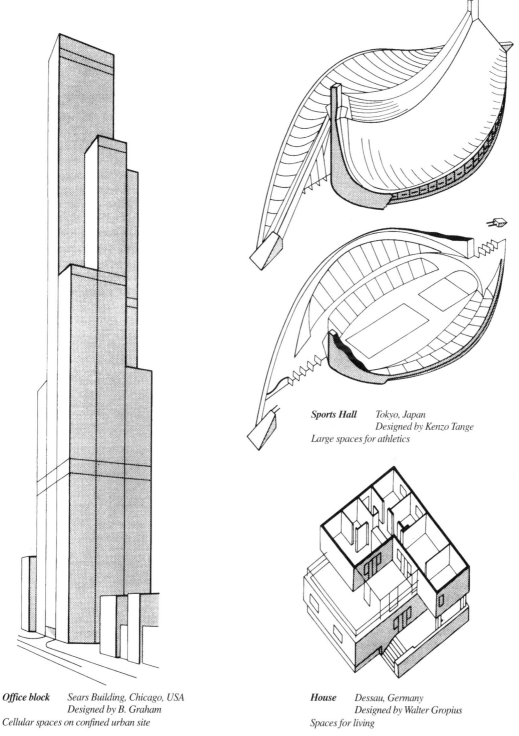

Sports Hall   *Tokyo, Japan*
Designed by Kenzo Tange
Large spaces for athletics

Office block   *Sears Building, Chicago, USA*
Designed by B. Graham
Cellular spaces on confined urban site

House   *Dessau, Germany*
Designed by Walter Gropius
Spaces for living

Figure 2.1   Effect of building function on appearance. (Adapted from material in *Great Architecture of the World*, published by Mitchell Beazley)

This serves in reinforcing the view that appearance and function are inseparable in a building, although the degree to which this may be recognised depends upon the skill of the designer. Even certain forms of 'decorative motif' may be based on a requirement to provide solar shading devices, or perhaps even form the structural tie between two separated flank walls. The appearance of a modern building should not rely on 'functional' or 'decorated functional' aspects alone. Properly controlled and sensitively located non-functional items (in terms of technical performance) – decorations, murals and sculptures – can be incorporated as an essential part of the overall aesthetic achieved by a building.

Some of the basic areas of consideration affecting the appearance of a building, both *externally and internally*, are as follows:

- The aesthetic objectives of the designer in terms of preferred form, shape, pattern, texture and colour, etc.
- The effects of location and siting on the design and construction methods adopted with particular reference to local planning guidelines, building byelaws, regulations and other relevant legislation.
- The design as part of the larger composition of the area – harmony with adjacent buildings and/or specific features, including landscaping.
- The 'viewing distances' applicable to the design.
- The use of a particular structural organisation.
- The use of materials that are suitable for particular applications which enhance, modify or even change the appearance of the design as it ages.
- The relationship of window and door openings, and the creation of a rhythm in the design.
- Architectural detailing used to reinforce the character required by the design and location, e.g. the use of particular types of window/door lintel construction, the creation of shade and shadows through location of components.
- The positioning of ductwork, service pipes, etc., and their contribution towards aesthetic character.
- The effects of maintenance on the initial design and subsequent use of a building (see section 2.3).

It is important that the aesthetic achieved by the smaller parts of a building is a reflection of the same design philosophy applied to its larger parts. The whole building, internally and externally (including approaches and landscaping), should display a similar character, taste, interest, wealth and aspiration. If sympathetically conceived and constructed, a homogeneous environment is created which is understandable to users, establishes interest, and develops taste for good design and construction generally.

2.3 Weathering and maintenance

More specific reference must be made about the necessity to anticipate the effects of future weathering and maintenance on the appearance of a building. Well designed and constructed, a building should accommodate the progress of time without causing a lessening of any functional requirement (see Chapter 3). Furthermore, the inevitable effects of weathering should make a positive rather than a negative contribution to the appearance of a building. This relies on carefully detailed constructional solutions which derive from a thorough understanding of the behaviour of materials to be used in a particular manner. Figure 2.2 illustrates a common design 'fault' which has marred the appearance of a building. The impermeable glass and metal frame surface within the opening between brick panels allows water to be caught, then to run over and clean the concrete beam immediately below. A lesser amount of rainwater will flow over the part of the beam occurring below the brick panels because of the greater

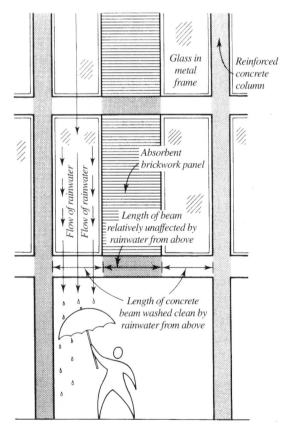

Figure 2.2   Effect of weathering on the appearance of a building.

absorption properties of the bricks. The result is a striped staining of the concrete beam, which changes the appearance of the building in a manner presumably never intended by the designer. The provision of an adequate sill and drip below the opening to prevent the free flow of water over the beam face would have helped to overcome this problem. Alternatively, a continuous gutter could have been provided along the top of the beam, or the beam faced with a material less susceptible to staining.

Further specific reading

Mitchell's Building Series

Internal Components	Section 2.3.1 Appearance of demountable partitions
	Section 3.3.1 Appearance of suspended ceilings
	Section 4.2.1 Appearance of raised floors
	Section 5.2 Choice of timbers
	Section 6.2.1 Appearance of doors
External Components	Section 3.2.1 Appearance of external glazing
	Section 4.2.1 Appearance of windows
	Section 7.2.1 Appearance of external doors
	Section 8.1 Introduction to roofings
Materials	Section 1.9 Appearance
	See also appearance subsections under every material chapter
Materials Technology	Chapter 2 Ceramics (introduction)
	Chapter 6 Metals (introduction)
	Chapter 9 Polymers (introduction)
	Chapter 11 Composites (introduction)
Structure and Fabric Part 1	Chapter 1 The nature of buildings and building
Finishes	Chapter 1 Introduction
	Chapter 2 Polymeric materials
	Chapter 4 Ceramic materials
	Chapter 6 Metals
	Chapter 7 Composites

Building Research Establishment

BRE Digest 45   *Design and appearance* Part 1
BRE Digest 46   *Design and appearance* Part 2
BRE Digest 269   *The selection of natural building stone*
BRE Digest 280   *Cleaning external surfaces of building*
BRE Digest 387   *Natural finishes for exterior wood*
BRE Digest 429   *Timbers: their natural durability and resistance to preservative treatment*
BRE Digest 446   *Assessing environmental impact of construction*
BRE Digest 448   *Cleaning buildings*
BRE Digest 449   *Cleaning exterior masonry*

3 Durability

CI/SfB (R8)

Apart from daily wear and tear by its users, a building is subjected to the constant influences of climate (wind, rain, snow, hail, sleet, sunlight), perhaps attack from vandals and vermin, or even damage by fire, explosions and structural movements. Both the inside and the outside of all buildings are therefore subject to forces which can cause deterioration during their life. Durability is the measure of the rate of deterioration resulting from these and other forces.

3.1 Changes in appearance

When related to external climatic or environmental factors, the durability aspects of a building are known as *weathering*. The action of frost, temperature variations, wind and rain on the materials of a building can cause *changes in appearance* by gradual erosion and/or the transportation of atmospheric pollutants which cause staining. Unless the building is carefully designed and detailed, these visual changes seldom enhance appearance and therefore produce disfiguration. The degree to which a building will suffer from erosion and/or staining depends upon many interacting variables. One involves the relationship between type and amount of atmospheric pollution with the exposure of a building to wind, rain, frost, snow and solar radiation. Their effects will be determined by the characteristics of the materials used in a building and include their capacity or otherwise for moisture absorption, as well as their surface profiles, orientation, texture and colour.

In an urban situation it is normal to see dark bands of staining below most horizontal projections (mouldings and sills, etc.) of a surface-permeable masonry wall. These projections provide a shield against direct rainfall or rainfall run-off from above, and dirt deposits are therefore left relatively undisturbed when compared with the lower regions of the wall (see also examples in Fig. 2.2). It is interesting to note that the design of projected mouldings for buildings of the past generally ensured an even weathering to the surrounding vertical surfaces. Nevertheless, although today's buildings are often devoid of mouldings and decorative devices, it is possible for designers to allow water to be guided down a specific route by means of ribbed projections or recessed grooves, and therefore to create 'controlled' areas of weathering which enhance rather than detract from the appearance of the building. Porous surfaces may accumulate dirt and dust, vertical surfaces stain according to roughness and absorptivity, surface colour affects visual density of staining, but carefully detailed designs can also use these properties to advantage.

It is first necessary to establish the degree of *climatic exposure* which is likely to affect a building (see Chapter 6). Generally, disfiguration will be slight where there is a moderate rainfall and a moderate rate of pollution, and constructional detailing needs to be far less 'bold' than for areas suffering greater amounts of rainfall and pollution. However, freak weather conditions can result from a group of tall buildings which create a weather pattern between them contrary to accepted predictions. Also, as buildings receive more direct rainfall on their upper storeys, high buildings are generally cleaner at the top and dirtier towards the lower storeys. This is caused by dust being washed down the face of the building and being retained at the point where the porosity of the surface absorbs (or constructional detailing collects and channels) a major part of the water. Low buildings, lacking exposure, will obtain an overall covering of pollution which will be unrelieved by washing. This is preferable to the random streaking to

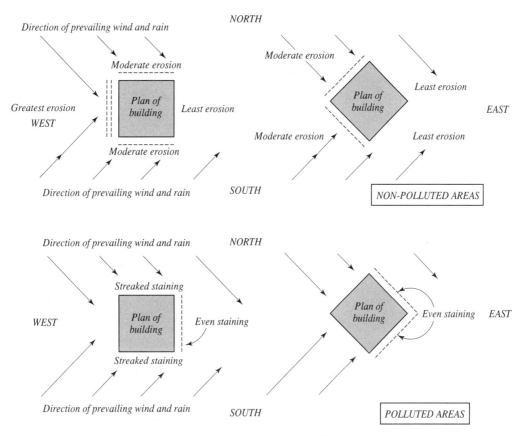

Figure 3.1   Effects of south-west and north-west rain-carrying winds on a building.

be seen on elevations which are subject to strong dust accumulation and which do not receive direct rainfall in sufficient quantity to provide overall cleansing.

The prevailing rain-carrying winds in the British Isles blow mainly from the south-west and the north-west quarters (Fig. 3.1). In non-polluted areas the rain will be clean and will wash the building down. In polluted areas the rain will absorb the dust in the air and thus will be dirty when striking the surface. It will also carry dirt from the upper surface to the lower surface and cause staining. South-west and north-west aspects of buildings will generally remain cleaner than the south-east and north-east aspects. North and north-east aspects are particularly liable to severe accumulation of dirt. Thus for a building to be cleaned the volume of clean rainwater must exceed the volume of dirty water passing over the surface. Snow has no beneficial cleaning effect.

3.2 Physical deterioration

Apart from causing the staining of a building, certain forms of weathering *also* result in chemical actions which cause physical deterioration by decay. From the time they are placed on a building site, all materials commence on a path of deterioration which could continue after a building has been constructed until a point is reached when they are not fulfilling a useful purpose. Again, the precise form of attack depends upon exposure conditions, and on the susceptibility of the materials used. (For further comment see section 3.1.) Corrosion, erosion, and disintegration of materials and construction details can follow from the effects of changes in moisture content, frost, sunlight, soil and groundwater action, atmospheric gases, electrolytic action, fungal and insect attack, or domestic and industrial wastes, etc. Critical conditions are influenced by the selection of materials appropriate for their design function, detailing for their application, and the skilfulness of their installation. These provide the 'control' on the extent to which deterioration and decay may be allowed over a given period of time.

The durability factors concerned with the effects of fire, explosions and structural movements will be mentioned under the appropriate performance requirement to follow. The problems of daily wear and tear by the users, and attack from vermin, vandals and burglars, also form an important part of the design criteria for the building. All involve the careful selection of appropriate materials and functional detailing as well as thoughtful overall planning which will lessen the likelihood of their occurrence.

3.3 Intended life span

Ideally, a chosen design and construction method should be able to resist all detrimental effects and provide prolonged durability. As economic considerations make this impossible, the selection of materials and assembly technique for a design should be made to ensure that their rate of deterioration will not impair the functional performance of a building, including appearance, during its *intended life span*. This period is difficult to quantify, so selection processes are generally influenced by an interrelationship between *initial costs* of materials and likely future *maintenance costs*. When deciding, it should be remembered that maintenance costs not only include labour and materials for renovation and for cleaning, but also sometimes financial losses resulting from the temporary curtailment of trade or business by the building owner or tenants while remedial work is being executed. Furthermore, where access for maintenance is difficult because of design detailing, consideration must be given to the cost of hiring scaffolding, special plant or even the provision of permanent gantries, mobile or otherwise. Gantries can form a dominant feature of the external or internal appearance of a building.

The degree to which subsequent maintenance of a building is possible, desirable or *necessary* provides the 'fine tuning' which assists in achieving an optimum design in terms of initial running and future costs. A designer must be expert in the choice of materials, or carry out necessary research into their physical and chemical properties, before devising construction techniques which will take full advantage of their potentialities. Continually rising costs are forcing serious consideration of designing buildings to give a *guaranteed, but limited life span* (limit state design). In this way, the life of a building is predetermined by quantifiable characteristics, including predictability of certain detrimental influences. For example, it would be unnecessarily costly to incorporate an earthquake-resisting structure in a building with an intended useful life of 20 years if the earthquake 'cycle' where the building is to be erected does not fall within this period. If available, the money saved by not using elaborate structural solutions can be effectively used on materials and constructional details with guaranteed performance requirements during the 20-year life of the building.

This approach can be extended to the statistical analysis of freak wind turbulences and amounts of rainfall in otherwise predictable climatic regions. However, financial savings in capital costs of construction methods (materials and technology) can only be made after thorough study and research have revealed the precise degree of durability achievable and therefore the life span of a building with minimal maintenance.

Further specific reading

Mitchell's Building Series

Environment and Services   Chapter 2 Moisture

Internal Components   Section 2.3.8 Durability and maintenance of demountable partitions
Section 3.3.6 Durability and maintenance of suspended ceilings
Section 4.2.6 Durability and maintenance of raised floors
Section 5.2 Choice of timbers
Section 6.2.2 Durability of doors

External Components   Section 1.2 Durability
Section 3.2.10 Durability and maintenance of external glazing
Section 4.2.8 Durability and maintenance of windows
Section 5.2 Performance requirements for rooflights and patent glazing
Section 7.2.2 Durability of external doors
Section 8.2.6 Durability and maintenance of roofings

Materials   Section 1.8 Deterioration
See also durability and deterioration subsections of each chapter

Materials Technology	Sections 3.6 and 3.9 Durability of concrete
	Section 4.5 Deterioration of brickwork
	Section 6.2 Corrosion susceptibility of metals
	Section 7.4 Steel corrosion
	Section 10.2 Durability of polymers
	Section 12.3 Timber decay
	Section 12.5 Corrosion and timber fixings
Finishes	Chapter 3 Applications of polymeric materials
	Chapter 5 Applications of ceramic materials
	Section 7.3 Flooring applications

Building Research Establishment

BRE Digest 45   *Design and appearance* Part 1
BRE Digest 46   *Design and appearance* Part 2
BRE Digest 144   *Asphalt and built-up felt roofings: durability*
BRE Digest 177   *Decay and conservation of stone masonry*
BRE Digest 217   *Wall cladding defects and their diagnosis*
BRE Digest 251   *Assessment of damage in low-rise buildings*
BRE Digest 268   *Common defects in low-rise traditional housing*
BRE Digest 280   *Cleaning external surfaces of building*
BRE Digest 304   *Preventing decay in external joinery*
BRE Digest 330   *Alkali-silica reactions in concrete*
BRE Digest 345   *Wet rots: recognition and control*
BRE Digest 387   *Natural finishes for exterior wood*
BRE Digest 390   *Wind around tall buildings*
BRE Digest 391   *Damage to roofs from aircraft wake vortices*
BRE Digest 403   *Damage to structures from ground borne vibration*
BRE Digest 405   *Carbonation of concrete and its effect on durability*
BRE Digest 406   *Wind actions on buildings and structures*
BRE Digest 415   *Reducing the risk of pest infestations in buildings*
BRE Digest 418   *Bird, bee and plant damage to buildings*
BRE Digest 420   *Selecting natural building stones*
BRE Digest 429   *Timbers: their natural durability and resistance to preservative treatment*
BRE Digest 434   *Corrosion of reinforcement in concrete*
BRE Digest 436   *Wind loading*

Building Regulations

Regulation 7   Materials and workmanship

4 Dimensional suitability

CI/SfB (F4)

The dimensional suitability of a construction method involves the consideration of two areas:

- The manner in which movement of materials causes dimensional variations in a building, or parts of a building during its life.
- Appropriate sizes for the parts of a building which suit the materials available to fulfil specific design functions, cost ratios, manufacturing processes and assembly techniques.

4.1 Movement

A building never remains inert; this is because changes in the environment and/or changes in loading cause dimensional changes in the building materials. Variations in moisture content and temperature produce movements in a building which tend to occur in relation to the stronger 'fixed points' in the building – between foundations and first floor, between top floor and roof, between partitions and main structure, or between panels and supporting frames (Fig. 4.1).

4.2 Irreversible and reversible movement

The moisture content of porous building materials can cause *irreversible* movement or *reversible* movement. Irreversible movement is generally associated with establishing a 'normal' or atmospheric moisture level in the materials of a component which have recently been manufactured. For example, clay bricks leaving a kiln will be very dry and will immediately begin to absorb moisture from the air which causes expansion. Conversely, calcium silicate bricks will be more saturated than normal bricks because they are cured by

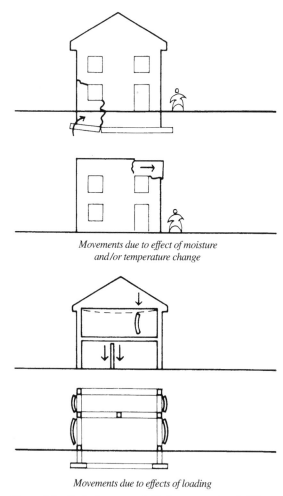

Movements due to effect of moisture and/or temperature change

Movements due to effects of loading

Figure 4.1 Typical movements likely to occur in a building.

autoclave processes and will immediately shrink after manufacture as their moisture content moves towards an equilibrium with that of the atmosphere. For this reason, newly manufactured bricks should not be immediately used for building walls as cracking will inevitably occur (Fig. 4.2). Reversible moisture movement occurs in materials which are in use and generally involves expansion on wetting and shrinkage on drying. These movements have both immediate and long-term effects on the fabric of a building and thoughtful detailing is essential if damage is to be avoided. Care must be taken to ensure movements are reduced to acceptable amounts by limiting the uninterrupted heights and lengths of components and elements. This can be achieved through a precise knowledge of the characteristics of the materials involved and the

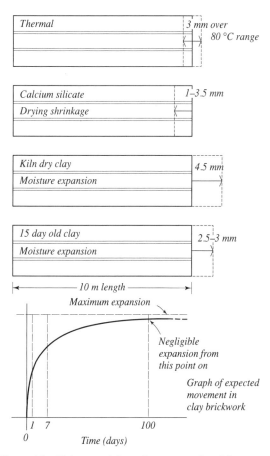

Figure 4.2   Moisture and thermal movements in calcium silicate and clay brick walls. (Based on data provided by the Brick Development Association)

incorporation of *movement joints* at centres beyond which excessive movement is likely to occur (Fig. 4.3). These movement joints should be positioned so as to take account of their visual effect on a building. It is important not to confuse changes in the size of materials due to absorption of moisture with the problems associated with moisture movement *through* materials. Movement through materials will be dealt with in Chapter 6.

Most building materials also expand to a greater or lesser extent with rises in temperature, and if they are restrained could induce considerable stress, producing cracking, bowing, buckling or other forms of deformation. Severe damage to walls can be caused by attempting to restrain beams and slabs – particularly where temperature ranges are likely to be great. Fortunately, dramatic failures of this nature are not very common. But daily (diurnal) temperature ranges are a frequent cause of damage to a building; this can occur immediately in the form of buckling metal sills, cracking glass, etc., or over a period by causing gaps in a weather-impermeable construction which allows the free penetration of moisture. The same care needed in constructional detailing and the provision of movement joints, which is required to limit moisture movements, is also necessary when considering thermal movements. In fact, both problems are often interrelated. In general terms, irreversible moisture movement in porous building materials is greater than reversible movement, and reversible moisture movement is usually less than movement due to temperature changes.

4.3 Softening and freezing

Detrimental effects resulting from temperature changes are also caused as a result of *softening* and of *freezing*. The majority of materials used for a building will not become softened by normal climatic temperature. However, those containing bituminous or coal tar pitch (e.g. asphalt and bituminous felts used for roofs, floor finishes, damp-proof courses, etc.) are liable to become more and more plastic as temperatures rise. This can produce indentation and perforation under load, or result in elongation causing their displacement. A bituminous damp-proof course can soften sufficiently for the load of a wall above to squeeze it outwards from its bedding and even upset the stability of the wall. This problem is most likely to occur on exposed south-facing walls.

Freezing causes a rather special form of thermal movement and, in this respect, is not dissimilar from the chemical attack described in Chapter 3. Sometimes

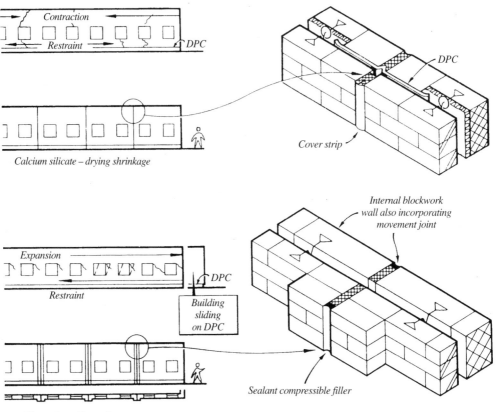

Figure 4.3   Moisture and thermal movements in brick walls.

water may penetrate into a structure and freeze along a junction between two materials. In forming ice lenses, the water expands by about 10 per cent and causes considerable damage. Water freezing in air pockets or other fissures in the mortar of brickwork can cause spalling of the joint and the adjoining arrises of brickwork. If the water freezes within the body of a material, a similar disintegration process will occur. A clear understanding is required about the ratio between amount of water absorbed and the volume and distribution of pore space available if this phenomenon is to be avoided. The presence of moisture in certain materials, even in very minute quantities, can produce chemical changes that lead to movement. The corrosion processes of iron and steel form a porous layer of rust, which becomes liable to expansion. This action can cause spalling of the concrete cover when the metal bars of reinforced concrete beams become rusted.

4.4 Shrinkage

By way of contrast, the chemical action for setting Portland cement produces a volumetric shrinkage. Constructional detailing should take this into account by allowing adequate gaps or tolerances between *in situ* concrete components and other materials. This is particularly true when detailing the junction between an *in situ* concrete frame and a brick infill panel, which is liable to expansion due to moisture absorption and climatic temperature increases (Fig. 4.4).

4.5 Loading

As most materials used for buildings are *elastic* to some degree, a certain amount of *plastic flow or creep* will occur over a period of many years, depending upon the type and amount of load or force to which they

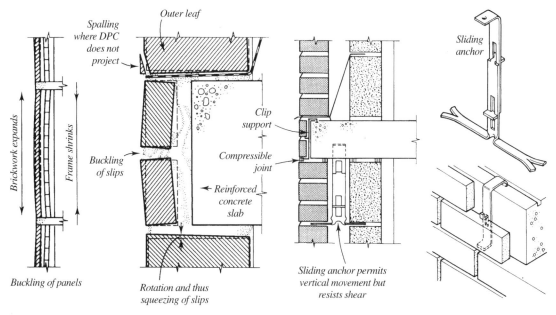

Figure 4.4   Movement joints to panel walls. (Based on data provided by the Brick Development Association)

are subjected. Regardless of this long-term movement, however, structural elements such as beams and columns of any material are likely to be subject to initial *deflection*. This will occur even as their superimposed loads accumulate during construction and will continue until full loading conditions are achieved. Construction methods should take into account the possibility of these movements by ensuring adequate tolerances between structural and non-structural parts of a building to avoid the effects of crushing and cracking. For example, the details for an internal partition of a framed building should ensure that the performance requirements (particularly strength and stability) are not negated by any deflection inherent in the nature of the materials from which beams and columns are formed (Fig. 4.5). If the tendency towards deflection or creep becomes too great in the external fabric of a building, secondary problems could develop from moisture penetration through cracks, etc. Similar problems may arise if a supporting or loading condition of a building is altered and the new conditions cannot be accommodated by the existing construction methods.

4.6 Appropriate sizes

Even the simplest form of building requires thousands of individual components for its construction (Fig. 4.6).

With traditional building practice, these products were completely unrelated in dimension and necessitated the use of skilled workers to scribe, cut, fill, lap and fit them together on the site. Such techniques were notoriously wasteful of both materials and labour. In order to make a building financially viable today, it is necessary that products are manufactured to sizes which coordinate with each other so they can be assembled on site without the need for alteration. This means that the role of the manufacturers becomes much more important as they must ensure the overall dimensions between various products coordinate, and ensure jointing methods between each allow connection with other related products. Products which can be easily assembled together using simple jointing methods are theoretically likely to be less reliant on exacting site skills to fulfil their intended functions. However, in practice, the ease and success of the joint again relies on adequate interpretation by the manufacturer of all performance requirements, labour skills and actual site conditions.

4.7 Dimensional coordination

Apart from providing savings in materials and labour, the use of *dimensionally coordinated* products for a building eases the processes of selection by allowing a

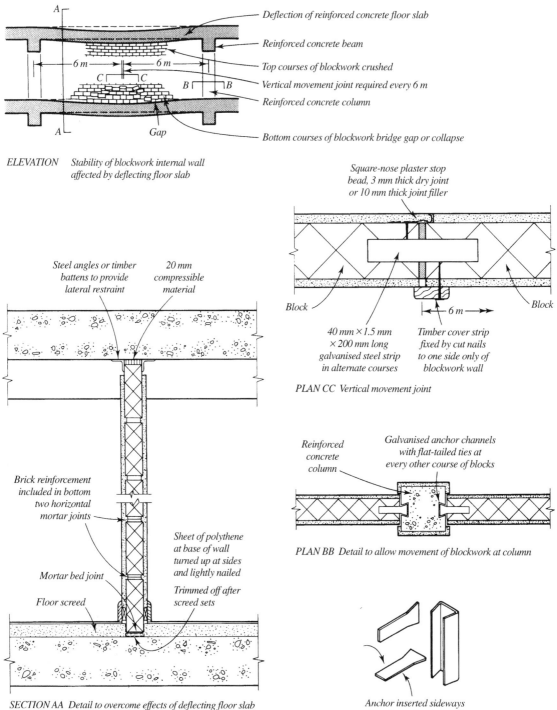

ELEVATION   Stability of blockwork internal wall affected by deflecting floor slab

Deflection of reinforced concrete floor slab

Reinforced concrete beam

Top courses of blockwork crushed

Vertical movement joint required every 6 m

Reinforced concrete column

Bottom courses of blockwork bridge gap or collapse

6 m   6 m

Gap

Square-nose plaster stop bead, 3 mm thick dry joint or 10 mm thick joint filler

Block   Block

6 m

40 mm × 1.5 mm × 200 mm long galvanised steel strip in alternate courses

Timber cover strip fixed by cut nails to one side only of blockwork wall

PLAN CC Vertical movement joint

Steel angles or timber battens to provide lateral restraint

20 mm compressible material

Brick reinforcement included in bottom two horizontal mortar joints

Sheet of polythene at base of wall turned up at sides and lightly nailed

Mortar bed joint

Trimmed off after screed sets

Floor screed

SECTION AA Detail to overcome effects of deflecting floor slab

Reinforced concrete column

Galvanised anchor channels with flat-tailed ties at every other course of blocks

PLAN BB Detail to allow movement of blockwork at column

Anchor inserted sideways into channel and turned to horizontal position

Figure 4.5   Deflection of reinforced concrete beams and the stability of non-load-bearing blockwork internal wall.

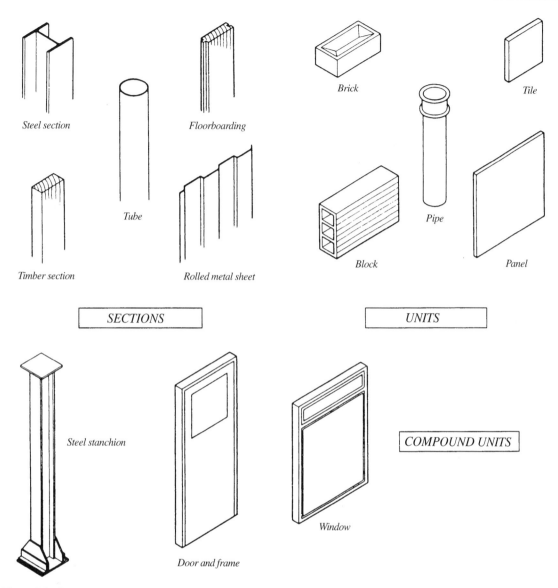

Figure 4.6 Component types.

greater range of similar items to be available (Fig. 4.7). In addition to the availability of home-based products, a designer can select from those abroad, providing they follow the same system of dimensional coordination (see section 15.6).

Many of the sizes of basic building materials used today derive through history and are directly related to the human scale (anthropometrics – see Chapter 12). The clay brick, for example, was dimensionally stand-ardised during medieval times relative to its weight, so as to enable easy manipulation by one hand, leaving the other hand free to operate a trowel. This standard size, now translated to 215 mm × 102.5 mm × 65 mm (BS 3921 : 1985 *Specification for clay bricks*), has given rise to the conventional aesthetic of a brick-built building. When used in large quantities, the bricks, together with their joints and the bonding method, determine the scale and proportion of the overall shape and thickness of the walls, as well as the size of window and door openings, etc. Provided the window and door components are manufactured to be fitted into an open-ing formed by the brick dimensions without adjustment,

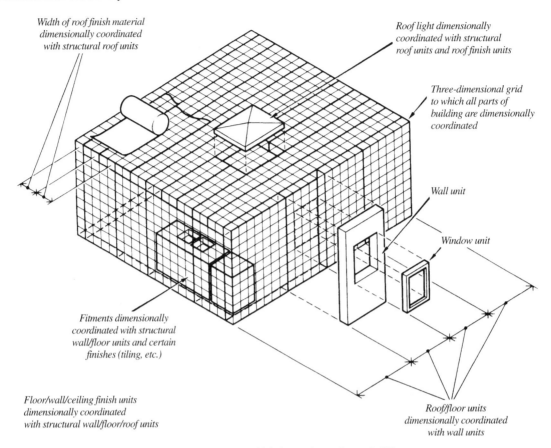

Width of roof finish material dimensionally coordinated with structural roof units

Roof light dimensionally coordinated with structural roof units and roof finish units

Three-dimensional grid to which all parts of building are dimensionally coordinated

Wall unit

Window unit

Fitments dimensionally coordinated with structural wall/floor units and certain finishes (tiling, etc.)

Floor/wall/ceiling finish units dimensionally coordinated with structural wall/floor/roof units

Roof/floor units dimensionally coordinated with wall units

Figure 4.7   Using dimensionally coordinated components which fit together to form a building.

the building shell can be constructed easily and economically. Similarly, internal partition units, floor joists, cupboards, floor and wall finishes should also be available in dimensions suitable to those of brickwork; see section 17.7.1.

Certain manufacturing processes and some materials cannot conveniently conform with common dimensional standards. For example, walls of stone, concrete or plastics will have different dimensional characteristics to walls of brick. It is necessary, therefore, that manufacturers should be given a guide on the likely range of dimensions for which particular components will be required, and the variations for which they should allow within this range. Accordingly, a range of overall dimensions of components related to building use, anthropometric requirements and manufacturing criteria has been recommended. Figure 4.8 illustrates the recommendations for the vertical dimensions used in the construction of a house. There are similar recommendations for the horizontal dimensions. In

order not to overstretch the resources of manufacturers and to further rationalise the available range of components, BS 6750 : 1986 *Specification for modular coordination in building* states that components should be manufactured in basic incremental sizes of 300 mm as first preference, or 100 mm as second preference, with 50 mm and 25 mm being allowed up to 300 mm (Fig. 4.9). For a more detailed explanation, see Chapter 1 of *Internal Components* in Mitchell's Building Series.

4.8 Modular coordination

Attempts have been made to persuade manufacturers to agree to manufacture components in standardised *modular* increments of 100 mm, which is the dimension most common in building products at home and abroad. This system is known as *modular coordination* and, when adopted, means that a building is designed within a three-dimensional framework of 100 mm cubes.

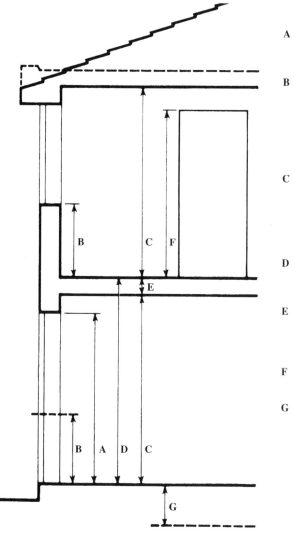

A Window head height
   2300 I
   2100 II

B Window sill height
   0   1000
   200   1100
   600   1200
   700   1400
   800   1800
   900   2100

C Floor-to-ceiling height
   2500
   2400
   2350
   2300
   2100 – garages only

D Floor-to-floor height
   2600 – standard for public sector housing
   2700

E Floor thickness
   200
   250
   300

F Door set height
   2100

G Change of level
   300   1700
   600   1800
   900   2000
   1200   2100
   1300   2300
   1400   2400
   1500

Figure 4.8 Recommended vertical controlling dimensions for housing.

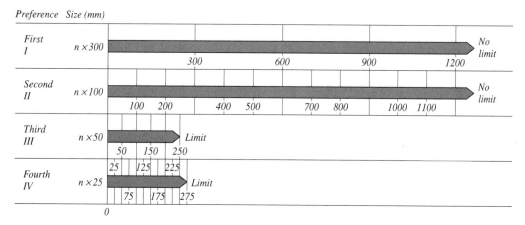

Figure 4.9 Preferred linear sizes for components used in a building.

A successful design relies on the certainty that a vast range of products will be available for the construction – a range which can be selected from almost any country adopting the dimensional system. Products requiring to have dimensions greater than the basic module are manufactured so as to be a *multiple of the basic module* (Fig. 4.10); or, if they are required to be smaller, are manufactured so that when placed together their combined dimensions suit either the basic module or a multiple thereof. Although the idea of modular

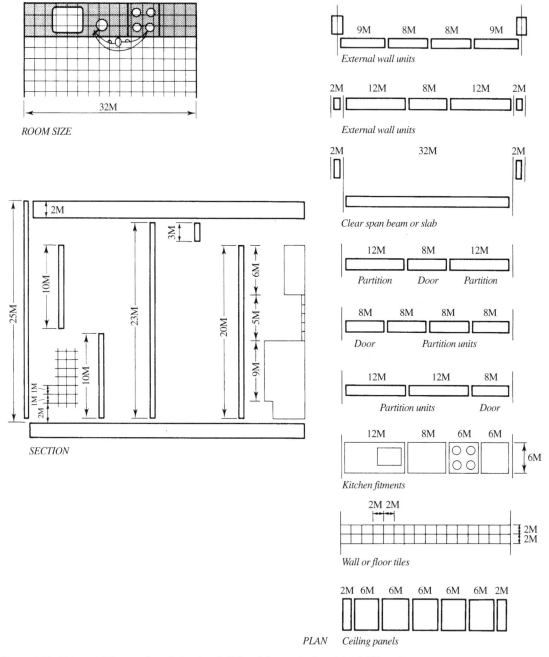

Figure 4.10   Using modular coordinated sizes in a building (M = standard module of 100 mm).

coordination has been in existence for a considerable time, many products will remain in imperial and unco-ordinated dimensions until it becomes cost viable for manufacturers to replace their machinery.

4.9 Jointing and tolerances

Whether modular or otherwise, components cannot be manufactured to precise dimensions to suit a regular three-dimensional grid. Allowance must be made for *jointing* the component, as well as for inaccuracies in the products due to *material properties* (shrinkage, expansion, twisting, bowing, etc.) and *manufacturing processes* (changes in mould shapes, effects on materials,

etc.). Furthermore, allowance must be made for *positioning* the components on site. This is particularly important for very large and heavy components which are manoeuvred into position by cranes, etc., and final location could vary by as much as 50 mm. Allowances of this nature are known as *tolerances*; if manufacturers supply products without these tolerances, cumulative errors in trying to place each component on a grid line would not only result in a building of incorrect overall dimensions, but also a great deal of frustration in the fixing of secondary items such as windows, or furniture and finishes within the building (Fig. 4.11). An important document covering the aspects of tolerances is BS 5606 : 1990 *Guide to accuracy in building*.

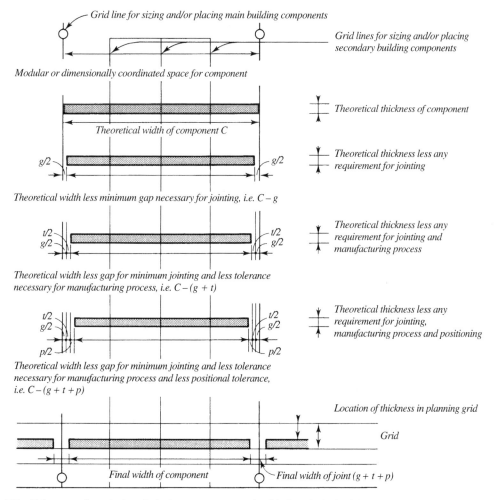

Figure 4.11 Tolerances allowed when designing components: a simplified method of arriving at the 'final dimension' of a component. More accurate but more complicated calculations are now available in BS 6954 : 1988 (1994) *Tolerances for building* Parts 1, 2 and 3.

Further specific reading

Mitchell's Building Series

Internal Components	Chapter 1 Component design
External Components	Chapter 2 Prefabricated building components
Materials	Section 1.7 Deformations Section 2.4 Natural defects of timber
Materials Technology	Chapter 3 Concrete (thermal movement) Chapter 4 Masonry construction (structural stability)
Structure and Fabric Part 1	Chapter 2 The production of buildings Chapter 3 Structural behaviour
Structure and Fabric Part 2	Section 3.3 Foundation design Chapter 4 Walls and piers (movement control) Chapter 5 Multi-storey structures (movement control) Chapter 6 Floor structures (movement control) Chapter 9 Roof structures (movement control)

Building Research Establishment

BRE Digest 157   *Calcium silicate brickwork*
BRE Digest 163   *Drying out buildings*
BRE Digest 199   *Getting good fit*
BRE Digest 223   *Wall cladding: designing to minimize defects due to inaccuracies and movements*
BRE Digest 227   *Estimation of thermal and moisture movements and stresses* Part 1
BRE Digest 228   *Estimation of thermal and moisture movements and stresses* Part 2
BRE Digest 229   *Estimation of thermal and moisture movements and stresses* Part 3
BRE Digest 234   *Accuracy in setting-out*
BRE Digest 357   *Shrinkage of natural aggregates in concrete*
BRE Digest 361   *Why do buildings crack?*
BRE Digest 389   *Concrete: cracking and corrosion of reinforcement*
BRE Digest 397   *Standardisation in support of European legislation: what does it mean for the UK construction industry?*

Building Regulations

A1   Loading
A2   Ground movement
A3   Disproportionate collapse

5 Strength and stability

CI/SfB (J)

The *strength* of a building refers to its capacity to carry loads without failure of the construction method; *stability* refers to the ability of a building to resist collapse, distortion, localised damage and movement.

5.1 Dead, live and wind loads

A building is required to resist loads imposed by gravity as well as other externally and internally applied forces: loads and forces from roofs, floors and walls must be transferred by load-carrying mechanisms to the supporting ground.

The loads and forces acting on a building are shown in Figs 5.1 to 5.3; they comprise the following:

(a) The weight of all the materials from which it is made (bricks, mortar, concrete, timber, plaster, glass, nails, screws, etc.). These weights are more or less constant during the life of a building and are called *dead loads*; they can be calculated from tables for weights of materials, etc.

(b) The weight of people using a building, and their furniture, goods, storage, etc. These weights are called *live loads* or *imposed loads* and, as they will vary, an average maximum load can be assumed from tables giving values applicable to the particular use of a building.

(c) Various forces may be applied to a building during its life such as those resulting from

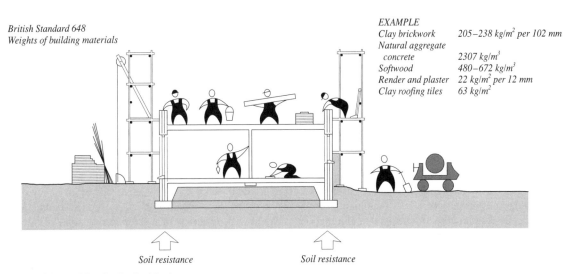

British Standard 648
Weights of building materials

EXAMPLE
Clay brickwork $205-238 \, kg/m^2 \, per \, 102 \, mm$
Natural aggregate
  concrete $2307 \, kg/m^3$
Softwood $480-672 \, kg/m^3$
Render and plaster $22 \, kg/m^2 \, per \, 12 \, mm$
Clay roofing tiles $63 \, kg/m^2$

Soil resistance   *Soil resistance*

Figure 5.1   Building loads: dead loads.

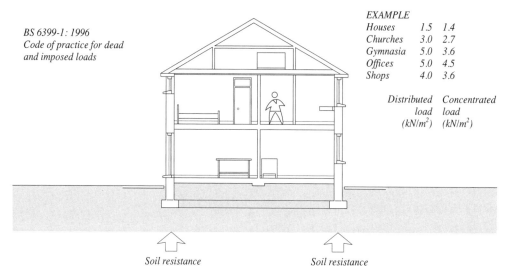

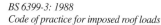

BS 6399-1: 1996
Code of practice for dead
and imposed loads

EXAMPLE

	Distributed load (kN/m²)	Concentrated load (kN/m²)
Houses	*1.5*	*1.4*
Churches	*3.0*	*2.7*
Gymnasia	*5.0*	*3.6*
Offices	*5.0*	*4.5*
Shops	*4.0*	*3.6*

Soil resistance   *Soil resistance*

Figure 5.2   Building loads: imposed loads.

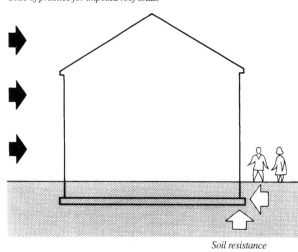

BS 6399-3: 1988
Code of practice for imposed roof loads

Soil resistance

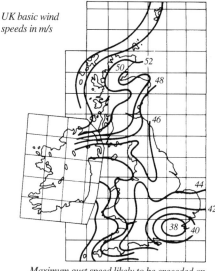

UK basic wind speeds in m/s

Maximum gust speed likely to be exceeded on the average once in ten years at 10 m above open ground level (Meteorological Office)

Figure 5.3   Building loads: wind loads.

wind, physical impact by people, machines, or explosion, and ground movements caused by changes in soil characteristics, earthquakes, mining subsidence, etc. The calculation of these forces is much more problematic and relies on adequate research and experience. Maximum *wind loads* (gusts) for various locations in the country have been tabulated and should be consulted before finalising the structural requirements for a building.

These loads and forces must be resisted by the supporting soil so that a building remains in equilibrium. A building can be visualised, therefore, as being 'squeezed' between the downward applied loading and the upward reactions of the supporting soil.

5.2 Structural organisation

There are four basic methods of structural organisation which can be employed in a building to ensure loads, forces and soil reactions act together in providing equilibrium to a building. The choice of method for a particular building is initially dictated by strength and characteristics of the soil providing support and analysis of the precise nature of all the structural influences. Structural influences are closely related to the function of the building and involve consideration of such factors as whether long or short spans are required, the height of the building, and the weight of materials necessary to fulfil other performance requirements, etc. Often, the juxtaposition of existing buildings provides positive guidelines or maybe limitations on the selection of structural form for a new building.

The structural organisation of a building forms one of the most important aspects which influence appearance and other functions. Since technological solutions are now available which make almost every structural organisation possible, an increasing burden of responsibility is being placed on designers to make rational decisions. The rigorous limitations imposed by simpler construction methods (materials and technology) no longer exist, and it is now feasible to develop a building with volumetric spaces (plan and height) greater than ever before; smaller, to complicated configuration; or to the same size using much less material.

Nevertheless, although perhaps interpreted with greater understanding, certain basic structural principles still remain. Expressed simply, the four basic methods involved in the use of construction methods resist the combined building loads by compression, tension or a combination of the two. These methods are defined by various terms, but here will be called: continuous structures, framed structures, panel structures and membrane structures.

5.3 Continuous structures

These are continuous supporting walls which transfer the combined loads and forces through their construction, mainly by direct *compression* (Fig. 5.4). Materials commonly used for this purpose are stone, horizontal timber logs, brick, block and concrete. In this respect, continuous supporting walls may be the oldest form of structural organisation.

Walls constructed from fairly small units such as bricks, blocks or stones rely on their strength by being laid in horizontal courses so that their vertical joints are staggered or *bonded* across the face of the wall (Fig. 5.5). In this way the compression loads which may initially affect individual, or a series of, bricks, blocks or stones can be successfully distributed through a greater volume of the wall. The units are held together by an adhesive mixture known as *mortar*, thereby completing the structural (and environmental)

enclosure. This mortar also serves the function of taking up any dimensional variations in the bricks or blocks so that they can be laid in more or less horizontal and vertical alignment. Most mortars today usually consist of water-activated binding mediums of cement and lime, and a fine aggregate filler such as sand, in the proportion of one part binder to three parts aggregate.

Laterally applied forces which could create *tension* in the wall are resisted by its preloaded condition. The weight of materials forming the wall, together with any loads they carry from floors or roof, combine in counteracting the tendency for horizontal movement or overturning as a result of horizontally applied forces, e.g. wind. Alternatively, where preloading is insufficient, the action of lateral forces can be resisted by the provision of *buttresses* at predetermined centres to resist the tendency for overturning (see Fig. 5.4). In this way a system of buttresses or piers can be used to provide stability to a long length of wall. Walls which are serrated or curved in plan will also be stronger than straight walls because they are more able to resist laterally applied forces.

Additional stability can also be provided by the floor(s) and roof of a building, provided there is adequate connection at the junction between horizontal and vertical elements. A wall which is laterally braced in this manner has the advantage of using considerably less material to support the same load as a thick unbraced wall (Fig. 5.6); see also section 5.7.

When it is required to provide openings (e.g. doors and windows) in a building using continuously supporting walls, it is necessary to use a beam or *lintel* made from a material or combination of materials capable of resisting both *compression* and *tension* forces resulting from the loads above. These loads must be transferred to the sides of the opening or *jambs*. Stone can be used for a lintel, but it provides only a limited resistance to tensile forces relative to its depth and therefore permits only small openings. Timber, steel, a combination of steel bars and concrete (reinforced concrete), or steel angles and bricks, can be suitable materials for lintel construction, although care must be taken to ensure their durability corresponds with the intended life of the building (see Chapter 3). Arches formed over openings use materials in direct compression only, in a similar manner to the wall itself. Therefore, arched openings are often considered by design purists to be more compatible with the aesthetic of this form of wall construction (Fig. 5.7).

5.4 Framed structures

These consist of a framework of timber, steel or reinforced concrete consisting of a regular system of

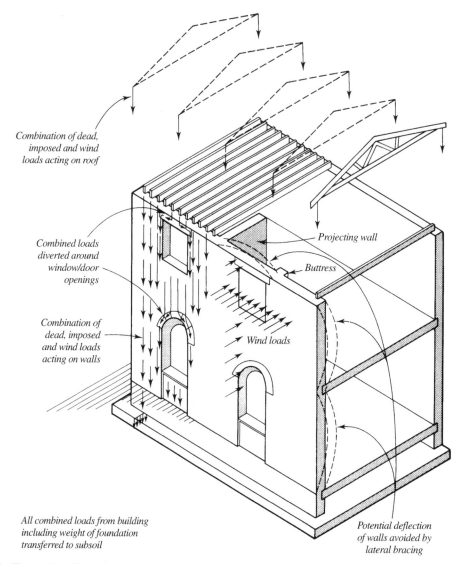

Combination of dead,
imposed and wind
loads acting on roof

Combined loads
diverted around
window/door
openings

Combination of
dead, imposed
and wind loads
acting on walls

Projecting wall

Buttress

Wind loads

All combined loads from building
including weight of foundation
transferred to subsoil

Potential deflection
of walls avoided by
lateral bracing

Figure 5.4   The transfer of loads in continuous structures.

horizontal beams and vertical columns (Fig. 5.8). The beams resist both compressive and tensile forces and transmit loads from the floors, roof and walls to the columns. The columns are required to resist mainly compressive forces; they transfer the beam loads (and the self-weight of beam and column) to the foundation and finally to the supporting soil. This obviously results in more concentrated loads being supported by the soil than for a similar weight of building using continuous supporting walls, unless special forms of foundations are used. The infill *panels* between the framework used to provide the external wall can be constructed of any suitable durable material which fulfils performance

requirements satisfactorily. If the wall material is positioned away from the framework so as to be externally or internally free of the columns and beams, it is known as *cladding*. Both panel and cladding walls are generally non-load-bearing, although in practice they must carry their own weight (unless suspended from above), resist the wind forces acting on their external face, perhaps provide support for internal fixtures, shelves, etc., and resist localised impact forces. However, depending upon the form of construction adopted, the resulting loads are usually transferred back to the supporting columns and beams by their fixing method. A structural framework and panels, or cladding walling, is an

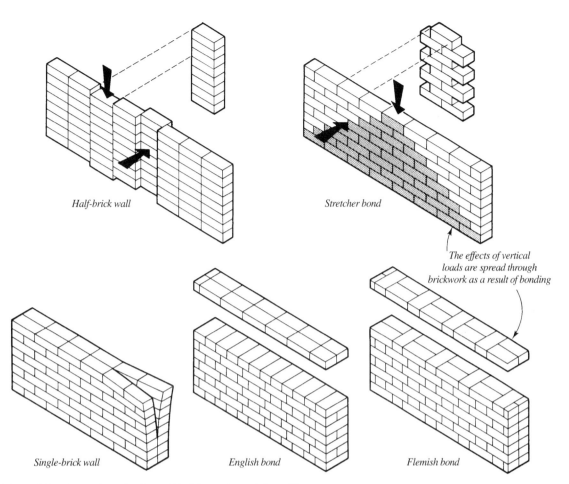

Figure 5.5 Effects of bonding in small building units used for walling.

Half-brick wall

Stretcher bond

The effects of vertical loads are spread through brickwork as a result of bonding

Single-brick wall

English bond

Flemish bond

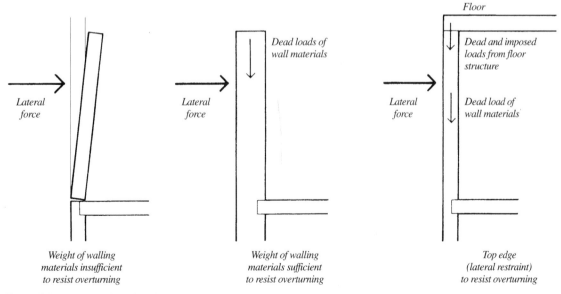

Figure 5.6 Lateral bracing of walls.

Lateral force

Lateral force

Dead loads of wall materials

Floor

Lateral force

Dead and imposed loads from floor structure

Dead load of wall materials

Weight of walling materials insufficient to resist overturning

Weight of walling materials sufficient to resist overturning

Top edge (lateral restraint) to resist overturning

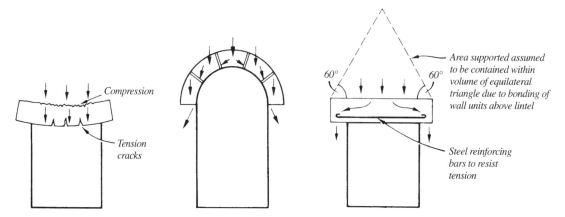

Figure 5.7   Reinforced concrete lintels and block arches.

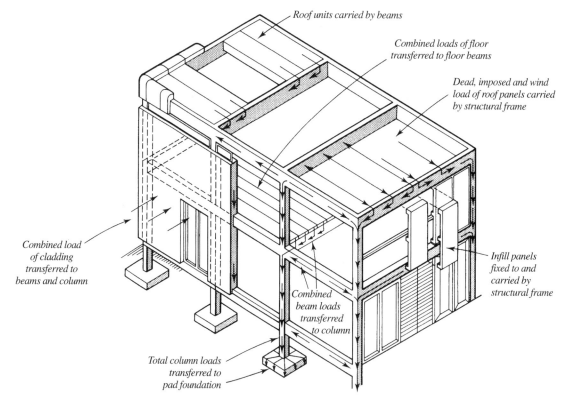

Figure 5.8   Transfer of loads in framed structures.

example of *composite construction*. Here the use of different materials to provide independent functions requires careful constructional detailing and skilled work to ensure an entirely successful enclosure.

The external appearance of a framed structure will vary according to the location of the beams and columns relative to the external wall. Figure 5.9 indicates the basic permutations. When it is not located within an enclosing panel wall, additional precautions may be necessary to protect the structural frame against possible detrimental effects resulting from an outbreak of fire (see section 9.3), and also from the effects of weathering when the frame is external to the wall. Although this may present no special problems for reinforced concrete – other than a slight increase in cross-sectional area – the use of timber and exposed steel frames requires special consideration. Table 5.1 provides a brief checklist of the structural materials used for a framed building.

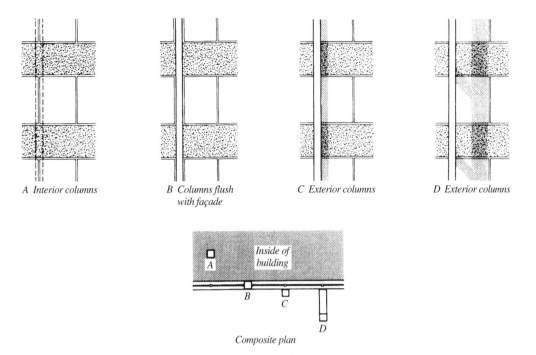

A Interior columns   *B Columns flush
with façade*   *C Exterior columns*   *D Exterior columns*

Composite plan

Figure 5.9   Position of structural frame (column) and its effect on a building's appearance. (Adapted from material in *Multi-storey Buildings in Steel* by Hart, Henn and Sontag)

Table 5.1   Comparison between timber, steel and reinforced concrete as structural framing

Criteria	Material		
	Timber	Steel	Reinforced concrete
Availability	Mostly imported as dimensionally coordinated sections	Most iron ore imported	UK supplies of aggregates and cement Steel is mostly imported
Conversion	Factory converted sections and components	Steelworks produce bulk sections and components manufactured in factory controlled conditions	Factories and steelworks convert into basic use materials (aggregates, cement, bar reinforcement)
	Preformed components manufactured to fine tolerances under factory conditions	Components manufactured to fine tolerances under factory conditions	Precast concrete components manufactured to fine tolerances under factory conditions
Site operations	Erected by skilled and semi-skilled operatives	Site erected by skilled operatives	*In situ* structures erected by semi-skilled and unskilled operatives
	Components relatively lightweight and to fine tolerances	Very accurate within small tolerances	Various shapes possible dependent upon potential of formwork material (up to 40% of total cost)
		Heavy sections used need crane	May need crane although *in situ* materials can be pumped to higher levels
			Precast components erected by skilled operatives
Site progress	Components relatively lightweight and can be quickly erected	Heavy components but can be erected quickly	Slow progress when necessary to wait for hardening before commencing next sequence or trade
	Foundation work less	Mistakes difficult to correct	
	Progress dependent upon weather conditions unless protection provided	Progress dependent upon weather conditions unless protection provided	Multi-storey structures can be formed near ground level and raised into position

Table 5.1 *(cont'd)*

Criteria	Material		
	Timber	Steel	Reinforced concrete
Site progress	Once erected floor/roof components can be placed and building sealed against weather	Once erected floor/roof components can be placed and building sealed against weather	Slow sealing against weather; design can help by using repeat formwork
Fire protection	Not considered to be good fire risk although designs can take account of 'sacrificial' sections to provide insulation (see text)	Not considered to be good fire risk and generally must be insulated for all but very small structures	Steel reinforcement insulated by concrete cover
		Possible to expose sections providing design permits shielding isolation or cooling (see text)	High inherent degree of fire protection and nil spread of flame
	Poor spread of flame characteristics, but can be improved by chemical treatment	Spread of flame characteristic depends on type of insulation provided	Fire resistance periods can be improved by selecting aggregate to reduce spalling
	Damaged timber irreparable	Damaged steel irreparable	Damaged reinforced concrete can be repaired
Adaptability	Easily adapted during construction	Difficult to adapt on site owing to precut lengths or difficulty in cutting	Can be adapted
	Extensions easy to provide	Extension can be achieved providing access to original sections available through fire protection	Difficult to provide extension of existing structure (steel reinforcement must be exposed) and separate new structure required
Maintenance	Finished with preservative stain, varnish or paint system	Needs regular maintenance if not encased or weathering steel	Self-finish quality depends on site skills and formwork quality
	Intumescent paint can be decorative and provides fire resistance	Intumescent paint can be decorative and provides fire protection	
	Designs must provide protection against insect and fungal attack		

(See page 18 regarding the use of *in situ* concrete framework with clay brick infill panel wall construction.)

5.5 Panel structures

These include preformed load-bearing panel construction for the walls, floors and roof which carry and transfer loads without the use of columns and, sometimes, beams (Fig. 5.10). This is similar to continuous supporting wall construction, but each panel is designed to resist its own imposed loads, as well as other performance requirements. They are generally more slender than most other forms of construction and are dimensionally coordinated so as to be interchangeable within their specific functional requirement. The main structural material of a panel is generally of steel or timber, and this can be faced with a suitable material (plywood, flat or profiled metal sheet) and incorporate thermal insulation to form a *sandwich construction*. (Certain forms of sandwich construction are also used for non-load-bearing panel cladding for a framed building.) Panels can incorporate window and door openings. The combined loads which this form of construction collects can be transferred to the supporting soil by continuous distribution or by concentrating them in a similar manner to that adopted for framed buildings.

5.6 Membrane structures

Thin non-structural membranes forming walls and roof (often combined in one place) are supported by tension and/or compression members (Figs 5.11 and 5.12). A typical example of this is a tent where the walls and roof are formed of canvas and the main structural support of timber or steel. Most permanent structures can be formed by columns, *compression members*, from which cables are suspended, *tension members*, which support a plastic membrane. Alternatively, a reinforced plastic or canvas membrane can be supported by air, as in inflatable structures. In this case the membrane is in tension because of the compression forces exerted by the air under pressure. Both these examples are suitable for a building where certain of the performance requirements discussed in Part A of this book do not form an essential part of a proposed building enclosure.

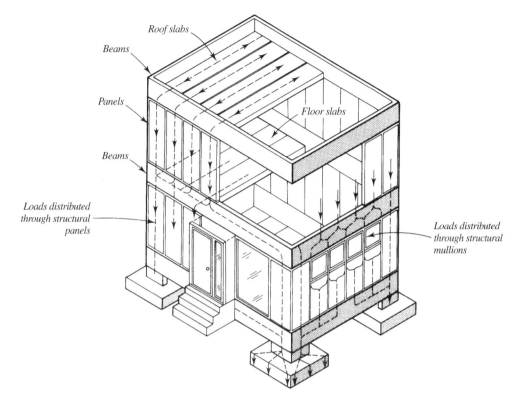

Figure 5.10 Transfer of loads in panel structures.

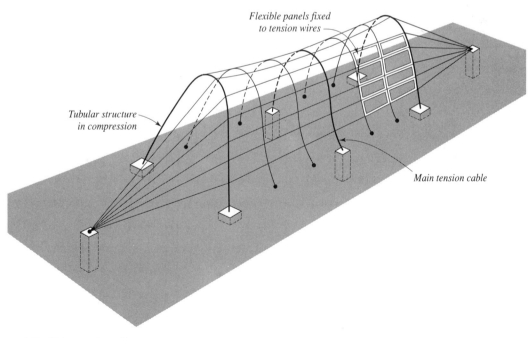

Figure 5.11 Using tension cables.

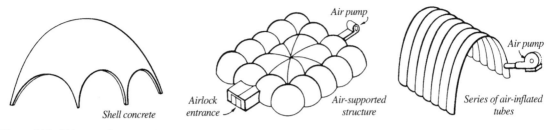

Figure 5.12   Using membrane structures.

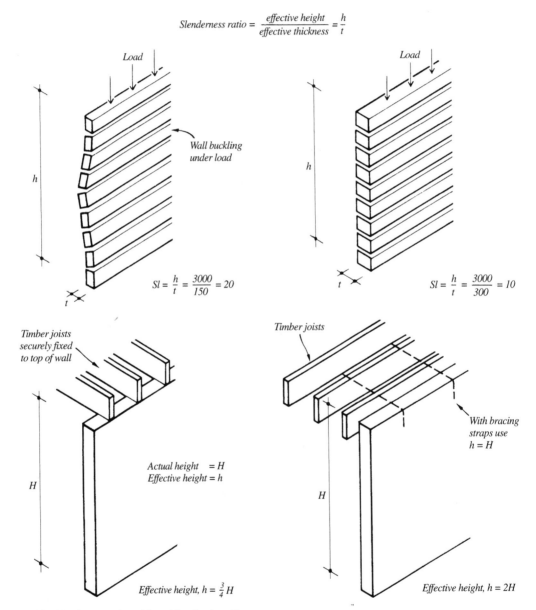

Figure 5.13   Slenderness ratio and lateral bracing in walls.

5.7 Slenderness ratio

Whenever the structural organisation of a building involves the resistance of vertical loads by a wall or column in compression, adequate *thickness* of sufficiently *strong* material must be employed in their construction to avoid *crushing*. A short wall or column can ultimately fail by crushing. But as height increases, ultimate failure is more likely to occur under decreasing loads by *buckling*. This form of failure results from lack of stiffness in a wall or column which causes bending to occur because, in practice, it is impossible to ensure that vertical loads act through the vertical centreline of their support. Very tall, thin walls or columns will buckle before crushing, short squat walls crush before buckling, and walls of intermediate proportions may fail by either method.

Obviously, the greater the height of a wall or column and the tendency towards buckling, the more critical becomes the relationship between *thickness and height*. This relationship is known as the *slenderness ratio*; as this ratio increases, so the load-carrying capacity of the wall or column decreases (Fig. 5.13). Because the stiffness of a wall or column can be increased by lateral bracing as described in section 5.3, for calculation purposes the dimensions used for the *effective* height and thickness can vary from the *actual* height and thickness in order to obtain a realistic slenderness ratio. The amount of variation depends on the degree and effectiveness of connection provided between the horizontal and vertical components. For example, where a wall is loaded by a floor construction which provides continuous lateral support (reinforced concrete slab or adequately connected timber joists), the height of the wall can be taken for calculation purposes to have *effective height* (h) equivalent to three-quarters of the *actual height* (H). This will give a slenderness ratio which either permits theoretically slightly less strong materials to be used than if no concession had been given, or permits the wall to be thinner and occupy less plan area. However, if the floor gives no lateral support whatsoever, the rules of calculation will double the actual height of the wall and vastly increase the slenderness ratio. A further *correction factor* may be applied to the slenderness ratio used for calculation purposes when applied loads are resolved eccentrically to the centre of the wall (Fig. 5.14). The design of columns is also subject to similar requirements regarding slenderness ratios and correction factors for eccentric loadings (Fig. 5.15).

5.8 Diagonal bracing

For frame buildings, the provision of effective lateral bracing will vary according to how well the materials employed will permit a rigid joint to be created between horizontal and vertical components. Very rigid joints can easily be created between beams and columns made of reinforced concrete, but it is more difficult for beams and columns made of steel and very hard when they are made of timber. When jointing techniques cannot provide sufficient lateral restraint, the structural frame can be made more rigid by inserting *diagonal bracing* in various locations around a building, or by using shear panels (Fig. 5.16).

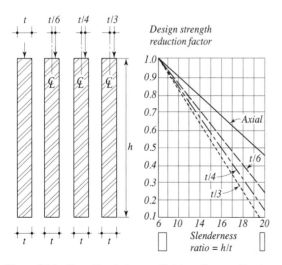

Figure 5.14 Correction factors applied to walls according to the eccentricity of loads. (Based on data provided by the Brick Development Association)

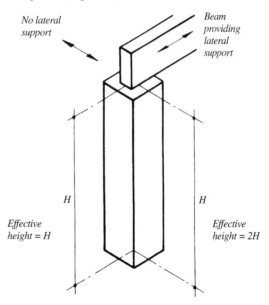

Figure 5.15 Slenderness ratio and lateral bracing in columns. (Based on data provided by the Brick Development Association)

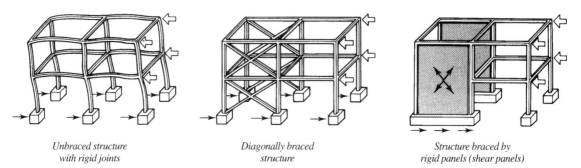

*Unbraced structure
with rigid joints*   *Diagonally braced
structure*   *Structure braced by
rigid panels (shear panels)*

Figure 5.16   Use of diagonal bracing in frame structures. (Adapted from material in *Multi-storey Buildings in Steel* by Hart, Henn and Sontag)

Further specific reading

Mitchell's Building Series

Internal Components	Section 2.3.3 Partition strength of demountable partitions Section 4.2.2 Loading on raised floors Section 6.2.6 Strength and stability of doors
External Components	Section 3.2.2 Strength of external glazing Section 7.2.6 Strength and security of external doors Chapter 8 Roofings
Materials	Section 1.2 Strength See also strength subsections in each chapter
Materials Technology	Strength and stability properties of various building components are considered in each chapter
Structure and Fabric Part 1	Chapter 1 The nature of buildings and building Chapter 3 Structural behaviour Chapters 4–10 Functional requirements sections or subsections
Structure and Fabric Part 2	Chapter 4 Walls and piers Chapter 5 Multi-storey structures Chapter 6 Floor structures Chapter 9 Roof structures

Building Research Establishment

BRE Digest 12   *Structural design in architecture*
BRE Digest 246   *Strength of brickwork and blockwork walls: design for vertical load*
BRE Digest 281   *Safety of large masonry walls*
BRE Digest 284   *Wind loads on canopy roofs*
BRE Digest 346   *The assessment of wind loads* Parts 1–8
BRE Digest 362   *Building mortar*
BRE Digest 390   *Wind around tall buildings*
BRE Digest 406   *Wind actions on buildings and structures*
BRE Digest 439   *Roof loads due to local drifting of snow*
BRE Digest 441   *Clay bricks and clay brick masonry* (2 Parts)

Building Regulations

A1   Loading

6 Weather exclusion

CI/SfB (H1)

Weather exclusion is concerned with methods of ensuring that wind and water (rain and snow) do not adversely affect the fabric of a building or its internal environment.

6.1 Wind and water penetration

Wind can cause direct physical damage by collapse or removal of parts of a building. It can cause dampness by driving moisture into or through a building fabric, and also excessive heat losses from the interior of a building by uncontrolled air changes.

Water penetration can produce rapid deterioration, as discussed in Chapter 2, and cause the fabric of a building to become moist enough to support life, including bacteria, moulds, mildew, other fungi, plants and insects. Saturated materials also permit the quick transferral of heat (water is a good conductor) and this, together with the other factors mentioned, will cause an uncomfortable, unhealthy and uneconomical building.

The sources of water likely to penetrate a building not only include those from rain and snow, but also those from moisture contained in soil or other material in immediate contact with the building fabric. For water to penetrate, there must be openings or passages in the building fabric through which it can pass, and a force to move it through these openings or passages. Without these two factors, the building fabric would remain in a watertight condition. Most buildings, however, have window and door openings, are made from lapped or jointed parts, or from porous materials ready to absorb moisture; wind currents and eddies are also normally present.

Condensation may also create moisture problems, considered in Chapter 8. Damage by flooding is beyond the scope of this book.

6.2 Exposure zones

Construction methods of earlier periods were generally capable of permitting a certain amount of wind and water to penetrate through to the interior of a building. Shapes for buildings were devised for particular climatic exposures which best provided an initial defence, and the constructional detailing endeavoured to provide a final barrier.

By way of progress, modern construction methods are expected to give almost total exclusion against wind and water penetration. With ever-changing fashions for building shapes – sometimes borrowed from areas vastly different in climatic influences – care must be taken to ensure that constructional methods are suitable for the exposure conditions dictated by the *specific* location and disposition of a building. That is to say, before considering form and construction method research must be carried out to reveal the degree to which a proposed building will be exposed to driving rain. In the United Kingdom, initial assessment can be obtained by reference to the *driving rain index* (DRI) for the particular location as published on maps by the Building Research Establishment (Fig. 6.1). Values are obtained by taking the mean annual wind speed in metres per second (m/s) and multiplying by the mean annual rainfall in millimetres. The product is divided by one thousand and the result is used to produce contour lines linking areas of similar annual driving rain index in m^2/s throughout the country.

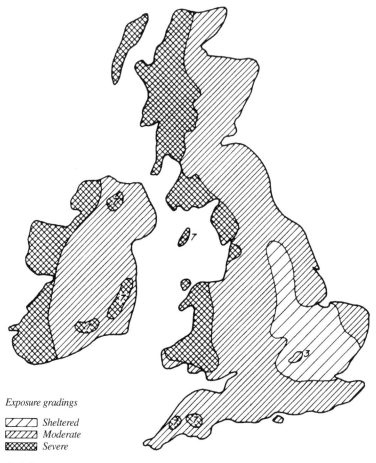

Figure 6.1 Driving rain index.

- *Sheltered exposure zone* refers to districts where the DRI is 3 or less.
- *Moderate exposure zone* refers to districts where the DRI is between 3 and 7.
- *Severe exposure zone* refers to districts where the DRI is 7 or more.

The value for a particular location within 8 km of the sea, or a large estuary, must be modified to the next zone above (sheltered to moderate, and moderate to severe) to take account of unusual exposure conditions. Furthermore, modifications may also be necessary to allow for local topography, special features which shelter the site or make it more exposed, roughness of terrain, height of proposed building and altitudes of the site above sea level. The proportion of driving rain from various directions within one particular location can be obtained by reference to *driving rain rose diagrams* (Fig. 6.2).

For greater accuracy, reference should be made to BS 8104 : 1992 *Code of practice for assessing exposure of walls to wind and driving rain*. This provides an alternative rose analysis for the extent of driving rain in various locations, to permit an exposure expectation for each face of a building. It is expressed in litres/m^2 per spell, where spell is the period that wind-driven rain occurs on a vertical face of a building. Figures for a wall annual index expressed in litres/m^2 per year can be derived from this data, as can local spell and local annual indices. This does not mean that a designer need adjust construction detailing for different elevations, but it will reveal which aspect of a building is most vulnerable to water penetration. This guidance is also helpful in assessing areas most compatible to lichens, mosses and other growths that can have a deteriorating effect on structure. The BS code provides a series of maps of the UK with superimposed rose diagrams allocating numerical values (Fig. 6.3) to specific areas.

One of the most important lessons to be learnt from driving rain indexes and roses is that design and construction details, of necessity, may vary from one

Scale of driving
rain index

Figure 6.2 Driving rain rose diagram.

Figure 6.3 Spell and annual rose values – source BS 8104, e.g. building with a south elevation = 8.

exposure zone to another; details suitable in a sheltered exposure zone would probably leak if simply transferred to a severe exposure zone without modification. For instance, in the severest of exposure structures should not have walls constructed with full cavity fill insulation. A minimum 50 mm air gap (see Fig. 6.5) is necessary to prevent dampness bridging the insulation.

6.3 Macro- and microclimates

However, there can exist a danger in using only meteorological climatic data of this nature for the final selection of appropriate material and constructional detailing. This data reveals the general climate, or *macroclimate*, liable to affect a building in a particular location. There is also a *microclimate* surrounding the immediate outer surface of a building, i.e. not more than 1 m from the surface of a building. This is created by specific environmental conditions arising from the precise form, location, juxtapositions and surface

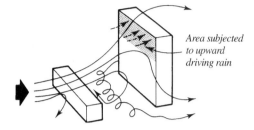

Figure 6.4 Alteration of general climate conditions as a result of building disposition and shape.

geometry of a building. A designer is expected to know when design ideas are liable to cause the microclimate to vary significantly from the macroclimate and make adjustments in materials and/or constructional detailing accordingly. Where there is no past experience, it may be necessary to make models of a building and its surroundings, test them under simulated environmental conditions and record the precise effects. For example, detailed analysis has revealed that tall buildings can receive more rainwater on their walls than on their roofs, especially on elevations facing the wind. Under these circumstances, rain is often driven *vertically up* the face of the building, making it necessary to use constructional details different from those considered suitable for lower buildings in the same exposure zone (Fig. 6.4).

6.4 Movement of water

Besides the numerous construction methods, there is a very wide range of materials and combinations of materials from which to choose when designing a building. Providing they are carefully matched to exposure conditions, all options can function with equal efficiency in controlling the *movement of water* from the exterior to the interior of a building. The precise way this is achieved will vary; it will depend upon the properties of the materials employed and the manner in which they are arranged to form the weather barrier. Figure 6.5 illustrates the three basic arrangements for the external fabric of a building where:

- walls initially act as a *permeable barrier* which permits subsequent evaporation of absorbed water;
- walls and roofs act as *semipermeable barriers* which permit a certain amount of water to penetrate until reaching a final barrier;
- walls and roofs act as an *impermeable barrier* which diverts water on contact.

When rain falls or is driven on *porous* building materials, such as most brick types, some stones, or blocks, it is absorbed then subsequently removed by natural evaporation. This process occurs when water adheres to the pores of the material. And if the adhesive force between the water molecules and the wall material is greater than the cohesive force between the molecules themselves, the water is drawn in by *capillary action*. A strong wind increases the rapidity of absorption, and only evaporation resulting from changes in the climatic conditions (rain ceases, temperature rises, air currents become warmer) will prevent the moisture penetrating through the thickness of the material. Earlier construction forms ensured that this thickness was sufficient to prevent water penetration by capillary action. Current economic trends and the need for energy conservation (saturated material loses about ten times more heat through it than when dry) have now firmly established construction methods using materials of thinner cross-sections. These are employed to provide an external initial weather check which is combined with materials

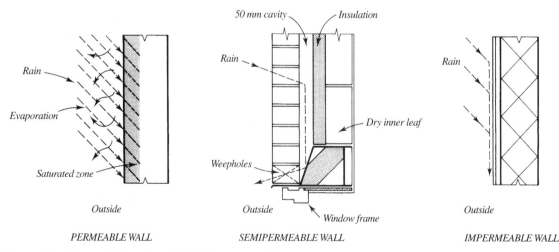

Figure 6.5 Movement of water in walls of different construction.

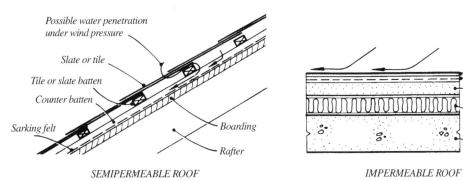

Figure 6.6 Movement of water in roofs of different construction.

used to provide *internal* thermal insulation. The external and internal components are separated by a water barrier (or air cavity) to interrupt the continuous flow of water from outside to inside. In this way, potentially wet areas are isolated from those which must remain permanently dry.

Constructional detailing must attempt to reduce the movement of water as much as possible. It is particularly important to ensure that the water barrier is incorporated in such a way that moisture is not trapped and kept in a position for a period of time liable to cause damage.

For similar reasons, when rain falls or is driven on building materials theoretically assumed entirely impervious, e.g. dense concrete, glass, metal, bituminous products or plastics, precautions must be taken to ensure quick and efficient run-off. The quantity of water must never be underestimated; on a glass wall of a building it can be as much as 5 litres per 10 m² of façade.

A typical flat roof construction to provide a weather-resisting barrier would consist of sealed lengths of multi-layer bituminous felt or a continuous homogeneous layer of asphalt (Fig. 6.6). A comparable pitch roof construction to resist the penetration of water under the influence of gravity would incorporate an outer surface finish of lapped tiles or slates backed by an impervious water barrier. Further comments about this form of construction are included in Chapter 17.

6.5 Joints

The need for joints arises because of the necessity to link, lap or bond materials together when providing the continuous and efficient weather enclosure for a building. Unless the many interrelated factors which influence their position and type are very carefully considered, they can form the weak link in the enclosure.

The first rule is to ensure that as much water as possible is kept away from this vulnerable point where two or more materials are brought together, each perhaps

having different characteristics and connected in some way with yet other types of materials. The designs for roofs and walls can provide shelter to their joints, or channel free-flowing water in predetermined directions to permit either collection or discharge to less damaging areas. The effects of wind-driven rainwater must always be taken into account in this respect, and boldly profiled upstands and overhangs at joint positions are desirable when it is difficult to form a continuous 'membrane type' seal between two building components.

The actual method of forming a joint will initially depend upon the physical and chemical properties of the materials involved. The comments made in Chapter 5 regarding *movements* and appropriate sizes are particularly relevant. The joint can be expected to behave in a similar manner to the surrounding surfaces by stopping water penetration at the outermost places, or by allowing water to be collected from its recesses and returned to the outside. Within these two extremes, there is a vast range of jointing possibilities (Fig. 6.7).

Some of the main joints necessary to provide a weather-resisting enclosure for a simple design of a timber-framed window in a brick/block cavity wall are illustrated in Fig. 6.8. These include the use of mortar

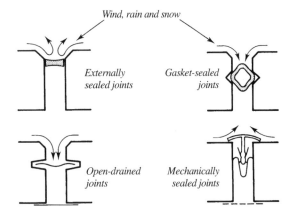

Figure 6.7 Types of jointing to resist the penetration of water.

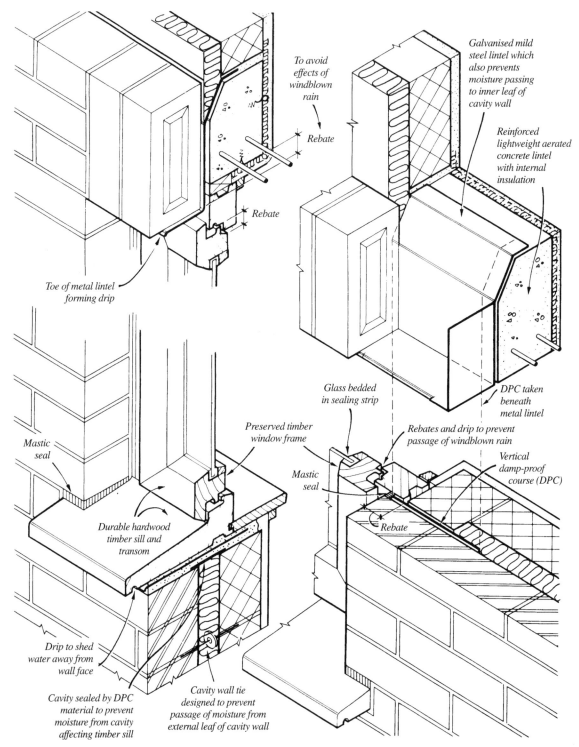

To avoid
effects of
windblown
rain

Rebate

Rebate

Galvanised mild
steel lintel which
also prevents
moisture passing
to inner leaf of
cavity wall

Reinforced
lightweight aerated
concrete lintel
with internal
insulation

Toe of metal lintel
forming drip

Mastic
seal

Glass bedded
in sealing strip

Preserved timber
window frame

Mastic
seal

Durable hardwood
timber sill and
transom

DPC taken
beneath
metal lintel

Rebates and drip to prevent
passage of windblown rain

Vertical
damp-proof
course (DPC)

Rebate

Drip to shed
water away from
wall face

Cavity sealed by DPC
material to prevent
moisture from cavity
affecting timber sill

Cavity wall tie
designed to prevent
passage of moisture from
external leaf of cavity wall

Figure 6.8 Construction method used to resist water and wind penetration through a window opening in a fully insulated brick/block masonry wall. *Note: this diagram is representative of many existing constructions, but the lintel and reveal details no longer satisfy new-build thermal insulation requirements for housing in the UK, see section 17.6.5.*

for the brickwork outer leaf of the wall, an impervious water barrier membrane (DPC) between brickwork and window frame, a seal between window frame and glass, and draught proofing between window sash and frame. The profiles between the fixed and opening parts of the window frame are also specially designed to reduce the movement of wind-borne water to the interior of the building. Draught-excluding devices can also be fitted in this gap to eliminate the flow of air and the possibility of heat loss. The relationship between the type of brick and the type of mortar used in the outer leaf of brickwork is also important, as indicated by Fig. 6.9.

One of the most important aspects of joints in a building involves their affect on appearance. The particular type of brick bond (stretcher, Flemish, Dutch, Quetta, etc.) and the width, profile and colour of mortar joint can have as much impact on the appearance of brickwork as the colour and shape of the bricks themselves. Similarly, the precise location and profile of the joints in preformed panel and *in situ* reinforced concrete walls will assist in determining not only the overall scale and

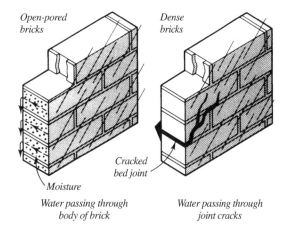

Figure 6.9 How water can penetrate brickwork.

proportion of a building, but also the pattern and rhythm of features on the façade. The surface texture created by the need to channel water away from widely-spaced joints of preformed wall units also helps in creating the particular character and expression of a building.

Further specific reading

Mitchell's Building Series

Environment and Services	Chapter 1 Environmental factors
	Chapter 2 Moisture
	Chapter 3 Ventilation and air quality
External Components	Section 3.2.3 Techniques for external glazing
	Section 4.2.4 Weather resistance of windows
	Section 7.2.3 Weather protection of external doors
	Section 8.2.1 Weather exclusion of roofings
Materials	Section 4.4 Properties of stone
	Chapter 6 Bricks and blocks (properties)
	Section 8.2 Dense concrete (properties)
Materials Technology	Chapter 1 Developing an attitude towards materials
Structure and Fabric Part 1	Chapter 5 Walls and piers
	Chapter 7 Roof structures
	Chapter 8 Floor structures
	Chapter 9 Fireplaces, flues and chimneys

Building Research Establishment

BRE Digest 54   *Damp-proofing solid floors*
BRE Digest 217   *Wall cladding defects and their diagnosis*
BRE Digest 304   *Preventing decay in external joinery*
BRE Digest 312   *Flat roof design: the technical options*
BRE Digest 346   *The assessment of wind loads* Parts 1–8
BRE Digest 350   *Climate and site development* Parts 1–3

BRE Digest 380   *Damp-proof courses*
BRE Digest 390   *Wind around tall buildings*
BRE Digest 406   *Wind action on buildings and structures*
BRE Digest 419   *Flat roof design: bituminous roofing membranes*
BRE Digest 428   *Protecting buildings against lightning*
BRE Digest 436   *Wind loading* Parts 1–3
BRE Report 262   *Thermal insulation: avoiding the risks*, 2nd edition

Building Regulations

A1   Loading
C2   Resistance to moisture
Regulation 7   Materials and workmanship

7 Sound control

CI/SfB (P)

The control of sound in a building must be considered from two aspects:

- The elimination or reduction of *unwanted sound* generated by sources within or outside a building (sound attenuation).
- The creation of good *listening conditions* within a building where speech and music need to be clear, unmarred by sound reverberation and echoes.

7.1 Unwanted sound

Figure 7.1 indicates the main ways by which sound can be transmitted into and through a building. This involves movement through air or other elastic media formed from solids, liquids or gases, *airborne sound*, or movement through a solid structure resulting from

an impact force, *impact sound*. Both are transmitted by direct paths from source to recipient or by indirect paths along adjoining elements. Transmission by indirect paths is known as *flanking transmission*. High levels of unwanted sound, or noise, can lead to a breakdown in people's mental health or even damage their hearing. Unwanted lower levels of sound are a nuisance and become a source of constant irritation, causing a loss of concentration.

7.2 Noise outside a building

Apart from industrial operations, *external noise* nuisance is most often caused by motor traffic and it is necessary for the designer to be familiar with the *noise climate* liable to affect the performance of a building. This

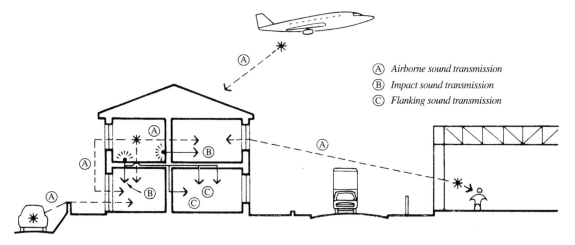

(A) Airborne sound transmission
(B) Impact sound transmission
(C) Flanking sound transmission

Figure 7.1 Unwanted sound.

is the range of sound levels achieved for 80 per cent of the time – the remaining 20 per cent being divided equally between sound levels occurring above and below the main range. The upper limit of the noise climate is called the ten per cent level ($L10$) and has become the unit used for specifying extreme exposure conditions to traffic noise which, when existing for 18 hours per day, permits government compensation or grants to be paid for control of sound levels in houses adjoining motorways, etc. Similar compensation is available for exposure to aircraft noise: this is assessed in terms of *noise and number index* (NNI) and takes into account the number of movements during the day and the loudness of each. Local authorities have powers and duties to control noise nuisance under the Control of Pollution Act 1974, which provide means of creating *noise abatement zones* for the long-term control of noise from fixed sources such as may exist in areas of mixed residential and industrial development. This act also provides the power to control noise on construction and demolition sites which, although usually short-lived, may inflict severe discomfort on normally peaceful neighbourhoods. Nevertheless, prevention is better than cure, and various documents exist which attempt to control the initial output of noise such as BS 5228 *Noise control on construction and open sites*; the Road Traffic Act 1988, Part II: *Construction and use of vehicles and equipment*; and the Health and Safety at Work, Etc., Act 1974, incorporating the Noise at Work Regulations 1989.

7.3 Noise inside a building

As indicated by Figure 7.1, the movement of either airborne or impact noises *inside* a building is a complex process involving transmission through walls and floors by direct and/or flanking paths. The relative weight and rigidity of a building fabric and the nature of construction affect the amount of transmission. Current building legislation attempts to define minimum standards for domestic construction methods which provide an acceptable degree of control. However, modern society requires the increasing use of sound-producing equipment for home entertainment devices and the frequent involvement of noisy household appliances. These requirements often conflict with the simultaneous trend towards lightweight materials and less homogeneous methods of assembly used for a building. Joints in constructions are particularly liable to cause weak links when considering sound control as a problem. The methods used to control the movement of sound within a building are similar to those adopted to control sound from external sources, although external sound is also reduced by weather exclusion measures, i.e. components having greater thicknesses and weights.

7.4 Frequency, intensity and loudness

When analysing the control of noise, it is useful to clarify the two basic factors, frequency and intensity, which initially influence the kind of sound received by the human ear.

Sound is normally created in the air when a surface is vibrated and sets up waves of alternating compression and rarefaction. The distance between adjacent centres of compression is known as the wavelength of the sound, which for human hearing varies from about 20 mm to 15 m. The number of complete movements or cycles from side to side made by the particles in the air (or any other elastic medium) during the passage of sound waves determines the *frequency* of the sound; this is usually quoted as the complete number of cycles made in a second, the number of hertz (Hz). The greater the number of cycles per second or hertz caused by the vibrations, the higher the pitch of the sound. People are most affected by frequencies between 500 Hz and 6000 Hz (Fig. 7.2). Similar waves can also be produced by air turbulence during explosive expansion of air or a combination of vibration and explosive expansion.

The *intensity* of sound is a measure of the acoustic energy used in its transmission through the air. It is calculated from:

$$I = P \div 4\pi r^2$$

where: I = Intensity at distance (W/m^2)
P = Power at sound source (W)
r = Distance from source (m)

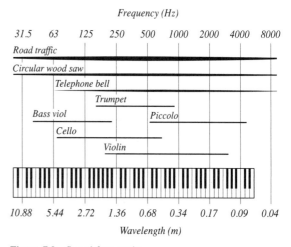

Figure 7.2 Sound frequencies.

Sound intensity level (SIL) is expressed in *decibels* (dB) conforming to a logarithmic scale (regular proportionate increments rather than equal increments) which closely approximates the way sound is heard. It also gives a manageable scale for a wide range of sounds (Table 7.1). SIL can be calculated from:

$$10 \log(I \div Io)$$

where: log = logarithm to the base 10
I = Intensity at distance (W/m^2)
Io = Intensity at the threshold of hearing
(taken as: $1 \times 10^{-12} \ W/m^2$)

The actual decibels produced by a particular external airborne sound will be reduced according to the amount and characteristics of the intervening space between source and recipient. There is a theoretical reduction of 6 dB each time the distance from the source is doubled.

In practice this amount may be modified by such factors as whether the source is a single point, a continuous line or an area origin; the source height; the amount reflected during its transmission; the effectiveness of screening devices provided by trees, other buildings, embankments, etc.; and meteorological conditions. In addition, the ability to hear a given sound within a building will depend to a considerable extent on the background noise generally existing within the interior. General room sounds created by radio, television and conversation often make traffic noises far less noticeable. Furthermore, the number of decibels created by a particular source does not necessarily provide an indication of how *loud* it sounds because the human ear is more sensitive to high frequency sounds than to low. Therefore, the *subjective* loudness of a noise is measured by a weighted scale known as dBA, which gives an overall intensity with a bias towards high frequency sounds. Other frequency weighted scales are also available (dBB, dBC).

Table 7.1 Typical sound intensity levels

Source of noise	Sound intensity (dB)
Four-engine jet aircraft at 100 m	120
Riveting of steel plate at 10 m	105
Pneumatic drill at 10 m	90
Circular wood saw at 10 m	80
Heavy road traffic at 10 m	75
Telephone bell at 10 m	65
Average male speech at 10 m	50
Whisper at 10 m	25
Threshold of hearing, 1000 Hz	0

7.5 Defensive measures

The first and obvious defence against the intrusion of unwanted airborne sound lies in placing as much distance as possible between the source and the recipient. For example, activities accommodated by the function of a building requiring a quiet environment (sleeping, studying, lecturing, etc.) can be placed remote from the external noisy distraction of motorways, sports stadiums and industrial applications. Further reduction can then be provided by the fabric of a building, although the precise nature of material and construction technique most suitable to reduce intrusive sound can be fairly complicated to assess. Nevertheless, data is available which specifies desirable sound levels within a building according to functions (Table 7.2), and construction methods can be selected which provide the necessary sound reduction to achieve these goals relative to the anticipated external noise environment. Similar considerations apply to the reduction of unwanted sounds which may occur inside a building.

Table 7.2 Acceptable intrusive noise levels in respect of broadband random frequency noise (e.g. road traffic)

Location	Noise level (dBA)
Banks	55
Churches	35
Cinemas	35
Classrooms	35
Concert halls	30
Conference rooms	30
Courtrooms	35
Council chambers	35
Department stores	55
Hospitals, wards	35
Hotels, bedrooms	35
Houses, living	45
Houses, sleeping	35
Lecture rooms	35
Libraries, loan	45
Libraries, reference	40
Music rooms	30
Offices, private	40
Offices, public	50
Radio studios	30
Restaurants	50
Recording studios	30
Shops	55
Telephoning, good	50
Telephoning, fair	55
Television studios	35
Theatres	30

These are optimum levels and may be exceeded by 5 dBA except in the critical situations shown in **bold**.

The amount of sound control or *sound attenuation** provided by certain construction methods has been established over the frequency range 100–3150 Hz (roughly corresponding to the lowest and highest frequencies normally experienced in a building). For this reason data for these forms of construction is useful when it is necessary to provide sound attenuation for broadband noise, but may be misleading when comparing dissimilar but adjoining methods of construction, or when noise is concentrated predominantly at selective frequencies, e.g. related to dBA scale.

For practical purposes, however, the airborne sound attenuation of a construction is controlled by four factors:

Mass or weight per unit area   Attenuation increases by approximately 5 dB for a doubling of weight or doubling of frequency.

Discontinuity   Elimination of direct sound paths where mass is insufficient by isolating those surfaces which receive the sound from those which surround the listener. The effect of a cavity between depends on its dimension in relation to the wavelength of sound to be controlled. Generally, a minimum practical gap of 50 mm is suitable for high frequencies, but a wider gap is necessary for low frequencies. Discontinuity must not incorporate any bridging along which sound may travel.

Stiffness   Elimination of vibration by sound waves. When incident sound waves have frequencies similar to those created by vibration of the construction, the sound attenuation will be less and unrelated to that theoretically provided by its mass.

Uniformity   Elimination of direct air paths through the construction. This applies not only to openings, holes and cracks, but also to materials which are not in themselves airtight. In the case of doors and windows, no matter how heavy the surrounding wall may be, the net sound attenuation will be limited to a maximum of about 7 dB *above* that of the door (Fig. 7.3).

The control of the effects of *impact* noise in a building requires different consideration from those given above. For example, the noise heard below a floor subject to impact noise will bear little relation to the airborne noise it causes in the room above. For the room below, weight has no advantage and the only defence is to prevent the transmission of the impact sound to the

* Avoid the term 'sound insulation' so as not to confuse it with 'thermal insulation'. Effective sound insulation requires a higher density of material, effective thermal insulation a lower density.

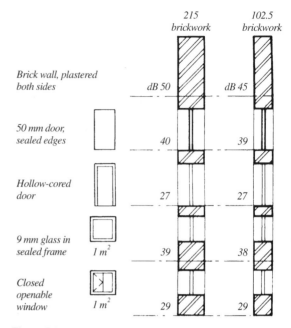

Figure 7.3   Sound attenuation: effects of openings in walls. (Based on data provided by the Brick Development Association)

structure by using a soft floor finish or a finish which is isolated from the structure.

7.6 Listening conditions

Sound within a room consists of two components: *direct*, which travels in a straight line through the air from the source to the recipient; and *reverberant*, which is the sum of all the sound reflections from the room surfaces (Fig. 7.4). As discussed in section 7.4, the direct noise decreases at the rate of 6 dB for each doubling of distance from the source, whereas the reverberant sound is theoretically constant throughout the room. This means that in a room containing a single source

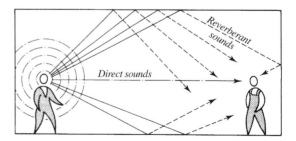

Figure 7.4   Direct sound and reflected (reverberant) sound.

there is a zone where the direct sound predominates; immediately beyond this another zone will occur where neither direct nor reverberant sound predominates, and then finally a zone where the reverberant component predominates. The sizes of these zones depend upon the dimensions of the room as well as the quantity and quality of the absorbent surfaces present.

7.7 Blurring and echoes

Reverberant sound can be extremely useful in large rooms such as auditoriums, where controlled reflections can send the sound levels to positions normally too distant to receive adequate sound from the direct path. However, focusing reflections in this way is a relatively skilled process, and if misjudged, reverberant sound and direct sound will be heard at slightly different intervals, thereby creating blurring (reflection arriving 1/30 to 1/15 second after direct sound) or echoes (more than 1/15 second discrepancy). Further explanation of the precise science used to achieve audibility and clarity of performance in auditoriums is beyond the scope of this book, and reference should be made to one of the many excellent monographs available on this subject for further clarification.

7.8 Sound absorption

It is important to differentiate between sound attenuation, as described earlier, and *sound absorption*. Attenuation is concerned with the passage of sound energy through a barrier; absorption with the sound reflection capabilities of the surfaces within a room or space. Sound produced by speech or music may be reflected several

times from walls, floors or ceiling before the residual energy in the sound waves is negligible. The continuance of sound during this period is known as reverberation, and the interval between the production of the sound and its decay to the point of inaudibility is known as the reverberation time. The sound-absorbing properties (absorption coefficient) of a surface are measured by the amount of sound reduction which occurs after waves strike the surface. Coefficients vary from 0 to 1, i.e. perfect reflection (hard surface) to total absorption (open window); see Table 7.3.

The sound levels in a room build up to a level which is determined by the absorption characteristics of the room. This can cause discomfort to the occupants, and increases the likelihood of noise being transferred to adjoining rooms where the original attenuation standards were sufficient. The provision of sound-absorbent materials on the surfaces of the room containing the sound source will eliminate these possibilities. If a room has mostly hard surfaces, no soft furnishings and few occupants to absorb sound effectively, it is usually not difficult to make a fourfold increase in absorption and obtain a reduction of 6 dB. However, care must be taken when selecting absorbent materials that the coefficient matches the frequency of the offending sound(s).

7.9 Sound reduction

A *sound reduction index* (SRI) may be used as a measure of the insulating effect of construction, against direct transmission of airborne sound. For consistency, tests are simulated in a laboratory where no flanking sound paths are possible. Insulation varies with frequency and

Table 7.3 Sound absorption coefficients

Item	Unit	Absorption coefficient at different frequencies		
		125 Hz	500 Hz	2000 Hz
Air	m^3	0	0	0.007
Audience (padded seats)	person	0.17	0.43	0.47
Seats (padded)	seat	0.08	0.16	0.19
Boarding or battens on solid wall	m^2	0.3	0.1	0.1
Brickwork	m^2	0.02	0.02	0.04
Woodblock, cork, lino, rubber floor	m^2	0.02	0.05	0.1
Floor tiles (hard)	m^2	0.03	0.03	0.05
Plaster	m^2	0.02	0.02	0.04
Window (5 mm)	m^2	0.2	0.1	0.05
Curtains (heavy)	m^2	0.1	0.4	0.5
Fibreboard with space behind	m^2	0.3	0.3	0.3
Ply panel over air space with absorbent lining	m^2	0.4	0.15	0.1
Suspended plasterboard ceiling	m^2	0.2	0.1	0.04

Table 7.4 Typical sound insulation values

Type of construction	SRI (Average dB in the range 100–3150 Hz
Brick/brick cavity wall, plastered	53
Brick/block cavity wall, plastered	49
Lightweight clad timber frame wall	38
215 mm brickwork, plastered	50
102.5 mm brickwork, plastered	45
50 mm dense concrete	40
100 mm dense concrete	45
300 mm lightweight concrete	42
12.5 mm plasterboard	25
Solid core door (25 kg/m^2)	26
Window, 4 mm single glazed	27
Window, 10 mm single glazed	31
Window, 4–12–4* double glazed	28
Window, 6–100–6* double window	37
Window, 10–200–10* double window	46
Roof (tiled) + 12.5 mm plasterboard ceiling	30
Roof (tiled) as above, with 100 mineral wool insulation	38
Flat roof, 100 mm reinforced concrete (230 kg/m^2)	48

* glass thickness–air space–glass thickness

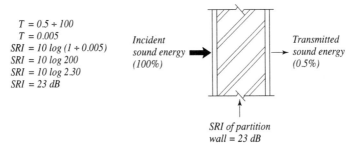

$$T = 0.5 \div 100$$
$$T = 0.005$$
$$SRI = 10 \log (1 \div 0.005)$$
$$SRI = 10 \log 200$$
$$SRI = 10 \log 2.30$$
$$SRI = 23 \, dB$$

Incident sound energy (100%)

Transmitted sound energy (0.5%)

SRI of partition wall = 23 dB

Figure 7.5 A wall transmitting 0.5 per cent of the sound energy incident on the exposed side at a given frequency.

the SRI is measured at octave* intervals between 100 and 3150 Hz. The arithmetical average is usually similar to the value at 500 Hz and this is generally a convenient figure for calculations and comparisons. Some typical values are shown in Table 7.4.

Accurate values for SRI can be obtained by calculation (see Fig. 7.5) from the laboratory data, using the following formula:

$$SRI = 10 \log(1 \div T)$$

Where: log = logarithm to the base ten
T = transmitted sound energy ÷ incident sound energy

Further details of laboratory analysis of sound insulation can be found in BS EN ISO 140: *Acoustics – measure-*

ments of sound insulation in buildings and of building elements.

7.10 Sound insulation regulations

The Building Regulations for England and Wales require resistance to the passage of sound:

- between dwellings (Approved Document E1) (see Table 7.5)
- within a dwelling (Approved Document E2)
- from external noise (Approved Document E3)

The regulations also have application to:

- common parts of buildings (Approved Document E4)
- acoustic conditions in schools (Approved Document E5)

It is not necessary to test every wall or floor in every new building on a site. Sample testing determined by the

* An octave is a range of frequencies between any frequency and double that frequency. For example, 500 Hz is one octave above 250 Hz. Octave bands for frequency analysis usually have frequency centres ranging between 31.5 and 8000 Hz (see Fig. 7.2).

Table 7.5 Insulation performance requirements for elements of construction which have a *separating* function (Building Regulations A.D. E1)

Situation	Airborne sound min. values (dB)		Impact sound max. values (dB)	
	Dwellings	Residential	Dwellings	Residential
Purpose built, i.e.				
new-build walls	45	43	–	–
new-build floors	45	45	62	62
Conversion/change of use/refurbishment				
walls	43	43	–	–
floors	43	43	64	64

building control authority is adequate to ensure quality control standards are achieved. The equipment used can be a compact hand-held sound level meter which converts variations in air pressure to variations in voltage. These variations are shown on a scale corresponding to decibels. The value indicated is the *root mean square* (RMS) of the signal, which is a type of uniform average, rather than extreme values.

The type of construction suitable for separating walls can be heavy or lightweight. Heavyweight walls reduce airborne sound by virtue of mass. Cavities can reduce sound by discontinuity or separation. Lightweight walls such as timber framing rely on a combination of mass, discontinuity and absorption of sound by a mineral fibre quilt. Table 7.6, Figures 7.6 and 7.7 provide a selection of standard constructions for separating walls and floors.

Table 7.6 Elements of construction to provide a separating function

Construction to include plastered room faces	Min. mass per unit area (kg/m²)
Solid walls:	
215 mm brickwork	375
215 mm concrete block	415
190 mm *in situ* concrete	415
Cavity walls:	
102 mm brickwork + 50 mm cavity	415
100 mm concrete blocks + 50 mm cavity	415
100 mm lightweight concrete blocks + 75 mm cavity	300
Timber frame wall:	
100 × 50 mm structural frame (×2), quilting of mineral wool in cavity (25 mm) or frames (25 mm each) + double layer of plasterboard (30 mm) each side	N/A
Floors:	
See Figure 7.7	

Solid masonry

Brick density 1610 kg/m³
Block density 1840 kg/m³
13 mm lightweight plaster,
min. mass per unit
area – 10 kg/m²

Cavity masonry

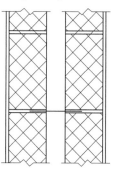

Concrete block
density 1990 kg/m³.
50 mm min. cavity
13 mm lightweight
plaster

Lightweight timber frame

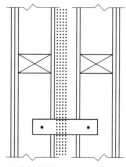

As Table 7.6.
Plywood sheathing in
cavity for structural
purposes
Plasterboard min. mass
per unit area – 9 kg/m²
40 × 3 mm galv. steel
continuity straps at 1.2 m

Figure 7.6 Separating walls.

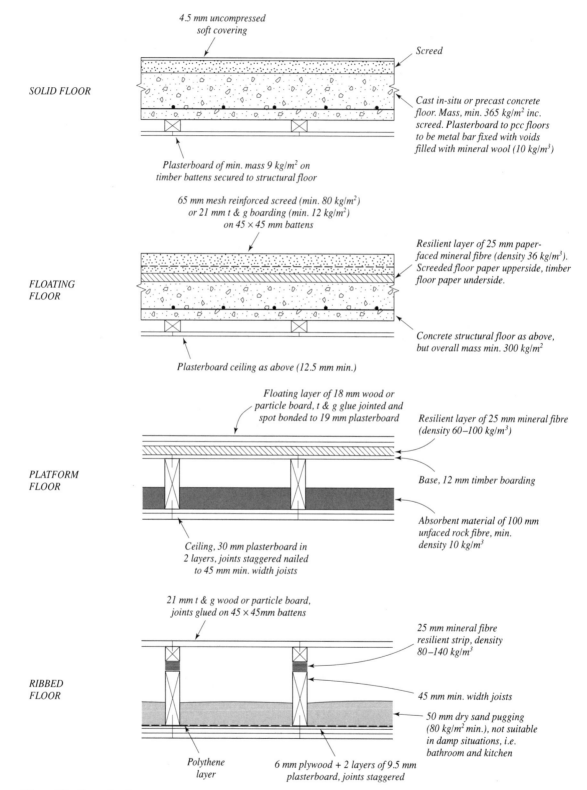

SOLID FLOOR

4.5 mm uncompressed
soft covering

Screed

Cast in-situ or precast concrete
floor. Mass, min. 365 kg/m² inc.
screed. Plasterboard to pcc floors
to be metal bar fixed with voids
filled with mineral wool (10 kg/m³)

Plasterboard of min. mass 9 kg/m² on
timber battens secured to structural floor

FLOATING
FLOOR

65 mm mesh reinforced screed (min. 80 kg/m²)
or 21 mm t & g boarding (min. 12 kg/m²)
on 45 × 45 mm battens

Resilient layer of 25 mm paper-
faced mineral fibre (density 36 kg/m³).
Screeded floor paper upperside, timber
floor paper underside.

Concrete structural floor as above,
but overall mass min. 300 kg/m²

Plasterboard ceiling as above (12.5 mm min.)

PLATFORM
FLOOR

Floating layer of 18 mm wood or
particle board, t & g glue jointed and
spot bonded to 19 mm plasterboard

Resilient layer of 25 mm mineral fibre
(density 60–100 kg/m³)

Base, 12 mm timber boarding

Absorbent material of 100 mm
unfaced rock fibre, min.
density 10 kg/m³

Ceiling, 30 mm plasterboard in
2 layers, joints staggered nailed
to 45 mm min. width joists

RIBBED
FLOOR

21 mm t & g wood or particle board,
joints glued on 45 × 45mm battens

25 mm mineral fibre
resilient strip, density
80–140 kg/m³

45 mm min. width joists

50 mm dry sand pugging
(80 kg/m² min.), not suitable
in damp situations, i.e.
bathroom and kitchen

Polythene
layer

6 mm plywood + 2 layers of 9.5 mm
plasterboard, joints staggered

Figure 7.7 Separating floors.

The minimum laboratory values for walls and floors *within* any new dwelling or new residential buildings, including those created by change of use is 40 dB (Building Regulations Approved Document E2). Some examples which should satisfy this requirement if constructed correctly are shown in Figure 7.8.

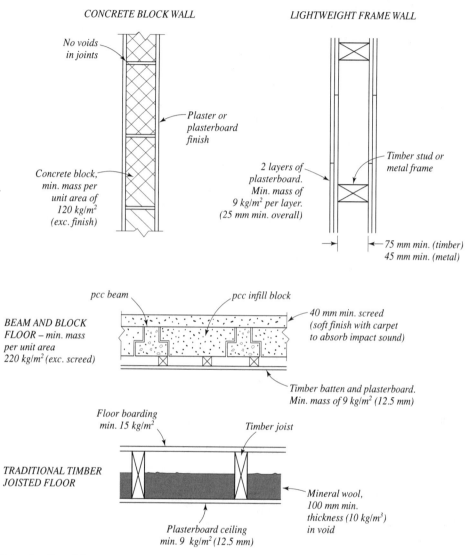

CONCRETE BLOCK WALL

No voids in joints

Plaster or plasterboard finish

Concrete block, min. mass per unit area of 120 kg/m² (exc. finish)

LIGHTWEIGHT FRAME WALL

Timber stud or metal frame

2 layers of plasterboard. Min. mass of 9 kg/m² per layer. (25 mm min. overall)

75 mm min. (timber) 45 mm min. (metal)

BEAM AND BLOCK FLOOR – *min. mass per unit area 220 kg/m² (exc. screed)*

pcc beam

pcc infill block

40 mm min. screed (soft finish with carpet to absorb impact sound)

Timber batten and plasterboard. Min. mass of 9 kg/m² (12.5 mm)

TRADITIONAL TIMBER JOISTED FLOOR

Floor boarding min. 15 kg/m²

Timber joist

Mineral wool, 100 mm min. thickness (10 kg/m³) in void

Plasterboard ceiling min. 9 kg/m² (12.5 mm)

Figure 7.8 Internal walls and floors.

Further specific reading

Mitchell's Building Series

Environment and Services	Chapter 6 Sound
Internal Components	Section 2.3.6 Sound control with demountable partitions
	Section 3.3.5 Sound control with suspended ceilings
	Section 4.2.5 Sound control with raised floors
	Section 6.2.4 Doors to limit sound
	Section 6.14 Sound-resisting doors
External Components	Section 3.2.7 Sound control with external glazing
	Section 4.2.7 Sound control with windows
Materials	Section 1.6 Acoustic properties
Materials Technology	Chapter 4 Masonry construction (sound insulation)
Structure and Fabric Part 1	Chapter 5 Walls and piers (functional requirements)
	Chapter 7 Roof structures (functional requirements)
	Chapter 8 Floor structures (functional requirements)
	Chapter 10 Stairs (functional requirements)

Building Research Establishment

BRE Digest 162	*Traffic noise and overheating in offices*
BRE Digest 192	*The acoustics of rooms for speech*
BRE Digest 193	*Loudspeaker systems for speech*
BRE Digest 293	*Improving the sound insulation of separating walls and floors*
BRE Digest 333	*Sound insulation of separating walls and floors* Part 1: Walls
BRE Digest 334	*Sound insulation of separating walls and floors* Part 2: Floors
BRE Digest 337	*Sound insulation: basic principles*
BRE Digest 338	*Insulation against external noise*
BRE Digest 347	*Sound insulation of lightweight dwellings*
BRE Digest 379	*Double glazing for heat and sound insulation*
BRE Digest 453	*Insulation glazing units*

Building Regulations

E1   Protection against sound from adjoining dwellings or buildings etc.
E2   Protection against sound within a dwelling etc.
E3   Reverberation in the common internal parts of buildings containing dwellings etc.
E4   Acoustic conditions in schools

8 Thermal comfort

CI/SfB(M)

A building must provide a satisfactory thermal environment for its occupants as well as for the mechanical systems it accommodates. Energy within the human body produces uninterrupted heat at varying rates in order to maintain an ideal temperature of 37 °C in the internal organs. However, heat is lost by the body through radiation, convection and evaporation from the skin and the lungs, and there is a continuous process of adjustment to ensure a thermal balance between heat produced and heat lost. It is particularly important that the brain temperature is maintained constant. The factors in the local environment which govern heat loss include not only air temperature but also air movement, relative humidity and the radiant temperatures from surrounding surfaces.

The fabric of clothing and of a building perform similar functions by maintaining temperature control through *passive means* which regulate natural flows of heat, air and moisture vapour. As a building involves a volume many times larger than that contained by clothing, and provides environmental conditions suitable for many occupants in different spaces, it must also provide *active means* for thermal comfort not unlike that achieved by the human body itself (Fig. 8.1).

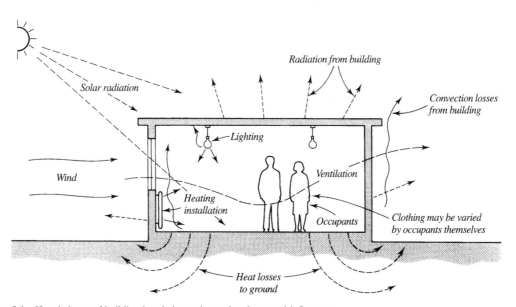

Figure 8.1   Heat balance of building in relation to internal and external influences.

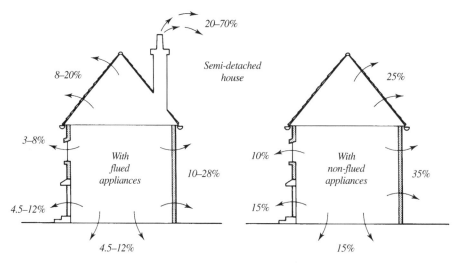

Figure 8.2   Potential percentage heat losses from domestic buildings.

8.1 Passive means

Simply stated, the creation of thermal comfort within a building by passive means involves the reduction of the rate of *heat losses* from the inside to the outside in colder climates, and the reduction of *heat gains* from the outside to the inside in warmer climates. In both cases the transfer of heat is regulated by the external fabric of a building which can provide varying degrees of thermal insulation. The type and amount necessary to achieve ideal conditions varies according to resource availability, climate of locality and the degree of exposure (Fig. 8.2).

In the United Kingdom the thermal performance of a building fabric is directly affected by such detailed criteria as seasonal changes and extremes of temperature; temperature differences between day and night (diurnal range); sky conditions (amount of sunlight and overshadowing); incoming and outgoing heat radiation; rainfall and its distribution; the effects of water absorption and repulsion on materials and their forms (weathering); air movements; and other special features influenced by location and orientation. A designer must interpret these requirements in conjunction with fashions in appearance for a building, together with the current legal requirements which endeavour to control atmospheric pollution and conserve the use of fuel for space heating by regulating the amount of heat loss from a building to the external environment.

8.2 Heat transfer

Heat may be lost from a building fabric to the external environment by *convection* (movement of heat by hot liquid or gas); *conduction* (transfer of heat through solids); and *radiation* (transfer of heat from a surface across a vacuum, gas- or liquid-filled space, or a transparent solid). Natural ventilation causes convection losses and takes the form of a stack effect, where warm air rises in a building, eventually escaping to be replaced by colder air. Although some degree of ventilation is desirable as one of the functions of a building (see Chapter 10), excessive heat losses should be minimised. Solid building materials used for a building fabric lose heat from warm to cold face by conduction. The ability of a material to conduct heat is known as its *conductivity* (λ) value, and is measured by the amount of heat flow (watts) per square metre of surface area for a temperature difference of 1 K per *metre thickness*, i.e. W m/m² K or W/m K. When comparing the insulation properties of a material to that of others, it is generally more convenient to use its *resistivity value* (*r*) because it takes no account of size or thickness. Resistivity is therefore expressed as the reciprocal of the conductivity, i.e. m K/W.

In order to calculate the actual *thermal resistance* (*R*) offered by a particular material of *known* thickness, the formula used is actual thickness (m) × m K/W, i.e. *R* has units of m² K/W. The higher the *R* value, the better the thermal resistance and the insulating performance of the material (Table 8.1). And the lower a material's density, the greater its properties of insulation.

It is unfortunate that generally these relationships of density are counter to the noise control and structural qualities required from a material. A solid monolithic wall may be up to three times the thickness required for structural purposes in order to provide an acceptable level of thermal insulation. If a dense structural

Table 8.1 Thermal conductivity and thermal resistances of some materials

Material	Conductivity (W/m K)	Resistance* (m² K/W)
Brickwork		
105 mm	0.84	0.125
220 mm	0.84	0.262
335 mm	0.84	0.399
Plaster		
15 mm hard	0.5	0.03
15 mm lightweight	0.16	0.09
10 mm plasterboard		0.06
Cavity		
Unventilated	–	0.18
With foil face		0.3
Behind tile hanging		0.12
Tile hanging	0.84	0.038
Expanded polystyrene		
25 mm	0.033	0.76
13 mm	0.033	0.39
Glass fibre		
50 mm	0.035	1.43
75 mm	0.035	2.14
100 mm	0.035	2.86
Aerated concrete		
100 mm	0.22	0.45
150 mm	0.22	0.68
Softwood, 100 mm	0.13	0.77
Weatherboarding, 20 mm	0.14	0.14
External surface	–	0.055
Internal surface	–	0.123

* The thermal resistance of a structural component is expressed as:

$$R = L \div \lambda$$

where: L = thickness of element (m)
λ = thermal conductivity (W/m K)

material is filled with pockets of air, its thermal insulation characteristics can be vastly improved while maintaining much of its strength characteristics. However, if these air pockets are allowed to become filled with water from exposure to rainfall or ground moisture, their insulation value will be cancelled altogether. This is because water is a much better conductor of heat than air; a saturated material could permit about ten times more heat to be transferred through it when compared with its 'dry' state.

Any heated material will radiate heat from its surface; bright metallic surfaces generally radiate least and dark surfaces the most. In this respect, radiation and absorption characteristics of materials correspond. Some construction methods incorporate air cavities which reduce the amount of heat transfer by conduction (and the passage of moisture) from inner warm materials to outer cold materials. Some heat, however, will be transferred across cavities by radiation, and the absorption

(and reflection) properties of the adjacent surfaces across the cavity will be significant to the thermal insulation value of the construction. (The performance of the cavity as an insulator will be impaired if convection currents take place.) Radiation losses generally depend on the *emissivity* – the rate of radiant heat emission – from the surface and values depend upon roughness of surface, the rate of air movement across it, its orientation or position, and the temperature of the air and other bodies facing it.

8.3 Thermal insulation values

It is unrealistic to rely on empirical rules to establish forms of constructions which provide satisfactory resistance to heat transfer *as well as* weather resistance, strength and stability and many of the other performance criteria. Moreover, current construction methods are generally less bulky than those used years ago, as greatly reduced thicknesses of materials are employed in combination with each other, each layer, leaf or skin perhaps fulfilling only one very specific function. The thermal comfort properties of these materials used in combination can only be assessed by calculating the amount of heat transfer from internal to external air (in cold climates). The thermal resistance properties of each layer must be taken into account, along with other factors relating to their surface texture and juxtapositions within the construction.

The internal to external thermal transmission rate of all the layers of a construction is known as a *U value* and is more accurately defined as the number of watts transmitted through one square metre of construction for each single degree Kelvin temperature difference between the air on each side of the construction, i.e. W/m² K. It is calculated by taking the reciprocal of the sum of all the thermal resistance values (*R* values of all materials used and any air cavities) as well as the internal and external surface resistances (Fig. 8.3).

As the wall shown in Figure 8.3 is consistent and continuous in its construction, i.e. the mortar and brickwork have similar density and thermal properties, the *U* value calculation is relatively straightforward. Therefore the thermal transmittance of this element is:

$$U = \frac{1}{R}$$

where, *R* is the sum of all the separate resistances for different materials or structural components in the element, i.e.

$$U = \frac{1}{Rsi + R_1 + R_2 + R_3 + Rso}$$

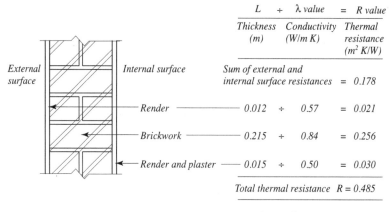

L	÷ λ value	= R value
Thickness (m)	Conductivity (W/m K)	Thermal resistance (m² K/W)

Sum of external and internal surface resistances = 0.178

Render — 0.012 ÷ 0.57 = 0.021

Brickwork — 0.215 ÷ 0.84 = 0.256

Render and plaster — 0.015 ÷ 0.50 = 0.030

Total thermal resistance R = 0.485

To calculate U value of element: $U = \dfrac{1}{R} = \dfrac{1}{0.485} = 2.062 \ W/m^2 \ K$

Figure 8.3 *U* value calculation for a one-brick-thick solid wall.

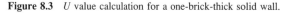

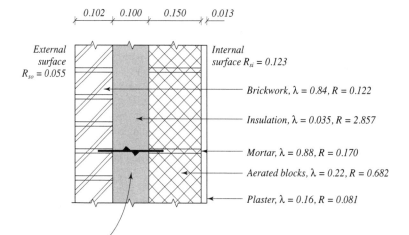

External surface $R_{so} = 0.055$

Internal surface $R_{si} = 0.123$

Brickwork, λ = 0.84, R = 0.122

Insulation, λ = 0.035, R = 2.857

Mortar, λ = 0.88, R = 0.170

Aerated blocks, λ = 0.22, R = 0.682

Plaster, λ = 0.16, R = 0.081

Wall ties – if butterfly type wire wall ties are used in narrower cavities (50–75 mm) at normal spacing*, no adjustment is made to calculations. However, vertical twist ties are usual in cavities over 75 mm wide and use of these ties requires an addition of 0.020 W/m² K to the calculated U value.

* 900 mm max. horizontally
750 mm max. horizontally where cavity >75 mm
450 mm max. vertically

Figure 8.4 Fully insulated cavity brick and block masonry wall.

where: U = thermal transmittance (W/m² K)
   Rsi = internal surface resistance (m² K/W)
R_1, R_2, R_3 = thermal resistance of structural
      components (m² K/W)
   Rso = external surface resistance (m² K/W)

Note: Ra, the resistance for an air space is included with the summation where a cavity or voids occur in the construction. Typical value is: 0.180 m² K/W.

Walls with differing materials and composition (Fig. 8.4) will require more detailed calculation to determine the *U* value. The *Proportional Area Method* allows for inconsistent construction such as the relatively dense mortar between lightweight concrete blocks. Here the mortar will have a much higher thermal conductivity and a thermal bridging effect in the wall. However, the *Combined Method* (BS EN ISO 6946: *Building components and building elements – thermal resistance and thermal transmittance – calculation method*) is now preferred. This method shows conformity with the Building Regulations by providing a *U* value which represents the average of the upper and lower thermal resistance (*R*) limits.

Proportional area method of *U* value calculation applied to the brick/block insulated cavity wall shown in Figure 8.4:

Standard block face size + mortar,
$$450 \times 225 \text{ mm} = 101\ 250 \text{ mm}^2$$
Standard block face/format size,
$$440 \times 215 \text{ mm} = \underline{94\ 600 \text{ mm}^2}$$
The face area of 10 mm deep mortar per block
$$= \quad 6\ 650 \text{ mm}^2$$

Proportional area calculations:
  Mortar $6\ 650 \div 101\ 250 = 0.066$ or 6.6%
  Block $94\ 600 \div 101\ 250 = 0.934$ or 93.4%

Thermal resistance (R) calculations:
  Outer leaf and insulation (all unbridged)
      $Rso = 0.055$
  Brickwork = 0.122
  Insulation = $\underline{2.857}$

$$3.034 \times 100\% = 3.034 \text{ m}^2 \text{ K/W}$$

  Inner leaf (unbridged part)
      $Rsi = 0.123$
    Plaster = 0.081
    Blocks = $\underline{0.682}$

$$0.886 \times 93.4\% = 0.828 \text{ m}^2 \text{ K/W}$$

  Inner leaf (bridged part)
      $Rsi = 0.123$
    Plaster = 0.081
    Mortar = $\underline{0.170}$

$$0.374 \times 6.6\% = 0.025 \text{ m}^2 \text{ K/W}$$

$U = 1 \div \Sigma R$

Therefore:

$$U = \frac{1}{3.034 + 0.828 + 0.025} = 0.257 \text{ W/m}^2 \text{ K}$$

0.020 W/m² K to be added for the use of vertical twist type wall ties in the wide cavity

Therefore, corrected *U* value = 0.277 W/m² K

Combined method of *U* value calculation applied to Figure 8.4:

Upper and lower thermal resistance (*R*) limits are calculated from the formula:

$$R = \frac{1}{\Sigma(Fx \div Rx)}$$

where: Fx = Fractional area of section x
       Rx = Total thermal resistance of section x
       Section x represents the different thermal paths

The wall (Fig. 8.4) can be seen to have two different thermal paths:

1. Through the concrete blocks.
2. Through the mortar between concrete blocks.

The remainder is consistent.

The upper limit (*R*) for the section containing concrete blocks is:

$$Rso = 0.055$$
$$\text{Brickwork} = 0.122$$
$$\text{Insulation} = 2.857$$
$$\text{Concrete blocks} = 0.682$$
$$\text{Plaster} = 0.081$$
$$Rsi = \underline{0.123}$$
$$3.920 \text{ m}^2 \text{ K/W}$$

The fractional area of this section containing blocks is 0.934

The upper limit (*R*) for the section containing mortar is:

$$Rso = 0.055$$
$$\text{Brickwork} = 0.122$$
$$\text{Insulation} = 2.857$$
$$\text{Mortar} = 0.170$$
$$\text{Plaster} = 0.081$$
$$Rsi = \underline{0.123}$$
$$3.408 \text{ m}^2 \text{ K/W}$$

The fractional area of this section containing mortar is 0.066

The upper limit of resistance (*R*) is:

$$R = \frac{1}{(0.934 \div 3.920) + (0.066 \div 3.408)}$$
$$= 3.881 \text{ m}^2 \text{ K/W}$$

The lower limit of resistance (R) is the summation of all the layers:

Rso	= 0.055
Brickwork	= 0.122
Insulation	= 2.857
Bridged layer	
$= 1 \div (0.934 \div 0.682) + (0.066 \div 0.170) = 0.569$	
Plaster	= 0.081
Rsi	= 0.123
	3.807 m² K/W

The total resistance (R) of the wall is taken as the average of the upper and lower limits:

$$(3.881 + 3.807) \div 2 = 3.844 \text{ m}^2 \text{ K/W}$$

$$U = 1 \div R$$
$$U = 1 \div 3.844 = 0.260 \text{ W/m}^2 \text{ K}$$

As before, 0.020 W/m² K is added to the calculation for the use of vertical twist type wall ties in the wide cavity. Therefore the corrected U value is 0.280 W/m² K.

The calculated U value for a particular construction is unlikely to provide an entirely accurate thermal transmittance rate because the heat flow conditions through the construction will vary with the amount of solar radiation, moisture conditions and the effects of prevailing winds. The quality of workmanship and supervision will also have an effect. Nevertheless, calculated U values are a reasonable guide for comparing the thermal insulation values of different forms of construction and as an indication as to whether or not the heat energy losses from a building are within the limitations of building legislation. The higher the U value, the poorer the thermal insulation properties of an element of construction. Also, a U value refers to the total constructional thickness of an element, whether consisting of a single layer of material or a combination of materials, with or without separating cavities.

The Building Regulations, Approved Document L: *Conservation of fuel and power*, aims to increase the energy efficiency of buildings and thereby to reduce the emission of burnt fuel gases into the atmosphere. These are generally known as greenhouse gases and they include:

- carbon dioxide
- chlorofluorocarbons
- hydrochlorofluorocarbons
- hydrofluorocarbons
- methane

- nitrous oxide
- perfluorocarbons
- sulphur hexafluoride

Carbon dioxide (CO_2) is the least powerful but it is the most prominent, accounting for about 80 per cent of all greenhouse gas emissions globally. Therefore, current thermal insulation and boiler efficiency calculations are designed specifically to reduce its contribution to the *greenhouse effect*. This term is widely used and includes changes to atmospheric conditions such as global warming and ozone depletion, leading to the physical effects of polar melting and rising sea levels.

Requirements for new buildings in the UK generally include an enclosing envelope of insulation, double glazing throughout and draught sealing of all doors, opening windows, loft hatches and other gaps at interfaces in the building fabric. The effectiveness of these provisions and the efficiency of hot water and heating systems can be calculated by a *standard assessment procedure* outlined in section 8.3.1. Standard U values for elements of construction are shown in Figure 8.5, but variations can apply, depending on which one of the three acceptable methods are used to attain compliance with the Building Regulations:

- elemental
- target U value
- carbon index

8.3.1 Standard assessment procedure (SAP)

Builders of new dwellings and conversions are required to determine by calculation an indication of expected fuel costs for hot water and heating. In addition to satisfying the Building Regulations, this benefits prospective new purchasers and tenants by providing comparisons between dwellings and a means to assess annual expenditure. SAP energy efficiency rating is expressed as a score on a scale from 1 to 120. Guidance for new dwellings is between 80 and 85 as shown in Table 8.2. SAP rating assessment calculations are quite extensive, therefore reference should be made to the appendices in Approved Document L of the Building Regulations where data and worksheets are provided.

8.3.2 Elemental method

The elemental method is the simplest to comprehend, requiring elements of construction to satisfy minimum standards of insulation. Conformity with the Building Regulations is usually achieved if construction of the external elements satisfies the following:

DWELLINGS

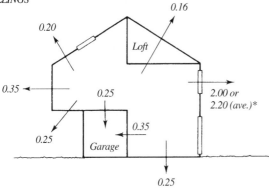

NON-DOMESTIC BUILDINGS

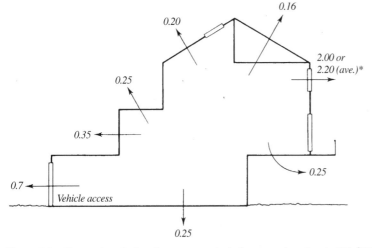

Notes:

1. SEDBUK min. values:
   Mains natural gas = 78%
   LPG   = 80%
   Oil   = 85%

2. Windows, doors and rooflights
   max. 25% of floor area,
   U (ave.) = 2.00 or 2.20*

3. Flat roof, U = 0.25 W/m² K max.

Notes:

1. WINDOWS AND PERSONNEL DOORS
   Residential: 30% exposed wall area
   Industrial: 15% exposed wall area
   Assembly places: 40% exposed wall area
   Rooflights: 20% exposed wall area

2. To provide some design flexibility,
   U values and glazed areas can be traded
   off, provided the heat loss rate does not
   exceed that of an equivalent compliant
   building and U values for specific
   elements do not exceed:
      roof, 0.35
      floor and wall, 0.7

* 2.00 W/m² K for wood or
  uPVC frames and 2.20 W/m² K
  for metal frames.

Figure 8.5   Elemental method performance standards for energy/heat loss in W/m² K.

Table 8.2   SAP energy rating guidance for new dwellings or conversions

Floor area of dwelling (m²)	SAP energy rating (min.)
<80	80
80–90	81
90–100	82
100–110	83
110–120	84
>120	85

- At least 200 mm total thickness of mineral wool insulation between and over roof ceiling joists.
- Full double glazing.
- Areas of doors, windows and rooflights, including components, not exceeding 25 per cent of the total floor area.

- Rigid insulation batts/boards of at least 50 mm thickness in ground floors (see Figs 17.26 and 17.27, and Table 8.3).
- Masonry wall construction with insulated cavity, full or part filled to suit required U value (e.g. Fig. 8.4), or
- Timber-framed wall construction (see Fig. 17.37) with 10 mm plywood cladding, 140 mm mineral wool insulation between framing and 25 mm (2 × 12.5 mm) plasterboard lining – U value approximately 0.3 W/m² K.

For floor insulation, high compressive strength mineral wool board and expanded polystyrene batts are generally available in thicknesses up to 100 mm. Less than 50 mm may satisfy the Building Regulations in certain circumstances as shown in Table 8.3. Tables of this type are provided in Approved Document L to the Building Regulations and in product manufacturers' catalogues.

Table 8.3 Example of mineral wool insulation (conductivity 0.02 W/m K) applied to solid concrete ground floors

U value	Perimeter/Area (P/A)							
	0.2	0.3	0.4	0.5	0.6	0.7	0.8	0.9
0.20	46	60	67	71	74	77	78	80 mm
0.25	26	40	47	51	54	57	58	60 mm
0.30	13	26	33	38	41	43	45	47 mm

Example: A room 10 m × 10 m on plan has a P/A of 0.4. If the required U value is 0.25, at least 47 mm of insulation should be included in the ground floor construction.

Thickness of insulation is determined by relating floor area to perimeter and required U value.

SEDBUK is an acronym for Seasonal Efficiency of a Domestic Boiler in the United Kingdom. Percentage values are defined in the government publication, *Standard Assessment Procedure for Energy Rating of Dwellings* (1998 edition). For example, using mains natural gas fuel, comparing an old boiler of say 10 years or more, with a new non-condensing boiler and a new condensing boiler, the SEDBUK values will be about 55 per cent, 75 per cent and 88 per cent respectively. (Note that a condensing boiler uses the primary heat exchange from burning fuel and the heat contained in the flue gases as a secondary means of heating water. Non-condensing boilers have only a primary heat exchange.)

8.3.3 Target U value method

$$\text{Target } U \text{ value} = 0.35 - [0.19(A_R/A_T) - 0.10(A_{GF}/A_T) + 0.413(A_F/A_T)]$$

where: A_R = Area of exposed roof
A_T = Total area of exposed elements
A_{GF} = Area of ground floor
A_F = Total area of floors (all storeys)

A worked example is shown using Figure 8.6 and Table 8.4.

Average U value = $78.1 \div 179 = 0.4363$ W/m^2 K
Target U value = $0.35 - [0.19(42/179) - 0.10(42/179) + 0.413(84/179)]$
Target U value = 0.4759 W/m^2 K

Table 8.4 Example target U value calculation (Fig. 8.6)

Element	Exposed area (m^2)	U value (W/m^2 K)	Heat loss (W/K)
Ground floor	42	0.25	10.5
Wall	78	0.35	27.3
Windows/doors	17	2.00	34.0
Roof/ceiling	42	0.15	6.3
Totals	**179**		**78.1**

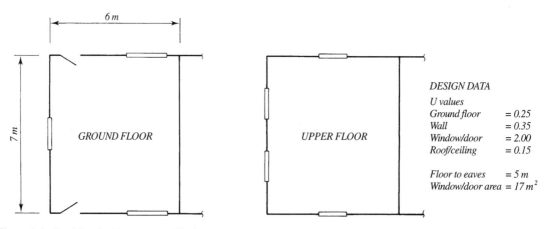

Figure 8.6 Semi-detached house: target U values.

As the average U value of the building does not exceed the target U value, the construction is acceptable. However, further adjustments to the target value may have some effect.

Target U value adjustments:

- Boiler rating. Where the SEDBUK value for the selected boiler differs (+ or −) from that given for the elemental method, the target U value should be multiplied by:

  Proposed boiler SEDBUK (%) ÷ Reference boiler SEDBUK (%)

e.g. Proposed oil fired boiler SEDBUK = 90%, reference boiler SEDBUK = 85% (see Fig. 8.5). So, 90 ÷ 85 = 1.059.

Applied to the previous example, the target U value will now be:

$0.4759 \times 1.059 = 0.5039$ W/m² K

Note: Where direct electric heating or a solid fuel boiler is used, the target U value is divided by 1.15 to compensate for the relative higher carbon emissions. Neither of these fuel systems should be used where the energy efficiency is established with the elemental method.

- Optional allowances for solar gain.
  1. Where metal framed windows are used, the target U value can be multiplied by 1.03 to take account of the greater area of exposed glass in this window type.
  2. If the glazed area to the south is greater than that to the north, the target U value can be eased by adding the following factor:

     $0.04[(A_S − A_N) ÷ A_{TG}]$

     where: A_S = Glazed area (including frame) facing south (+ or − 30°)
     A_N = Glazed area (including frame) facing north (+ or − 30°)
     A_{TG} = Total glazed areas

     e.g. $A_S = 5$ m², $A_N = 4$ m² and $A_{TG} = 17$ m²

     $0.04[(5 − 4) ÷ 17] = 0.0023$

     Revised target U value is now $0.5039 + 0.0023 = 0.5062$ W/m² K

8.3.4 Carbon index method

The objective is to create new dwellings with a similar energy performance to that achieved with the elemental method, whilst providing more flexibility in their design. The assessment criteria is based on calculated carbon dioxide emissions in kilograms or tonnes per year relative to dwelling floor area. The following formulae may be used:

$$CF = CO_2 ÷ (TFA + 45)$$
$$CI = 17.7 − (9 \log._{10} CF)$$

where: CF = Carbon factor
CO_2 = Carbon dioxide emission in kg/year
TFA = Dwelling total floor area in m²
CI = Carbon index
$\log._{10}$ = logarithm to the base of ten

e.g. A dwelling of 180 m² total floor area producing CO_2 emissions of 2500 kg/year:

$$CF = 2500 ÷ (180 + 45) = 11.11$$
$$CI = 17.7 − (9 \log._{10} 11.11) = 8.29 \ (8.3)$$

The carbon index is represented between 0 and 10, rounded to one decimal place. All new dwellings should have a value of at least 8.0. SAP worksheets from Approved Document L can be used to determine the annual CO_2 emissions.

8.4 Condensation and interstitial condensation

The consistency of construction is a very important factor, as illustrated in Figures 8.7 and 8.8. Here most of the construction is of layered form but is 'bridged' at intervals by a material or an air space providing less thermal resistance. This results in variations in thermal transmittance values along the construction and the occurrence of *cold bridging*. The slightly cooler internal surface at the point of the *cold bridge* provides a dew-point temperature at which the warm internal air cools, causing droplets of water (condensation) as well as dust to be deposited locally, i.e. *pattern staining*.

In practice, it is also necessary to prevent the occurrence of *condensation* at the surface or within material(s) used for a building fabric (Fig. 8.9). This takes place when atmospheric temperature (dry bulb) falls below the dew-point temperature – a property that depends upon the water vapour content of the air and therefore upon the vapour pressure. The amount of water vapour contained in the atmosphere will fall according to temperature and relative humidity (ratio of vapour pressure present to completely saturated air).

Surface condensation is likely to occur when air containing a given amount of water vapour is cooled by coming into contact with a cold plane. High relative humidity, more than 80 per cent, will cause mould

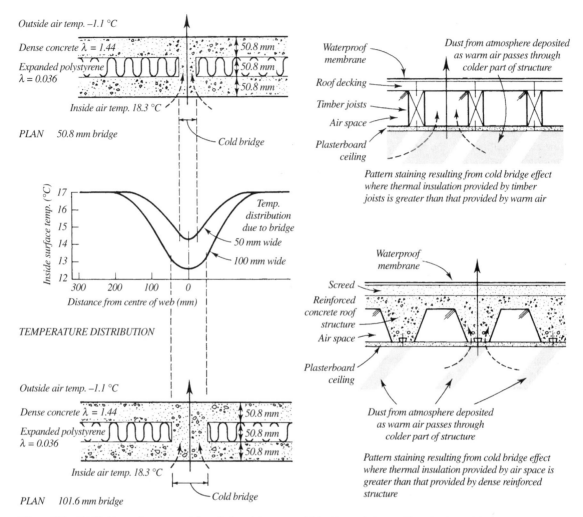

Figure 8.7　Cold bridge and pattern staining. (Adapted from material in *Materials for Buildings* by Lyall Addleson)

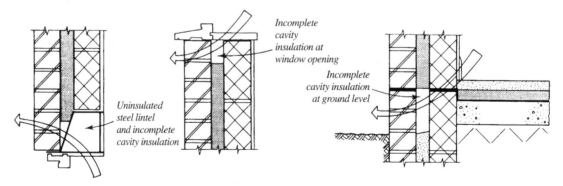

Figure 8.8　Cold bridging by incomplete cavity insulation.

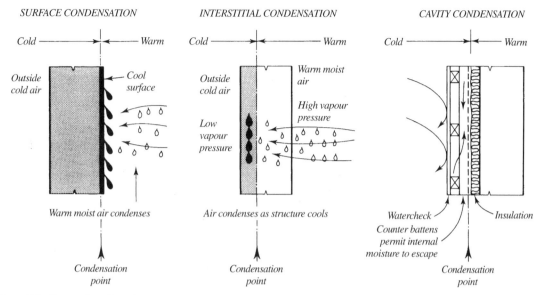

Figure 8.9 Forms of condensation.

growths on the surface of organic materials or droplets of water on non-organic surfaces. This risk can be reduced by keeping internal surfaces at a higher temperature than the dew-point of the internal air, which involves carefully balancing the building fabric insulation *and* internal heating and ventilating requirements. Simply to increase the insulation value of the building fabric may only move the dew-point of the air *into* the body of the material and produce *interstitial condensation*, causing a loss of insulation and deterioration.

One solution to the damage caused by surface and interstitial condensation is to use a construction method which makes the condensation or dew-point coincide with a *cavity* which separates external weathering layers from internal insulating layers (see section 6.4). This cavity should be ventilated, and the moisture formed by condensation (together with any which penetrates the external weathering layers) must be adequately collected and diverted to the outside. When a cavity cannot be incorporated in the correct position relative to the point of condensation, then a *vapour control layer or membrane* must be used. This consists of a thin impermeable sheet (e.g. plastics or reinforced aluminium foil) and should be placed on the *warm side* of the insulating material to prevent transference of moist air from the warm interior of a room into the fabric of the building (see Figs 17.37 and 17.61). Care must be taken to ensure that the sheets are adequately lapped and folded at their edges to prevent moisture penetration. A vapour control layer must also be incorporated for composite construction methods employing

different layers of materials providing varying thermal and vapour resistance properties. The problems associated with the practical application of vapour control layers are discussed in section 17.9.6.

Modern forms of construction and living patterns have greatly increased the risk of condensation, not only upon internal surfaces, but also within wall, floor and roof constructions. Traditional construction allowed moisture vapour to pass more easily through its fabric and through air gaps around doors and windows. There was often less moisture vapour to be vented, owing to lower temperature requirements and greater ventilation rates at moisture sources (air bricks and chimney flues, etc.). See also section 10.4 for a summary of current legislation governing the ventilation of rooms in domestic buildings.

8.5 Thermal capacity

The selection processes for a building fabric to provide satisfactory thermal comfort conditions must include consideration of the form of space heating to be employed. When the heat source remains virtually constant, it may be considered expedient to adopt a building fabric which will store heat or have a *high thermal capacity*. For example, masonry walling material will gradually build up a reservoir of heat while it is being warmed. When the heat source ceases for a short period (i.e. overnight) or when the external temperature drops below normal, the stored heat will be slowly given back. At any event, the internal wall surfaces will feel relatively warm and comfortable, and the possibility of condensation occurring will be less likely, except during the period when the heat source is initially resumed. In hot climates, very thick and heavy construction buffers the effect of very high external daytime temperatures on the internal climate of the building.

When intermittent heat sources are used, or other marked fluctuations from steady temperatures occur, dense construction will be slow to warm up, and lower surface temperatures may be present for some time. This will give rise to discomfort for the occupants and condensation. In rooms used occasionally, and there-fore requiring only intermittent heating for comfort, lining the interior surfaces with material of *low thermal capacity*, incorporating a vapour control layer, will reduce the amount of heat required and produce a quick thermal response. Alternatively, a construction method of entirely low thermal capacity can be adopted (timber walling incorporating vapour control layer and thermal insulation in spaces between structural members), and often this is the chosen form where the need to conserve fuel resources is paramount. In both cases – internally insulated dense construction or lightweight construction – less heat will be stored and the room will cool more rapidly when the heat source is curtailed. One acceptable compromise is to adopt a dense construction with thermal insulation on the *external* face. In this way, once a wall has been warmed, there will be a long period before it is finally dissipated completely, during which time the heat source may be resumed again.

8.6 Heat gains

Unwanted solar heat gain into a building can be a source of considerable thermal discomfort and interrupt the working of normal methods of space heating. However, heat gains of this nature can be modified by careful adjustment of the amount of glazed areas (either fixed or openable); the type of glass used for windows; build-ing and configuration; thermal character of the building fabric; surface colours and texture; and the degree of absorption permitted by exposed materials. The usual methods adapted are shading or screening devices, in-creasing the thermal capacity of the building fabric, and adopting a reflective outer surface. Nevertheless, with special reference to colder climates like the United Kingdom, the beneficial physiological and biological effects upon humans, animals and plants should be considered before deciding to eliminate solar heat gains altogether.

Besides solar sources, considerable heat gains can be encountered when the function of a building requires the use of large amounts of energy. Office lighting sys-tems and mechanical ventilation plant, exhibition dis-play cases and lamps, and even people crowded together in supermarkets, cinemas, swimming-pools and disco halls can all create a considerable amount of heat. In large buildings the heat generated in this way may sometimes mean that cooling or ventilation plant is needed instead of heating plant, even during the cold winter season. The designer of such a building must ensure the correct *thermal balance* between heat require-ment, heat inputs or gains and the required insulation standards for the building fabric. Considerable financial benefits may also be obtained if the heat gains can be transformed and stored for later use; see also section 8.3.

8.7 Active means

To arrive at a suitable method of space heating for a building requires examination of the fuel to be used as a source of heat, the methods of distributing the heat source to the heat emitters, and the appropriate means of heat output. Detailed analysis of these factors is beyond the scope of this book, but it will be realised from the immediately preceding paragraphs that they exert a profound influence on the thermal comfort standards achieved by a building.

With few exceptions, any form of heat emission can be powered by any fuel. The choice currently available includes fossil fuels (wood, coal, gas and oil); electri-city; and, to a lesser extent, renewable resources (solar, animal wastes, geothermal, tidal, wind, wave, and ambi-ent energy from light fittings, mechanical plant and even human beings). Fossil fuels are being consumed at an increasingly rapid rate to keep up with comfort standards for today. In order to regulate consumption of these finite resources and to control the atmospheric pollution from their combustion, the following items now have high priority:

- Increasing building insulation standards
- Developing more efficient fuel combustion plant
- Introducing energy management systems
- Promoting schemes and systems to renew and reuse energy

Wood is a renewable resource, but as materials for building purposes are also being depleted, it is becom-ing an extremely valuable commodity which is far too important to burn. Electricity is generated mainly from fossil fuels, and for this reason *ideally* should not be used as a direct heating source because its production by this means is both inefficient and expensive.

The increasing use of nuclear reactors for the produc-tion of electricity, using naturally occurring resources,

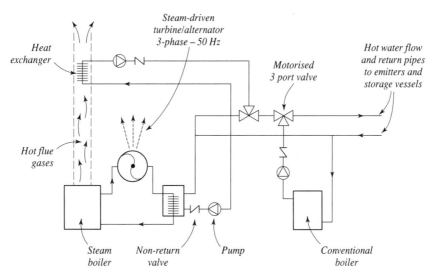

Figure 8.10 Principles of a CHP installation.

certainly makes a more viable heating source. Developments in nuclear fusion (there is still an ample supply of uranium) could make electricity the major future source for thermal comfort, provided conversion to a useful form can be accomplished in safety and without danger of radiation or pollution to people and the countryside. Fear of these happening is one of the reasons for increasing interest in the use of certain renewable resources and ambient alternative energy sources for generating electricity, e.g. wind- and wave-powered turbines.

Combined heat and power (CHP) systems have enjoyed far more success. Size of plant can vary from small units applied to hotels, schools, etc. to larger-scaled district heating/energy systems. The principle shown in Figure 8.10 uses the surplus energy in flue gases and cooling water from conventional oil- or gas-fired electricity generators, directed through heat exchangers for use with hot water storage and heating.

One of the most important decisions, as far as user comfort is concerned, is the choice of the form of *emitter* (radiator, convectors, ducted warm air, etc.); see Figures 8.11 and 8.12. Figures 11.1 and 11.2 show installation principles. The selection of emitter is influenced by the response of a building fabric to changes in external air temperature, to temperature variations caused by the sun, and to the heat gains from changing occupancy, cooking or the use of other heat-producing equipment. If discomfort is to be avoided, either from underheating or overheating, the response of the heating system must be equal to or shorter than the response of the building fabric. The response of the fabric is

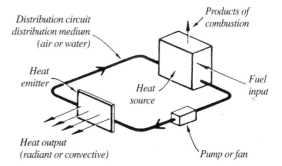

Figure 8.11 Circuit diagram for heater.

determined by its mass, the degree and position of insulation, the reflectivity of external surfaces where exposed to the sun, and the area and orientation of windows. Furthermore, controls have a vital role to play in achieving economy of operation; the generally available controls are not yet able to provide immediate response to changes in temperature, so there may be a time lapse, creating discomfort.

8.7.1 Space heating and hot water controls – Building Regulations

The regulations are intended for systems with centralised boiler heat energy sources and not individual solid fuel, gas or electric heaters. Electric storage heaters are included, with an expectation that units have an automatic charge control mechanism (thermostat) to regulate the energy consumption relative to room temperature.

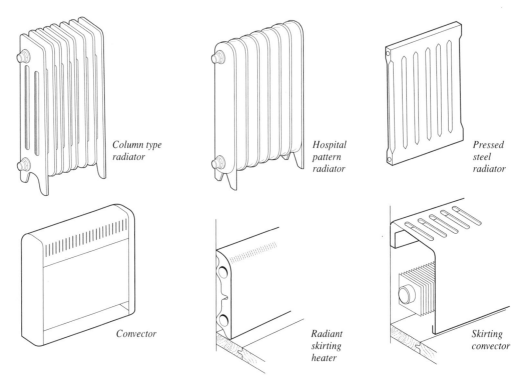

Figure 8.12   Heat emitters.

Conventional gas- or oil-fired boiler systems of hot water or warm air circulation satisfy the regulations if they contain the following:

● zone control
● timing control
● boiler control

Zone control – a means of controlling individual room temperatures where different heating needs are appropriate. This may be achieved by fitting thermostatic radiator valves to each emitter, or by using room thermostats to control zoned circuits. For instance, kitchen and workroom temperatures need not be the same as in living areas. Some examples with minimum air change rates are listed in Table 8.5. Large dwellings should be zoned into maximum floor areas of 150 m² with each zone having separate time controls.

Timing control – systems should be capable of being programmed to provide both hot water and space heating independently at predetermined times. This applies specifically to gas- and oil-fired installations, and solid fuel boiler systems with forced air draught electric fans. Separate timing is unnecessary for boilers with an instantaneous draw-off facility, e.g.

combination boilers and natural draught solid fuel boilers.

Boiler control – gas- and oil-fired boilers should be controlled within the programmed cycle by room thermostats for space heating and a storage cylinder thermostat for hot water. Where thermostatic radiator valves are deployed, overall control from a room thermostat should disconnect the boiler and pump when there is no demand for heat. A thermostatic pipeline switch would be suitable to control stored

Table 8.5   Internal design temperatures and air infiltration rates

Location	Temperature °C	Air changes per hour
Bathroom	22	2.0
Bedroom	18	1.0
Bed/sitting room	21	1.5
Dining room	21	1.5
Hall/landing	18	1.5
Kitchen	18	2.0*
Living room	21	1.5
WC	18	2.0

* Note: See Building Regulations, Approved Document F1: *Means of ventilation*, and section 10.4 in this book.

hot water. Additionally, manufacturers provide a manually controllable working thermostat within a boiler and a limit (high temperature) thermostat as a supplementary safety cut-out on boilers applied to larger non-domestic installations.

Additional requirements for domestic hot water storage systems include:

- Storage vessels to have a minimum of 35 mm factory applied polyurethane foam insulation (minimum density = 30 kg/m³).
- Insulated primary pipework with pumped circulation.

See also:

- Building Regulations, Approved Document G3: *Hot water storage*.
- Building Regulations, Approved Document L1: *Conservation of fuel and power in dwellings*.
- Water Manufacturers' Association: *Performance specification for thermal stores – 1999*.

Additional requirements for hot water storage systems in non-domestic buildings include:

- Particular attention to correct sizing of plant to avoid energy wastage.
- Supplementary use of solar power.
- Minimising number and length of hot water draw-off pipes (see Fig. 11.1).

See also: Building Regulations, Approved Document L2: *Conservation of fuel and power in buildings other than dwellings*.

8.8 Solar energy

Notwithstanding section 8.6, the reuse of solar energy is becoming increasingly important as an alternative means of both space and water heating, and various forms of construction have been devised which 'capture' this free resource. For example, a glazed conservatory on the south side of a building will create an accumulation of heat which can be absorbed into an interior wall of high thermal capacity (Fig. 8.13). The heat stored in this manner will be emitted into the interior spaces at night or during other periods of no sunshine. The amount of heat absorbed by an internal wall of this nature can be regulated by louvres opening in the conservatory so that excessive heat gains will not occur and cause discomfort. This system is known as a *passive* solar energy resource.

Active solar systems have been the subject of considerable research over the past 50 years. Basic systems consist of an exposed glass collector panel, behind which are located pipes containing water circulated through a heat exchanger in a storage vessel – see Figure 8.14. More recent developments are far more sophisticated and include collectors inside clear glass vacuum-sealed cylinders, as well as power generation from photovoltaic fuel cells. In the UK, the water heated by solar means can provide a useful supplement to hot water and space heating systems, whereas in many parts of the world it is the principal energy resource due to the high levels of solar radiation.

The potential for solar energy is considerable as solar radiation can be quite effective even on cloudy

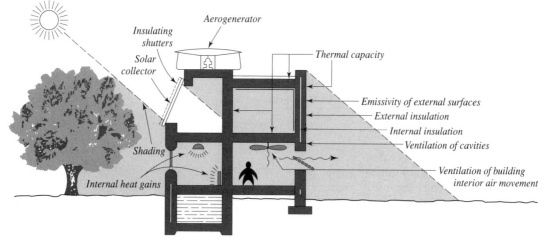

Figure 8.13   Using solar energy.

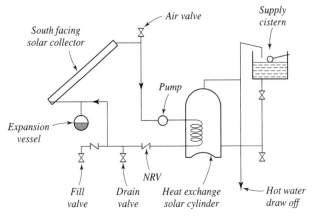

Figure 8.14   Principle of solar-powered hot water system.

days. In the UK, the average amount of solar radiation on a south-facing roof inclined at an optimum angle of about 40°, is around 1000 kW/m². To date, there has been some reluctance to accept these systems in Britain, as the capital outlay may take several years to recoup in fuel savings. Also, the exposed panels can be perceived as a visual intrusion on the appearance of a building.

Further specific reading

Mitchell's Building Series

Environment and Services	Chapter 5 Heat
	Chapter 7 Thermal installations
Internal Components	Section 6.2.4 Doors to limit thermal effect
External Components	Section 3.2.5 Thermal control with external glazing
	Section 4.2.5 Thermal control with windows
	Section 7.2.4 Thermal control with external doors
	Section 8.2.3 Thermal control with roofings
Materials	Section 1.5 Thermal properties and insulation
Materials Technology	Chapter 4 Masonry construction (thermal performance)
Structure and Fabric Part 1	Chapter 5 Walls and piers
	Chapter 7 Roof structures
	Chapter 8 Floor structures

Building Research Establishment

BRE Digest 108   *Standard U-values*
BRE Digest 145   *Heat losses through ground floors*
BRE Digest 162   *Traffic noise and overheating in offices*
BRE Digest 180   *Condensation in roofs*
BRE Digest 232   *Energy conservation in artificial light*
BRE Digest 236   *Cavity insulation*
BRE Digest 254   *Reliability and performance of solar collector systems*
BRE Digest 270   *Condensation in insulated domestic roofs*
BRE Digest 297   *Surface condensation and mould growth in traditional-built dwellings*

BRE Digest 302   *Building overseas in warm climates*
BRE Digest 306   *Domestic draughtproofing: ventilation considerations*
BRE Digest 319   *Domestic draughtproofing: materials, costs and benefits*
BRE Digest 324   *Flat roof design: thermal insulation*
BRE Digest 336   *Swimming pool roofs: minimising the risk of condensation using warm-deck roofing*
BRE Digest 339   *Condensing boilers*
BRE Digest 355   *Energy efficiency in dwellings*
BRE Digest 369   *Interstitial condensation and fabric degradation*
BRE Digest 370   *Control of lichens, moulds and similar growths*
BRE Digest 377   *Selecting windows by performance*
BRE Digest 379   *Double glazing for heat and sound insulation*
BRE Digest 404   *PVC-U windows*
BRE Digest 438   *Photovoltaics: Integration into buildings*
BRE Digest 453   *Insulating glazing units*
BRE Report 262   *Thermal insulation: avoiding risks*
BRE Information Paper 21/92   *Spillage of flue gases from open flued combustion appliances*
BRE Information Paper 3/90   *The 'U' value of ground floors: application to building regulations*
BRE Information Paper 7/93   *The 'U' value of solid ground floors with edge insulation*
BRE Information Paper 12/94   *Assessing condensation risk and heat loss at thermal bridges*
BRE Information Paper 14/94   *'U' values for basements*
BRE Information Paper 15/94   *Energy efficiency in new housing*

Building Regulations

G3   Hot water storage
L1   Conservation of fuel and power in dwellings
L2   Conservation of fuel and power in buildings other than dwellings

9 Fire protection

CI/SfB (K)

It is ironic that, although fire is used in the manufacture of most materials and can provide the thermal conditions required in a building, it can also be highly destructive to a building and its occupants. Death or injury by fire is particularly horrifying and incidents naturally receive special concern in the minds of most building occupiers. Account must also be taken of the financial loss of a building and its contents when considering the full effects of devastation by fire. Losses can be in excess of £500 million per annum when taking account of the current inflationary trends in the cost of material and labour.

The design and construction method employed for a building must therefore safeguard occupiers from death or injury and also minimise the amount of destruction. These goals can be achieved through an understanding of the nature of fire and its effects on materials and construction used in a building; methods of containing a fire and limiting its spread; methods of ensuring the occupants of a building being attacked by fire can escape to safety; and methods of controlling a fire once it has started.

9.1 Combustibility

Fire is a chemical action resulting in heat, light, and flame (a glowing mass of gas), accompanied by the emission of sound. To start, a fire needs a combustible substance, oxygen, and a source of heat such as from a flame, friction, sparks, glowing embers or concentrated solar rays. Once a fire has started, heat is usually produced faster than it is dissipated to its surrounding, therefore the temperature rises with time. The fire will burn out once the fuel is removed, become smothered if oxygen is not available, or die if the heat is removed

by water, for example. Before this happens, however, a fire is likely to ignite nearby material, causing the fire to spread through a building, and perhaps to adjoining buildings by processes which include radiation.

In the context of fire protection, materials used in a building fall into two broad categories: *combustible* and *non-combustible*. To determine the extent of combustibility or non-combustibility of a material or application, BS 476: *Fire tests on building materials and structures*, should be consulted. It is produced in several parts to cover the extensive and varied testing procedures for fire propagation, ignitability of materials, resistance to fire, etc. It will generally be found that inorganic materials are non-combustible, e.g. stone, brick, concrete or steel, whereas organic materials are combustible, e.g. timber and its by-products (fibreboard, plywood, particle board, etc.) as well as petrochemical products, including plastics.

9.2 Fire resistance

In situations where it is imperative that a building must not contribute fuel to a fire, only non-combustible materials should be employed. However, even if practical, the sole use of non-combustible materials will not necessarily avoid the spread of a fire generated by the burning contents of a building. Avoidance of this requires the parts of the building – materials and construction used for elements (walls, floors, etc.) – to have *fire resistance*. This is the term used to describe the ability of an element of building construction to fulfil its assigned function in the event of a fire without permitting the transfer of the fire from one area to another. BS 476-21 and 22: *Methods for determination of the fire resistance of load-bearing and non-load-bearing*

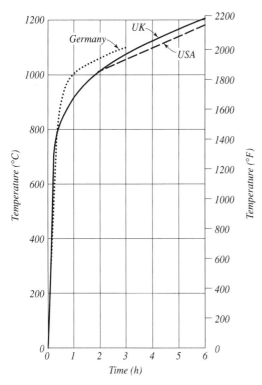

Figure 9.1   Time–temperature curves used in different countries.

elements of construction, respectively, establish a time period during which elements can be expected to perform this function. A sample must be subjected to a simulated building fire which, research has indicated, rises in temperature according to duration; see the time–temperature curve in Figure 9.1. The tested element is then rated according to the time (0.5, 1, 2, 3, 4, 5 or 6 hours) it is able to fulfil three performance criteria. These criteria approximate to those necessary to ensure the reasonable safety of occupants and contents in a building, including time to discover the fire and make a safe escape. The performance criteria (Fig. 9.2) applied in the tests are as follows:

Resistance to collapse (previously referred to as *stability*)   A *load-bearing* element of construction must support its full load during a fire for a specified period. This minimum period varies from 15 minutes up to 2 hours, depending on which part of a building it relates to and the purpose grouping or type of building it applies to. See tables in Appendix A of Approved Document B to the Building Regulations.

Integrity   Structural resistance to the passage of flames and hot gases. Failure occurs when cracks or other openings form, through which flame or hot gases can pass; this would cause combustion on the side of the element remote from the fire.

Insulation   The ability of the construction to resist fire-transmitted heat. Failure occurs when the temperature on the side of the element remote from the fire is increased *generally* by more than 140 °C, or *at any point* by more than 180 °C above the initial temperature.

The period of fire resistance suitable for particular elements of a building depends on the functions to be accommodated and the volume, height and floor areas of the spaces involved (Table 9.1). In addition to the Building Regulations, more stringent requirements are likely to be determined by the Loss Prevention Council (representing insurers) and recommendations in BS 5306: *Fire extinguishing installations and equipment on premises*.

Combustibility cannot be equated with fire resistance: one is a characteristic of a material, the other relates to the performance of an element as a whole. For example, an external wall of fibre cement profiled sheeting (non-combustible) on a mild steel frame (non-combustible) has no notional period of fire resistance because a construction of this nature would rapidly permit fire to spread by heat transfer, i.e. it would not meet the insulation requirement of fire resistance. Conversely, an external wall of timber cladding (combustible) on timber studs (combustible) with a plasterboard internal lining (combustible because of paper sheathing) will provide full fire resistance for a period of half an hour. However,

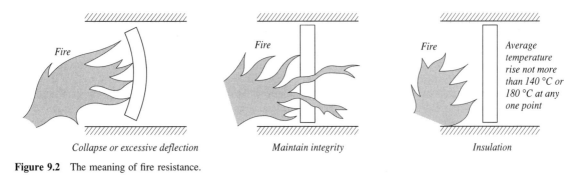

Collapse or excessive deflection   Maintain integrity   Insulation

Figure 9.2   The meaning of fire resistance.

Table 9.1 Fire resistance for elements of construction in buildings of various purpose groups

Building type	Height (m) to top storey floor	Compartment floor area (m²) per storey, max.		Compartment volume, max. (m³)		Fire resistance (mins)
		Multi-storey	Single storey	Multi-storey	Single storey	
Storage:						
non-sprinklered	<18			20 000	no limit	90
	18–30			4 000	no limit	120
sprinklered	<18			40 000	no limit	60
	>18			8 000	no limit	90
	>30			8 000	no limit	120
Industrial:						
non-sprinklered	<18	7 000	no limit			90
	18–30	2 000	no limit			120
sprinklered	<18	14 000	no limit			60
	>18	4 000	no limit			90
	>30	4 000	no limit			120
Shops:						
non-sprinklered	<18	2 000	2 000			60
	<30	2 000	2 000			90
sprinklered	<18	4 000	no limit			60
	<30	4 000	no limit			60
	>30	4 000	no limit			120
Offices:						
non-sprinklered	<5	no limit	no limit			30
	<18	no limit	no limit			60
	<30	no limit	no limit			90
sprinklered	<18	no limit	no limit			30
	<30	no limit	no limit			60
	>30	no limit	no limit			120
Dwellings:						
houses	<5					30
	<18					60
flats and maisonettes	<5					30
	<18					60
	<30					90
	>30					120

there are cases where legislation requires both fire resistance and non-combustibility, e.g. external walls of a building located in close proximity to a boundary.

9.3 Spread of fire

The *spread of fire* within a building and from one building to another can most effectively be restricted by the identification and isolation of potential hazards. Therefore, the first defence involves siting. Having established that potential fire risks exist by the nature of the functions accommodated, it is necessary to select the appropriate siting for a building relative to safety of nearby properties. An extreme example might be that of a building accommodating particularly dangerous fire hazards (e.g. manufacture and/or storage of highly flammable chemicals). This should be located in a remote part of the countryside, away from properties likely to be damaged by heat radiation caused as a result of the fire.

Building legislation exists (Building Regulations, Approved Document B: *Fire safety*) to restrict fire spread by stipulating periods of fire resistance and construction methods appropriate to the function of a building, as discussed earlier. It also limits the use of certain construction methods according to their distance from other properties and/or boundaries. For example, only a nominal amount (0.1 m²) of unprotected combustible material (such as timber cladding) is permitted on an external wall of a small residential building if this is within 1 m of its boundary. The unprotected area can increase relatively between distances of 1 m and 6 m; over 6 m the amount is unrestricted. Windows, doors and other openings in external walls are also carefully controlled since, unless special forms are used, they do little in stopping the spread of fire from inside or into a building. Regulations restrict the positioning, size and amount of these openings according to function (fire risk of a building as well as location of the external wall relative to other properties and/or boundaries).

Whereas the general approach on walls and floors is to 'contain' the fire, there is no legislative requirement for the external surfaces of a roof to have fire resistance. Instead reference is made to test procedures in BS 476-3, which grades the suitability of specific roof coverings according to their ability to resist external penetration by fire and their resistance to surface spread of flame.

The spread of fire within a building can similarly be restricted by the special segregation of particular fire hazards. Also, fire-resisting *compartments* or 'cells' can be used to limit the spread of a fire through a building. The precise location, enclosed volume and period of fire resistance required again depends on the nature of the activities accommodated. The height of a building and the ease with which its occupants can escape, and/or fire-fighting can be successfully carried out, are also deciding factors (Fig. 9.3). Structural organisation (see

section 5.2) is also important in this respect. Whereas continuous masonry wall construction with reinforced concrete floors uses materials capable of providing a high standard of fire resistance, the 'infill' type of constructions needed for framed buildings must be carefully selected to provide adequate fire resistance as well as fulfilling other performance requirements, including compatibility and continuity with the support system.

One aspect of the fire-resisting requirements is that the structural organisation of a building should not collapse or deform before the occupants can escape safely. A collapse will be of major importance when considering the spread of fire and also fire-fighting (see section 9.7), and framed structural organisation may need special consideration relative to the materials employed. As long as sufficient insulating cover of concrete is provided to the steel bars of reinforced concrete columns and beams, adequate defence against the untimely collapse or deformation is reasonably easy to achieve. However, steel columns and beams (Fig. 9.4) can present bigger problems since, depending on the precise

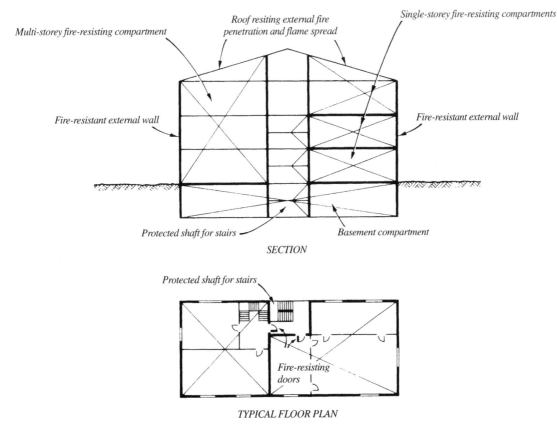

Figure 9.3 Using compartments to limit the spread of fire in a building. (Adapted from material published in the *Architects' Journal*)

FIRE PROTECTION CONCEALS STEELWORK

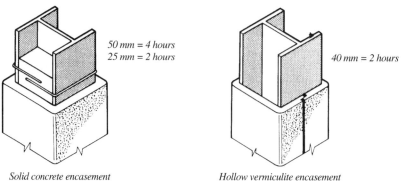

50 mm = 4 hours
25 mm = 2 hours

40 mm = 2 hours

Solid concrete encasement

Hollow vermiculite encasement
on expanded metal backing

FIRE PROTECTION FOLLOWS PROFILE OF STEELWORK

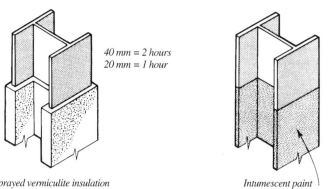

40 mm = 2 hours
20 mm = 1 hour

Up to 1 hour
depending on
applications

Sprayed vermiculite insulation

Intumescent paint

EXPOSED STEELWORK

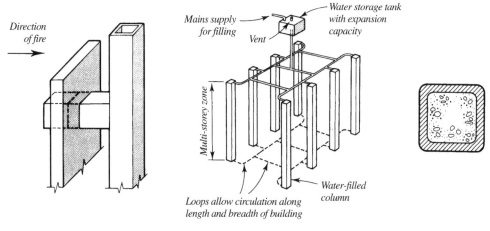

Direction
of fire

Mains supply
for filling

Water storage tank
with expansion
capacity

Vent

Multi-storey zone

Loops allow circulation along
length and breadth of building

Water-filled
column

Heat-shielded sections

Water-filled sections

Concrete-filled hollow sections

Figure 9.4   Methods of providing fire resistance to steel columns.

physical composition, their ultimate strength is about 50 per cent when subjected to a temperature of 550 °C. This can be achieved within 15 minutes in a BS 476 test fire, as already indicated in Figure 9.1. Unless special design techniques are employed which keep temperature below this level for at least the required fire-resisting period of a building, structural steel sections must be insulated with a solid or a hollow casing. This protection must also be compatible with the prevailing environmental and functional conditions. External casings should be weather and impact resistant, durable and of acceptable appearance to maintain their effective life in a building. Similarly, internal fire-insulating casings may be required to withstand knocks from trolleys, to provide fixings for fitments and space for service pipes or cables, as well as to be aesthetically suitable.

Interestingly, timber columns and beams can provide greater fire resistance than unprotected steel. Figure 9.5 indicates the use of sacrificial timber, which provides effective insulation to structural sections subjected to a fire.

Apart from normal fire-resisting requirements (including doors, etc.), care must be taken in the selection of finishes which will not contribute too much to the spread or growth of fire. To insist on the use of non-combustible materials in all circumstances would be too onerous and restrictive. For this reason, the surface spread of flame test set out in BS 476-7 was developed to classify the relative risk of various combustible materials, and these can be used in certain positions, though the area may be limited. Building legislation has introduced a further surface spread of flame classification that covers not only non-combustible materials but also materials with a surface finish giving low fire-propagation properties (BS 476-6), which therefore contribute little to the growth of a fire.

9.4 Means of escape

The occupants of a building must be provided with clearly-defined and safe escape routes in the event of a fire. These routes must be kept clear of obstruction, be easy to manoeuvre and, very importantly, should be free of the effects of flames, heat and smoke (Fig. 9.6). For this reason escape routes need special consideration regarding safety from fire; access points must be shielded by lobbies and fire-resisting (and smoke-resisting) doors; walls and floors should have sufficient fire resistance to allow escape and subsequent access for fire-fighters to tackle the fire. Sometimes it may be necessary to provide positive air pressure in the escape route to ensure smoke is forced back into the body of the building, making exit easier, or an extractor ventilation unit can be provided to remove smoke. Internal surface finishes must not contribute to the fire or provide hazard to people escaping, and escape areas must be adequately lit, sometimes by emergency lights powered from a separate generator.

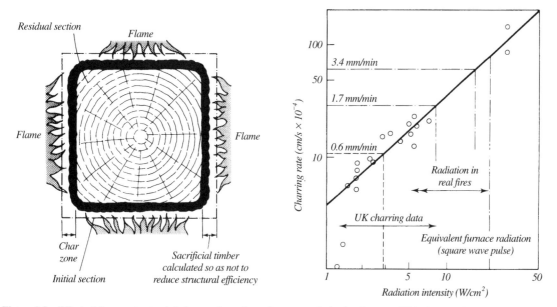

Figure 9.5 Effect of fire on structural timber sections. Once the structural size has been established, extra thickness can provide insulation. The thickness required for a specific period of fire protection for a particular type of timber can be derived from its charring rate. (Adapted from material in *Fire Research Note 896* published by the BRE)

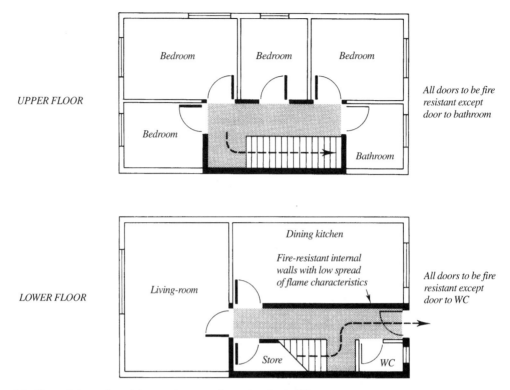

UPPER FLOOR

Bedroom   *Bedroom*   *Bedroom*

All doors to be fire
resistant except
door to bathroom

Bedroom

Bathroom

LOWER FLOOR

Dining kitchen

Fire-resistant internal
walls with low spread
of flame characteristics

All doors to be fire
resistant except
door to WC

Living-room

Store   *WC*

Figure 9.6   Fire-resisting walls and doors assist escape from a burning building.

Tragedies can occur unless means of escape from a building are correctly designed and sited. In the event of a fire it should be possible to evacuate a building in reasonable time ($2\frac{1}{2}$ minutes is considered normal for everyone to reach a place of safety, except for special premises such as hospitals). Specific fire safety requirements for hospitals are contained in Health Technical Memorandum (HTM) No. 81: *Fire precautions in new hospitals*. This document satisfies the objectives set out in the Building Regulations, Approved Document B. The width, the size of treads and risers, and heights of handrails for staircases used for escape purposes must also conform to similar safety requirements. Escape-route planning should therefore be related to the use of the building, the number of occupants, the risks involved, and to the heights of floors above ground and the shapes and dimensions of floors. The risk of persons being trapped or overcome by the effects of a fire is greatest in multi-storey buildings, and access routes and standing positions for fire-brigade appliances are critical factors in the design of escape routes. For an appreciation of the minimum dimensions and location of escape routes in a variety of building types, refer to BS 5588: *Fire precautions in design, construction and use of buildings*.

9.5 Fire alarms

Dwellings should be provided with an automatic fire detection and alarm system to the recommendations of BS 5839: *Fire detection and alarm systems in buildings*, i.e.

- BS 5839-1: *Code of practice for system design, installation and servicing*, or
- BS 5839-6: *Code of practice for the design and installation of fire detection and alarm systems in dwellings*.

An alternative is a mains electricity operated smoke alarm, conforming to BS 5446: *Components of automatic fire alarm systems for residential premises*. Specifically:

- BS 5446-1: *Specification for self-contained smoke alarms and point-type smoke detectors*.

Note: These mains operated detectors should also have a battery back-up secondary power supply.

A large house, i.e. premises having a floor area greater than 200 m^2 in any one storey, and having more than three storeys (a storey includes ground floor and basement floors) should be fitted with a BS 5839-1 system. A large house of no more than three storeys may be fitted with a BS 5839-6 system.

Smoke alarms should be ceiling mounted at least 300 mm from light fittings, with provision of at least one per storey. Specific requirements for loft conversions include an alarm linked to operate the signal in other detectors. Other preferred locations include:

- Circulation spaces between bedrooms.
- Circulation spaces, no further than 7.5 m from any door to a habitable room.
- Kitchens.
- Living rooms.

For buildings other than dwellings, consideration must be given to the type of occupancy and purpose of the building. Whilst the basic requirements of BS 5839 alarm systems will suit some buildings, others such as large shopping units may be best cleared by trained staff, rather than an alarm system which could cause panic. Consultation on specific requirements for non-domestic buildings should be undertaken with the local building control authority and the area fire authority, to determine optimum warning systems and procedures in event of a fire. See also:

- Fire Precautions Act.
- Fire Precautions (Workplace) Regulations.
- BS 5839-8: *Code of practice for the design, installation and servicing of voice alarm systems.*
- BS 5588: *Fire precautions in the design, construction and use of buildings.*

9.6 Fire (resistant) doors

Fire doors have a performance expectation in excess of normal means of access requirements. They are required to have sufficient *integrity* (see section 9.2) in a fire, for a predetermined period which complements the element of construction in which they are located. BS 476-22: *Methods for determination of the fire resistance of non-load-bearing elements of construction*, categorises door integrity for a period of minutes. For example, an FD 30 classification indicates a Fire Door which can resist fire penetration for 30 minutes. If an S is added to the classification, e.g. FD 30S, this indicates that the door has the added facility to resist smoke leakage. Reference to the Building Regulations, Approved Document B, Appendix 3 Table B1 provides a listing of fire door locations. These are primarily in compartment walls, escape routes and enclosures.

All fire doors should be fitted with an automatic closing device. Some exception is permitted if self-closing would be considered a hindrance. In these circumstances the door can have a fusible link to hold the door open (unless the doorway is an escape route), an automatic release mechanism actuated by a fire detection system, or a door closure delay device. Hardware and ironmongery should satisfy the requirements of:

- BS 476-31: *Methods of measuring smoke penetration through door sets and shutter assemblies.* Note: A door set refers to a door and its lining or frame, as an assembly.
- Association of Builders' Hardware Manufacturers: *Code of Practice. Hardware essential to the optimum performance of fire-resisting timber door sets.*

A colour-coded plastic plug is fitted into the door hanging style to indicate the fire resistance, e.g. a blue core with a white background indicates FD 20 with no intumescent strip in the frame, or FD 30 with an intumescent strip in the frame. An FD 60 door with two intumescent strips in the frame has a colour-coded plug with a red core and blue background. Refer to BS 8214: *Code of practice for fire door assemblies with non-metallic leaves.*

An intumescent strip can be fitted into a recess in the frame as shown in Figure 9.7. At temperatures of about 150 °C the seal expands to prevent the passage of fire and smoke, but it does not prevent the door being moved physically. An alternative for smoke sealing only is a brush strip seal which also functions as a draught seal.

9.7 Fire-fighting

The techniques so far described for fire protection can be classified as *passive measures* as they are an inbuilt

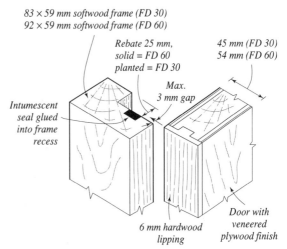

Figure 9.7 Fire door and frame.

feature of a building. *Active measures* of fire protection incorporate fire-fighting techniques which can be inbuilt (sprinklers, dry/wet risers, alarm systems, fusible links to doors, shutters, fire-fighters' lifts, etc.) or be brought to a building in distress (fire-fighters and appliances). The siting of a building must allow easy access for fire-fighters and their appliances in an emergency. Hard-standing areas must be included in the landscaping and located at prescribed distances from a building to facilitate fire-fighting activities: see Building Regulations, Approved Document B5.

9.8 Fire insurance

The influences associated with the insurance of a building against damage or loss through the occurrence of a fire are important aspects of initial design considerations. Fire insurers often look for higher standards of fire protection than are stipulated in the clauses of fire legislation documents or recommendations given directly by the fire authority. For this purpose the insurance industry has published its own *Code of Practice for the Construction of Buildings*, produced by the Loss Prevention Council. It complements the Approved Documents and British/European Standards supporting the Building Regulations and generally requires higher performance standards. Failure to consider this publication and any other specific insurance requirement at an early stage of design, or during the selection processes for appropriate construction methods, may result in high premiums for fire insurance or the adoption of costly modifications.

Further specific reading

Mitchell's Building Series

Environment and Services	Chapter 18 Firefighting equipment
Internal Components	Section 2.3.4 Fire precautions with demountable partitions
	Section 2.3.5 Fire resistance of demountable partitions
	Section 2.4 Building Regulations and fire resistance
	Section 3.3.4 Fire precautions with suspended ceilings
	Section 4.2.4 Fire precautions with raised floors
	Section 5.1.5 Thermal and fire resistance
	Section 6.2.5 Fire doors
	Section 6.9 Fire-resisting doors
	Section 6.10 Sliding fire doors
External Components	Section 3.2.6 Fire precautions with external glazing
	Section 4.2.6 Fire precautions with windows
	Section 5.2 Performance requirements for rooflights and patent glazing
	Section 7.2.5 Fire precautions with external doors
	Section 8.2.5 Fire precautions with roofings
Materials	Section 1.8 Deterioration
Materials Technology	Section 3.7 Mechanical behaviour of concrete (fire resistance)
	Section 4.1 Fire parameters in masonry construction
	Section 5.2 Design of glazing – fire
	Chapter 12 Timber (fire protection)
Structure & Fabric Part 1	Chapter 5 Walls and piers (functional requirements)
	Chapter 6 Framed structures (functional requirements)
	Chapter 7 Roof structures (functional requirements)
	Chapter 8 Floor structures (functional requirements)
	Chapter 9 Fireplaces, flues and chimneys (functional requirements)
	Chapter 10 Stairs (functional requirements)
Structure & Fabric Part 2	Chapter 10 Fire protection

Building Research Establishment

BRE Digest 208   *Increasing the fire resistance of timber floors*
BRE Digest 225   *Fire terminology*
BRE Digest 230   *Fire performance of walls and linings*
BRE Digest 285   *Fires in furniture*
BRE Digest 288   *Dust explosions*
BRE Digest 294   *Fire risk from combustible cavity insulation*
BRE Digest 300   *Toxic effects of fires*
BRE Digest 317   *Fire resistant steel structures: free standing blockwork-filled columns and stanchions*
BRE Digest 320   *Fire doors*
BRE Digest 367   *Fire modelling*
BRE Digest 385   *Sprinkler protection of storages*
BRE Digest 388   *Human behaviour in fire*
BRE Digest 396   *Smoke control in buildings*

Building Regulations

B1   Means of warning and escape
B2   Internal fire spread (linings)
B3   Internal fire spread (structure)
B4   External fire spread
B5   Access and facilities for the fire service

10 Lighting and ventilation

CI/SfB (N) + (L2)

Together with the provision of thermal comfort and sound control, lighting and ventilation provide the initial environmental aspects of a building which ensure the physiological and psychological well-being of its occupants. However, the advancement of technology has led to some degree of complacency on the part of designers concerning the influence which the provision of acceptable lighting and ventilating standards has on the overall appearance and construction method of a building. Apart from purely functional requirements, they provide the principal means of creating aesthetic atmosphere and character. Nevertheless, whereas these functions once derived purely through an interrelationship between architectural form and construction method, it is becoming increasingly possible to design a building where fashion can be *made* to function by the use of artificial devices.

In all but the simplest form of building, often it may be considered normal to make 'corrections' in the lighting levels not achieved from the designed building by the use of electric light systems. Similarly, air-conditioning apparatus can be made to compensate for the lack of sufficient natural ventilating openings (windows, chimney openings, etc.). Indeed, artificial systems may even be oversized to make the thermal environment acceptable because of ill-considered decisions about siting, orientation or the amount of glazing in a building envelope.

10.1 Standards

Although the size, position and amount of window openings are necessarily controlled by current needs to conserve the use of energy and to ensure safety from the spread of fire or the intrusion of unwanted sound, a sensible balance must be achieved between these aims and acceptable lighting/ventilating standards. This can be accomplished by detailed analysis of each requirement, careful design decisions and adoption of suitable construction methods. Artificial lighting is required for certain activities to assist concentration without eye strain, and also after natural light periods. A building with a deep plan form required by virtue of optimum function, e.g. large office floor spaces, warrants continuous artificial light sources. In this case a suitably designed system can convert the otherwise wasted heat, generated by the lamp, into useful back-up or supplementary space heating for a building. The residue energy can be similarly used from artificial ventilating systems required by a large building, or a building with the external envelope entirely sealed against noise or extremes of climate.

10.2 Natural lighting

Among its many other vital functions, the sun provides the sources of *natural light* which create the first psychological connection between the inside and the outside of a building. The influence of the sun in creating shaded areas affecting the vegetation and human enjoyment of spaces is a critical factor when considering the dimensions and shape of a building and its distance from others. The use of 'natural' coloured daylight to illuminate interior spaces can create interesting effects, caused by variations in intensity during the day influencing the shading of planes and the hue and depth of coloured surfaces.

Daylight is admitted into a building through 'holes' in external fabric (windows, rooflights, etc.), which in adverse climates generally incorporate glass or an alternative transparent material to control the effects of heat loss and/or inclement weather on the interior spaces. The amount of light received inside a building is usually only a small fraction of that received outside – because of modifications imposed by the size and position of openings – and will also constantly vary, owing to the influences imposed on the 'whole sky' illumination level by clouds, buildings and/or other reflecting planes. Therefore, it is impracticable to express interior daylighting in terms of the illumination actually obtainable inside a building at any one time, for within a few minutes that figure is liable to change with corresponding changes in the luminance of the sky.

For practical purposes, use is made of the *daylight factor*. This is a percentage ratio of the instantaneous illumination level at a reference point inside a room to that occurring simultaneously outside in an unobstructed position. Typical daylight factors are indicated in Table 10.1. A simple rule of thumb can also be used to approximate the daylight factor:

$$D = 0.1 \times P$$

where: D = Daylight factor
P = Percentage glazing to floor area

e.g. given a room of 100 m² floor area with 20 m² of glazing:

$$D = 0.1 \times (20 \div 100) \times (100 \div 1) = 2\%$$

This can be more usefully represented in calculation of the natural *illuminance* (see section 10.3 and Table 10.2) at the reference point inside a building by applying the following formula:

$$D = (Ei \div Eo) \times 100$$

where: D = Daylight factor
Ei = Illuminance at reference point in building
Eo = Illuminance at the reference point if the room was unobstructed

Both factors of E are measured in lux (lumens per square metre), with Eo taken as a standard 5000 lux

Table 10.1 Typical recommended minimum daylight factors for rooms with side lighting only

Building type	Location	Daylight factor* (%)
Dwellings	Living rooms (over $\frac{1}{2}$ depth, but for minimum area 8 m²)	1
	Bedrooms (over $\frac{3}{4}$ depth, but for minimum area 6 m²)	0.5
	Kitchens (over $\frac{1}{2}$ depth, but for minimum area 5 m²)	2
Offices and banks	General offices, counters, accounting book areas, public areas	2
	Typing tables, business machines, manually operated computers	4
Drawing offices	General	2
	Drawing boards	6
Assembly and concert halls	Foyers, auditoriums, stairs (on treads)	1
	Corridors (on floors)	0.5
Churches	Body of church	1
	Chancel, choir, pulpit	1.5
	Altars, communion tables (depending on lighting emphasis required)	3–6
	Vestries	2
Libraries	Shelves (on vertical surfaces of book spines), reading tables (additional lighting on book stacks)	1
Art galleries and museums	General	1
	On pictures (but special provision for conservation where required)	6 (max.)
Schools and colleges	Assembly and teaching areas	2
	Art rooms	4
	Laboratories (benches)	3
	Staff rooms, common rooms	1
Hospitals	Wards	1
	Reception rooms, waiting rooms	2
	Pharmacies	3
Sports halls	General	2
Swimming pools	Pool surfaces	2
	Surrounding floor areas	1

* The minimum daylight factors recommended do not necessarily apply to the whole area of the interior. Unless otherwise stated, the values are for a horizontal reference plane at table or desk height (0.850 m above floor level).

for unobstructed sky in the UK. So, transposing the formula to make Ei the subject:

$$Ei = (D \times Eo) \div 100$$
$$Ei = (2 \times 5000) \div 100 = 100 \text{ lux}$$

A comparison of illumination levels is shown in Table 10.2.

Daylight reading of a reference point in a room can be made up of three components: *sky component*, or the light received directly from the sky; *externally reflected component*, which is the light received after reflection from the ground, building or other external surface; and *internally reflected component*, which is the light received after being reflected from the surfaces inside a building. The design of a building must take into account these three factors if the 'correct' amount of daylight is an essential factor in its function and if the design and construction method are closely related. Figure 10.1 indicates various arrangements to achieve acceptable daylight factors within a building for specific visual functions. The arrangement of the windows and other openings in the walls provides the main architectural character of a building, generally called *fenestration*.

No account is taken of the influence of direct sunlight, but various methods of calculating daylight factors have been devised for overcast sky conditions. Modifications to values can be made for glazing materials other than clear glass, dirt on the glass and reductions caused by the window framing.

If windows or skylights are within the normal field of vision inside a building, they are likely to be distractingly bright compared with other things occupants may wish to study. To reduce this apparent brightness, or *glare*, the openings should generally be placed away from interior focal points. Glare can also be reduced by reducing the brightness of the light source (tinted glass, louvres or curtaining) while increasing the brightness of the interior spaces by better light distribution techniques, such as the use of lighter colours for surfaces, or in extreme conditions by supplementary artificial lighting.

Although exerting a very pleasing influence, brightening interior colours and providing both psychological and physical warmth, direct sunlight in a building can cause intensive glare, overheating (see Chapter 8) and fading of surface colours. For this reason, sunlight used to illuminate a building is also often diffused or reflected to reduce its intensity. Shading and reflecting devices include trees, vines, overhangs, awnings, louvres, blinds, shades and curtains. Overhead shading devices (*brise soleil*) block or filter direct sunlight, allowing only reflected light from the sky and ground to enter a building.

Louvres and blinds are capable of converting direct sunlight into a softer, reflected light.

10.3 Artificial lighting

The chief drawback of daylighting is its inconsistency, especially its total unavailability after dusk and before sunrise. Artificial lighting can be instantly and constantly available, is easy to manipulate and can be controlled by the occupants of a building. However, daylighting and artificial lighting should be regarded as complementary. Artificial lighting is used mainly for *night-time illumination* and as a *daytime supplement* when daylighting alone is insufficient.

An acceptable balance of brightness within a building can be accomplished by an integration between the design of natural daylight sources and *artificial supplementary lighting* (Fig. 10.2) to provide the combined level of light appropriate to a specific visual task. During daylight hours natural light should appear dominant wherever possible. However, quite apart from artificial light sources supplementing lighting levels, the use of artificial lighting in a building could lead to more flexible internal planning arrangements and to the incorporation of fewer or smaller windows. Thus daytime supplementary artificial lighting schemes directly affect the appearance of a building and its economy of construction. Against this must be levied the probability of greater energy usage, although reference has already been made to the effects which artificial lighting installations have upon the heating load for a building, and the possible economic advantages obtained by the recycling of heat generated by lamps, etc.

The objective of lighting design is to achieve an appropriate brightness or *luminance* for a visual task to be performed. When establishing desired luminance levels, account must be taken of the appearance (position, colour, shape and texture) of all wall, ceiling and floor surfaces, as well as the selection of suitable light fittings not only to light the task to be performed, but also to provide appropriate amounts of reflected light. *Luminance* should not be confused with *illuminance*. Illuminance is the measure of light failing on a surface (lumens per square metre or lux), whereas luminance refers to light reflected from it or emitted by it (candela per square metre or alternatively apostilbil-luminance × reflection factor).

Table 10.2 lists illumination levels suitable for a range of situations: the quality of these levels could be influenced by glare and an acceptable limiting index is also shown. The glare index is calculated by considering the light source location, the luminances of the source, the effect of surroundings and the size of the source.

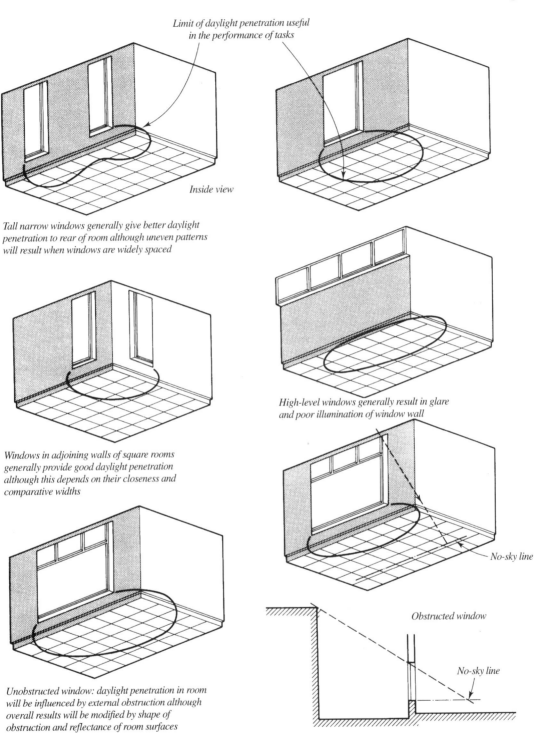

Figure 10.1 Effects of window shape and position on penetration and distribution of daylight. (Adapted from material in *Windows: Performance, Design and Installation* by Beckett and Godfrey, published by CLS/RIBA)

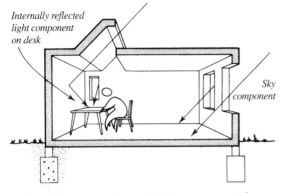

Space designed to provide natural lighting to area remote from external wall

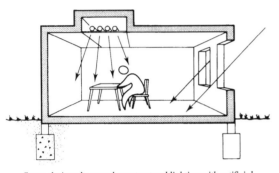

Space designed to supplement natural lighting with artificial lighting to area remote from external wall

Figure 10.2   Artificial supplementary lighting.

Table 10.2   Illumination levels and limiting glare indices for various functions

Location	Illuminance (lux or lm/m²)	Limiting glare index
Entrance hall	150	22
Stairs	150	22
Corridors	100	22
Outdoor entrances	30	22
Casual assembly work	200	25
Rough/heavy work	300	28
Medium assembly work	500	25
Fine assembly work	1000	22
Precision work	1500	16
General office work	500	19
Computer room	750	16
Drawing office	750	16
Filing room	300	22
Shop counter	500	22
Supermarket	500	22
Classroom	300	16
Laboratory	500	16
Public house bar	150	22
Restaurant	100	22
Kitchen	500	22
Dwellings		
Living room	50	N/A
Reading room	150	N/A
Study	300	N/A
Kitchen	300	N/A
Bedroom	50	N/A
Hall/landing	150	N/A
Library		
Reading area	200	19
Tables	600	16
Counter	600	16

N/A = not applicable.

Glare indices for artificial light range from about 10 for a shaded light fitting having low output to about 30 for an unshaded lamp.

As seen from Figure 10.2, various basic decisions have to be made concerning lighting objectives and whether the system involves daylight, electric light or a combined system. With electric or combined systems, further decisions must be taken concerning the way light is distributed by particular fittings, and upon their positions relative to each other as well as in relation to the surface to be illuminated. As with daylighting, light-coloured and highly reflective room surfaces help to provide more illumination from the same amount of energy source.

For artificial lighting there is also a problem concerning the way the internal colours of a building may be changed depending upon the way they are affected by the light source. Designers must always ensure that particular light fittings not only provide the correct level of illumination in the required direction, but also that the light (energy) source allows the desired colour rendition of the objects to be illuminated. In this respect, the reflectance value of the surrounding surfaces and the contrast created between them also play an important role by reducing the effects of glare. Particular attention should be paid when two or more sources are visible together, such as daylight and supplementary artificial light, or tungsten (incandescent) and fluorescent fittings (see also section 12.6).

The ease with which the maintenance and cleaning of artificial lights can be carried out will depend upon design of fittings and how they are incorporated into a building. Generally, fittings, lamps and auxiliary gear should be readily accessible, and it is an advantage if fittings can be easily removed for replacement and servicing. Access for servicing, whether by reaching, ladders, demountable towers, winches, catwalks or external access from the roof, will depend upon the fixing height, space allocation and the general structural/constructional design details adopted for a building. However, maintenance of light systems must not be considered in isolation since it generally forms part of similar needs for other services, equipment and perhaps even of window-cleaning procedures (see section 12.6).

A building of special importance is often floodlit for prestige or security. And there is often a need for emergency or safety lighting in public buildings, e.g. theatres, cinemas and department stores. This is generally supplied from an independent energy source and could involve the use of gas, batteries or an automatic-start diesel generator. This type of lighting must again form an essential part of the design of a building and its construction method.

10.4 Natural ventilation

The simplest ventilation system in a building uses external air as its source, the wind as its motive force, and openings in the external enclosure for fresh air intake on the windward side and stale air extraction on the leeward side (Fig. 10.3). In a tightly constructed building, air infiltration is slow, but in a building with loose-fitting components, doors and windows, air movements will be excessive and cause draughts and high heat losses. The idea, therefore, is to create a naturally ventilated building using correctly fitting components of sizes and configurations which provide the optimum amount of air changes for the occupants, according to their activities, and also permit a minimum amount of heat loss. Tables 10.3 and 10.4 indicate the desirable

Table 10.3 Recommended minimum rates of fresh-air supply to buildings for human habitation

Type of space	Recommended m^3/h per person*
Factory Open-plan office Shops Department store Supermarket Theatre	18–30
Cafeteria Dance hall Hotel bedroom Laboratories Private offices Residential	30–43
Cocktail bar Function room Luxury residential Restaurant/commercial dining room	43–65
Boardroom Executive office Conference room	65–90

Type of space	Recommended m^3/h per m^2 of floor area
Corridors	5
Domestic kitchen	36
Commercial kitchen	72
Sanitary accommodation	36

* To convert to air changes per hour, divide by room volume and multiply by the number of occupants, e.g. a function room of 200 m^3 volume designed to accommodate 20 people requires

$$65/200 \times 20 = 6.5 \text{ air changes per hour}$$

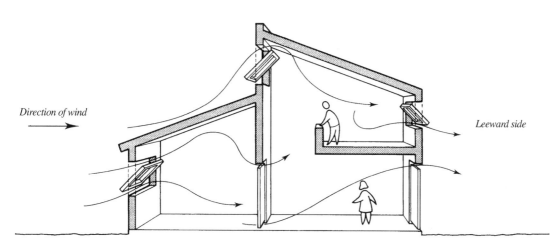

Direction of wind

Leeward side

Figure 10.3 Natural ventilation.

Table 10.4　Approximate air change rates

Accommodation	Air changes per hour
Offices, above ground	2–6
Offices, below ground	10–20
Factories, large and open	1–4
Factories/industrial units	6–8
Workshops with unhealthy fumes	20–30
Fabric manufacturing/processing	10–30
Kitchens, above ground	20–40
Kitchens, below ground	40–60
Public lavatories	6–12
Boiler accommodation/plant rooms	10–15
Foundries	8–15
Laboratories	10–12
Hospital operating theatres	<20
Hospital treatment rooms	<10
Restaurants	10–15
Smoking rooms	10–15
Storage/warehousing	1–2
Assembly halls	3–6
Classrooms	2–4
Domestic habitable rooms	~1
Lobbies/corridors	3–4
Libraries	2–4

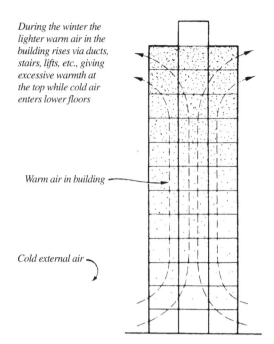

During the winter the lighter warm air in the building rises via ducts, stairs, lifts, etc., giving excessive warmth at the top while cold air enters lower floors

Warm air in building

Cold external air

Figure 10.4　Stack effect in a naturally ventilated tall building.

minimum fresh-air requirements for persons taking part in various activities. These figures are also known as *ventilation rates* or *air change rates*.

In practice these rates may be very difficult to achieve by natural methods of ventilation because airflow will be governed by areas of openings, the degree to which their use can be controlled by obstruction within a building restricting air movements, and by the pressure differences causing the flow. Also, the recommended quantities of air indicated in Table 10.3 will need some adjustment relative to the likely presence of offensive fumes or smells (including tobacco smoke), as well as for the moisture content of the ventilating air (relative humidity). It is usually considered that relative humidities of 30–70 per cent are acceptable as healthy, and increased ventilation rates will be required to reduce higher levels.

For natural ventilation, windows are used to control the volume, velocity and direction of airflow, and they are designed so as to provide openings capable of adjustment. Without wind forces, airflow through a building is simply produced by the migration of air from a high pressure zone to a low pressure zone by convection currents created through the difference in density between warmer air and cooler air. Unless employing sealed-duct devices, fuel-burning plant such as fireplaces, boilers or furnaces draw oxygen from the external air via interior spaces, and produce a *stack effect* which will also assist ventilation (Fig. 10.4). This technique often avoided the build-up of air in an interior space containing a large amount of water vapour and thus reduced the possibility of condensation.

However, the need to conserve energy and reduce room heat losses has given rise to a technology which endeavours to reduce air movements in an interior space to a minimum compatible with the functions to be carried out. The spaces created by joints between components are now reduced, finer tolerances are possible and, where gaps are inevitable, rubber or synthetic seals are used to ensure small amounts of air circulation from the exterior to the interior of a building. This often means that fresh air ducts or grilles now have to be provided to allow sufficient air for both combustion of fuel and the well-being of the occupants in a building.

Building legislation (Building Regulations Approved Document F) requires open spaces outside ventilating openings in domestic buildings to ensure an adequate volume of air for ventilation. Nevertheless, when a building is subjected to high velocity winds which are likely to cause excessive ventilation or draughts, as well as high heat losses, care should be taken to locate openings in the building exclusively on the leeward side, or to protect them by shielding devices such as fences or trees.

With contemporary construction practice, it is essential to provide controlled ventilation in occupied

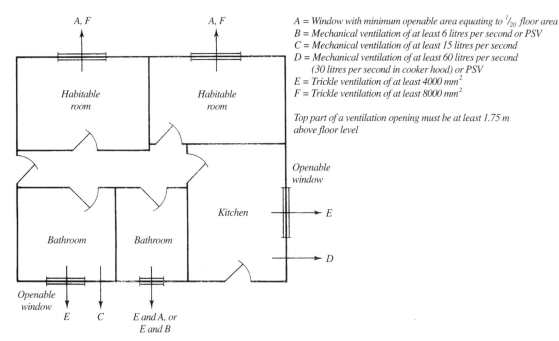

A = Window with minimum openable area equating to $^1/_{20}$ floor area
B = Mechanical ventilation of at least 6 litres per second or PSV
C = Mechanical ventilation of at least 15 litres per second
D = Mechanical ventilation of at least 60 litres per second
    (30 litres per second in cooker hood) or PSV
E = Trickle ventilation of at least 4000 mm²
F = Trickle ventilation of at least 8000 mm²

Top part of a ventilation opening must be at least 1.75 m
above floor level

Figure 10.5   Ventilation of dwellings.

Table 10.5   Building Regulations requirements for domestic ventilation

Accommodation	Ventilation provision		Extract rate (l/s)
	Rapid	Background* (mm²)	
Bathroom (with or without WC)	Openable window	4000	15 or PSV†
Habitable room	Openable window at least equivalent to 1/20 floor area	8000	–
Kitchen	Openable window	4000	30 in cooker hood 60 elsewhere or PSV
Sanitary accommodation (if separate from bathroom)	Openable window at least equivalent to 1/20 floor area, or mechanical extract of 6 l/s or PSV	4000	–
Utility room	Openable window	4000	30 or PSV
Common space	Ventilation opening of 1/50 floor area or extract fan providing at least one air change per hour		–

* Background ventilation may be by trickle vents set in window frames.
† PSV = passive stack ventilation.

rooms (Fig. 10.5). Table 10.5 indicates rapid and background means to achieve 0.5–1.0 volume air change per hour, sufficient to complement current high standards of insulation and prevent condensation. This may be by trickle ventilators (Fig. 10.6), fanned extracts or passive stack ventilation (Fig. 10.7). Where kitchens, utility rooms, bathrooms or lavatories are located without an external wall an extract fan is effected with the lighting switch. The fan has a time-delay mechanism

to provide an automatic 15-minute overrun facility. An air inlet of, or equivalent to, a 10 mm gap under the door must also be provided.

Passive stack ventilation (PSV) combines with trickle ventilators to create air movement by the stack effect principle (see Fig. 10.4), i.e. warm air rises and gains velocity in the small vertical (or almost vertical) ducts from kitchens and bathrooms, to be replaced by cool fresh air drawn in through ventilation grilles in the

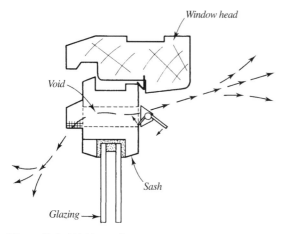

Figure 10.6   Trickle ventilator.

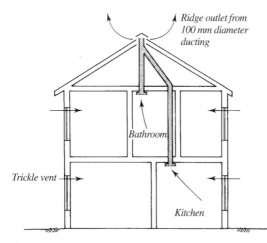

Figure 10.7   Passive stack ventilation.

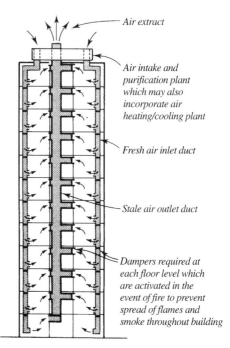

Figure 10.8   Artificial ventilation provides clean air uniformly throughout a building.

window frames. Plastic drainpipes or flexible wire reinforced tubes, 100 mm in diameter, are adequate for the application shown in Figure 10.7. A mechanically assisted PSV system may be installed where several internal rooms (no external walls) would otherwise each require an extract fan, e.g. kitchens and bathrooms in a block of flats. The ducted PSV system is linked to each room to provide permanent ventilation and only one extract fan is positioned at the duct outlet. Activation of the fan is from each compartment light switch and inlet air is through a ventilation gap under the door as previously described.

10.5 Artificial ventilation

Where a reliable and positive flow of air for ventilation is required, fans can be installed in a building to *extract* stale air, which is immediately replaced by fresh air from the outside flowing through gaps around window and door frames, or through purposely designed grilles. Fans in more elaborate ventilation schemes are connected to systems of ductwork for better air distribution throughout a building. Separate duct systems can be installed to pull away stale air from those used to distribute clean air. Often such systems are coupled with heating and cooling plant in such a way that the clean air is distributed at the selected temperature for thermal comfort air-conditioning (Fig. 10.8).

Artificial ventilation systems are usually employed for internally located rooms; crowded rooms where natural ventilation is insufficient; special rooms needing closely controlled humidity and/or freedom from any dust (e.g. computer rooms and operating theatres); and where polluted air is required to be either removed or prevented from entering internal spaces.

Certain high buildings will need artificial air movement control to ensure a balanced thermal environment because of the otherwise exaggerated 'stack effect' causing the highest parts to be hot as a result of rapidly rising warm air.

The selection, design and integration of artificial ventilation and especially air-conditioning systems into a building require specialist knowledge, techniques and skills. It is essential that the implications of a chosen scheme are realised at a very early stage in the design of a building. The effects of plant, exposed or concealed ductwork, additional fire protection to prevent the spread of fire through ductwork, etc., suspended ceilings, and additional sound control must all be carefully considered as they may have a profound effect on the appearance of the building and other technical aspects, including construction method. The integration of plant and ductwork within the dimensional discipline of the structure of a building requires particular attention.

Further specific reading

Mitchell's Building Series

Environment and Services	Chapter 3 Ventilation and air quality
	Chapter 4 Daylighting
	Chapter 7 Thermal installations
	Chapter 8 Electric lighting
Internal Components	Section 3.8 Integrated service systems and suspended ceilings
External Components	Chapter 3 External glazing
	Section 4.2.2 Lighting from windows
	Section 4.2.3 Ventilation from windows
	Chapter 5 Rooflights and patent glazing
	Chapter 6 Structural glazing

Building Research Establishment

BRE Digest 162 *Traffic noise and overheating in offices*
BRE Digest 170 *Ventilation of internal bathrooms and WCs in dwellings*
BRE Digest 232 *Energy conservation in artificial lighting*
BRE Digest 256 *Office lighting for good visual task conditions*
BRE Digest 272 *Lighting controls and daylight use*
BRE Digest 306 *Domestic draughtproofing: ventilation considerations*
BRE Digest 309 *Estimating daylight in buildings* Part 1
BRE Digest 310 *Estimating daylight in buildings* Part 2
BRE Digest 319 *Domestic draughtproofing: materials costs and benefits*
BRE Digest 398 *Continuous mechanical ventilation in dwellings: design, installation and operation*
BRE Digest 399 *Natural ventilation in non-domestic buildings*
BRE Report 162 *Background ventilation of dwellings: a review*
BRE Report 265 *Minimising air infiltration in office buildings*
BRE Report 288 *Designing buildings for daylight*
BRE Report 345 *Environmental design guide for naturally ventilated and daylit offices*
BRE Report 415 *Office lighting*
BRE Information Paper 13/94 *Passive stack ventilation systems: design and installation*

Building Regulations

F1   Means of ventilation

11 Sanitation and drainage

CI/SfB (U46)

Some of the essential activities taking place within and around a building are liable to encourage the growth of bacteria, insects and vermin, which could cause pollution, disease and foul smells. It is therefore necessary to control carefully the conditions most favourable to the development of these unwanted infestations: the provisions for *drinking-water, food preparation and washing*, and the generation of *waste products, refuse and dirt*. As far as the building designer is concerned, this involves a careful analysis of suitable water supply and storage systems, and the effective methods of waste and refuse removal. Decisions in these areas must be closely related to the selection of materials least subject to contamination, and the provision in design for efficient cleaning and freedom from deterioration.

11.1 Drinking-water, food preparation and washing

In the United Kingdom, the supply of water to a building is subject to statutory undertakings, mostly controlled by the provisions of the Water Acts 1945 and 1973, the Water Industry Act 1991 in England and Wales, and the Water Act 1980 for Scotland. These are supplemented by the Water Supply (Water Fittings) Regulations 1999 and BS 6700, 1997: *Specification for design, installation, testing and maintenance of services supplying water for domestic use within buildings and their curtilages*. They seek to empower the water authorities to protect drinking-water supplies, effect more efficient use of water, enforce prevention of waste, undue consumption, and misuse or contamination of water. Clean and potable water is therefore generally in ample supply even during times of drought and, as the same source is used for all purposes in a building, installations are of simple design (Fig. 11.1).

The principal mechanical requirements for the system are that all *pipework* used should be non-corrodible, capable of being tightly jointed, and resistant to deformation and mechanical damage; the *layout* of pipework should provide the minimum resistance to water flow and should be protected from freezing; and *forming, supporting* and *connecting* techniques to appliance should reduce the possibility of noise generation and transmission. The water supply system or *plumbing* allows the water not intended for human consumption to be stored in a replenishable cistern, incorporating its own feed system to sanitary appliances such as WCs, bidets, basins and baths; washdown points; and the hot water system supplied via another storage vessel and cylinder.

The provision of separate water storage facilities, isolated from the mains, reduces the risk of misuse and contamination of the water within the rising main. Misuse could adversely affect the supply source, as well as the water to be used for direct human consumption (drinking and food preparation) within a building. The storage cistern also ensures a continued supply of water in the event of the mains supply being cut off for a short period due to damage or maintenance work. By installing the storage tank at high level in a building, sufficient supply pressure can be ensured. However, if the pressure of the water in the mains supply is insufficient or likely to fluctuate dramatically owing to demand, the mains water must be pumped to the required level for a building to be adequately serviced.

It is usual to provide a cold water storage cistern as this assists in reducing the size of the pipe used for the mains supply. To avoid the possibility of pollution, there must be a clear horizontal gap between the outlet

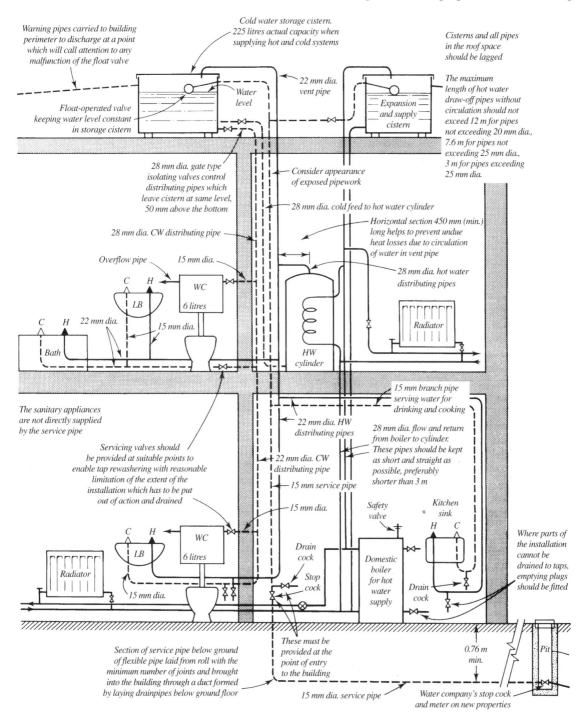

Warning pipes carried to building perimeter to discharge at a point which will call attention to any malfunction of the float valve

Cold water storage cistern. 225 litres actual capacity when supplying hot and cold systems

Cisterns and all pipes in the roof space should be lagged

Float-operated valve keeping water level constant in storage cistern

Water level

22 mm dia. vent pipe

Expansion and supply cistern

The maximum length of hot water draw-off pipes without circulation should not exceed 12 m for pipes not exceeding 20 mm dia., 7.6 m for pipes not exceeding 25 mm dia., 3 m for pipes exceeding 25 mm dia.

28 mm dia. gate type isolating valves control distributing pipes which leave cistern at same level, 50 mm above the bottom

Consider appearance of exposed pipework

28 mm dia. cold feed to hot water cylinder

28 mm dia. CW distributing pipe

Horizontal section 450 mm (min.) long helps to prevent undue heat losses due to circulation of water in vent pipe

Overflow pipe   *15 mm dia.*

28 mm dia. hot water distributing pipes

C *H*
LB
WC
6 litres

C *H* *22 mm dia.*
Bath

15 mm dia.

Radiator

HW cylinder

The sanitary appliances are not directly supplied by the service pipe

15 mm branch pipe serving water for drinking and cooking

22 mm dia. HW distributing pipes

28 mm dia. flow and return from boiler to cylinder. These pipes should be kept as short and straight as possible, preferably shorter than 3 m

Servicing valves should be provided at suitable points to enable tap rewashering with reasonable limitation of the extent of the installation which has to be put out of action and drained

22 mm dia. CW distributing pipe

15 mm service pipe

15 mm dia.

Safety valve

Kitchen sink

C *H*
LB
WC
6 litres

Radiator

15 mm dia.

Drain cock

Stop cock

Domestic boiler for hot water supply

H *C*

Drain cock

Where parts of the installation cannot be drained to taps, emptying plugs should be fitted

Section of service pipe below ground of flexible pipe laid from roll with the minimum number of joints and brought into the building through a duct formed by laying drainpipes below ground floor

These must be provided at the point of entry to the building

15 mm dia. service pipe

0.76 m min.

Pit

Water company's stop cock and meter on new properties

Figure 11.1 Typical domestic water supply system.

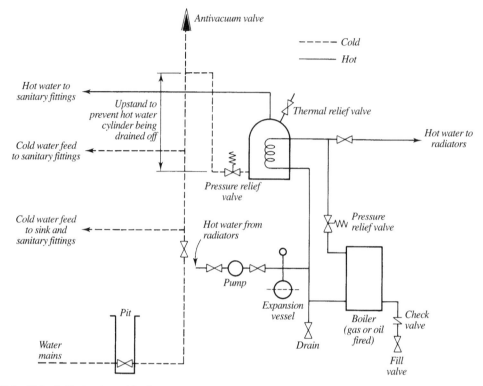

Figure 11.2   Mains-fed hot water and heating system.

of the supply pipe and the top surface of the stored water; the gap is maintained by siting an overflow pipe at a distance *below* the supply pipe. Regulation of the flow of water is achieved by a float valve, usually in the form of a hollow plastic ball (see Fig. 11.1).

A similar gap is provided between the outlet of a tap on sanitary appliances and the maximum permitted height of water (controlled by an overflow) which the appliances are likely to contain, e.g. the drinking-water supply to a kitchen sink is isolated from the hot water supply in this manner. However, as a result of research based on systems in America and some European countries, some UK local authorities now allow taps to most sanitary appliances to be fed directly from the mains without involving water storage cisterns. Figure 11.2 indicates such a plumbing system, and pollution is prevented by using an antivacuum valve which maintains a positive pressure in the system to prevent any possibility of water backflow (back syphonage). Sealed hot water heating systems are currently used in the United Kingdom, but the diagram also indicates a sealed system for the hot water supply. The latter is similar to that which is widely employed in other EU countries and has now been introduced in the United Kingdom.

In addition to the mechanical and technical criteria, it is very important that detailed consideration is given to the *appearance* of a water supply system within a building. Pipework, ancillary devices and appliances must be carefully integrated as an essential part of the design of a building. No plumbing system should 'occur' in a building as an afterthought; if adequately considered during the early stages of design, the system can enhance the appearance of purely practical areas, or be successfully concealed in other areas by incorporation with other services within structural elements, e.g. wall, floor and roof construction.

11.2 Waste products, refuse and dirt

Depending upon its precise function, usually a building must also provide installations which facilitate the disposal of water-borne organic waste matter, including human excrement (sewage); permit the collection of rainwater otherwise liable to cause some form of deterioration; and allow other organic and non-organic waste (refuse) to be removed.

Sanitary fittings (WCs, bidets, urinals, baths, showers, basins and sinks) which receive the sewage are designed on the basis of specific anthropometric data for efficient function; they also flush away bacteria, and prevent foul smells. They are manufactured from non-porous, smooth, durable and easily cleaned materials, and they incorporate a water-filled trap to prevent gases escaping into the building from the pipework beyond. Further precautions against smells and other forms of pollution are provided by building designs which promote good ventilation in rooms used for bathrooms, laboratories, operating theatres, abattoirs, etc., where sanitary appliances are situated; by ensuring adequate space around a building; and by its correct location within a particular environment.

11.3 Drainage systems

The sanitary fittings discharge sewage along a network of gas and watertight pipes which together form the *drainage system* of a building. This system conveys both solids and liquids to a treatment plant in the local authority sewage works or, if this plant is not available, to a *cesspool* or *septic tank* (Fig. 11.3) and filter bed in close proximity. Pipes which convey liquids only are generally called *waste pipes*, and those which convey solid matter and liquids are called *soil pipes*.

Efficient and economic design considerations usually mean that the waste and soil pipe system in a building takes one of two forms. The simplest occurs when sanitary appliances can be fairly closely grouped around vertical discharge stacks (soil and waste pipes) which then convey sewage by the most direct route to underground drains (Fig. 11.4). Such installations are found in many types of buildings, including blocks of flats, where the activities to be accommodated allow vertically repetitive planning arrangements. It is now usual to combine the soil and the waste pipe into a *centralised*

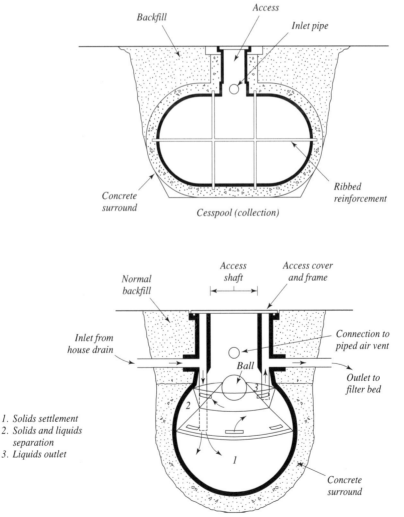

Figure 11.3 Plastic cesspool and septic tank set below ground in concreted surround.

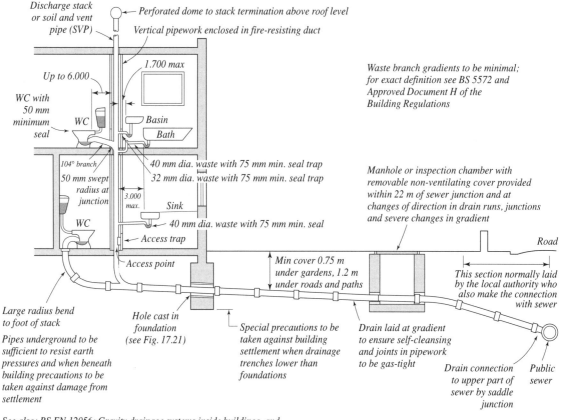

Figure 11.4 Typical centralised single stack sanitary installation and drainage for small building.

single discharge stack system; the overall dimensions of this pipe are dictated by the amount and frequency of solid material it must transfer to the drains. Great care must be taken when designing the system to ensure that the water seals or traps in the sanitary fittings are not prevented from reforming after use. Seals may not reform because of *induced syphonage*, which occurs when the seal is sucked away by the force of water discharging from branch pipes on other floors as it passes down the vertical discharge stack. The rush of water down the stack absorbs air from the branch pipe of the fittings below and this causes the external air pressure on the seal to force the water out of the trap. This syphonic action can be prevented by ventilation pipes (antisyphon pipes) on the drain side of the sanitary fitting to equalise air pressure within the main stack. Alternatively, special antisyphon resealing traps can be used, which let air into the system without complete loss of trap seal. Nowadays it is more usual to design the whole drainage system with careful regard to sizes, lengths, slopes and positioning of pipework. Design rules are based upon

empirical and scientific analysis of water flow and air pressure characteristics. A variation suitable for repetitive domestic installations uses an *air admittance valve* at the top of the discharge stack. It does not allow foul air to escape and therefore can be located within the roof space to reduce installation and roofing costs. It also eliminates the unsightly stack projecting through the roof slope. However, every fifth dwelling must have a conventional stack to ventilate the sewer.

When the activities within a building are complex and it is not possible to group sanitary fittings around a centralised stack – as in hospitals, schools and certain office layouts – a satisfactory drainage system can be accomplished by grouping fittings in 'islands', and by providing extensive horizontal pipework connection to strategically located vertical soil and/or waste stacks. This system is expensive and more complicated than the centralised system because the necessary proliferation of horizontal pipes not only makes concealment by 'false' ceilings necessary, it also means that further syphonage problems must be overcome. Overlong horizontal lengths of pipework are liable to cause the discharged water from a sanitary fitting to remove the trap seal by *self-syphonage*. This occurs when a horizontal pipe flows full and prevents a build-up of positive pressure on the drain side of the trap until the last of the water has been drawn away. Fortunately, baths and large flat-bottomed sinks rarely suffer self-syphonage; this is because the last of the discharging water moves very slowly, allowing the seal to resettle, and the waste pipe of a WC is too large to flow at full bore. For other sanitary fittings with extensive horizontal lengths of pipework, larger sizes of pipe and greater depth of seal must be incorporated, and also probably an elaborate ventilating system.

11.4 Rainwater collection

A drainage system for the removal of waste should be planned in conjunction with the drainage system required for the collection of rainwater. This involves collection of rainwater falling on a building *and* perhaps around a building. The collected rainwater must then be conveyed in a similar manner as waste water to a local authority *surface water drain*. If this drain is not available, rainwater can be taken to a purpose-built *soakaway* situated away from areas likely to be detrimentally affected by excessive amounts of water, or to a conveniently located *water course* or a *storage vessel* for subsequent use as a water supply (Fig. 11.5).

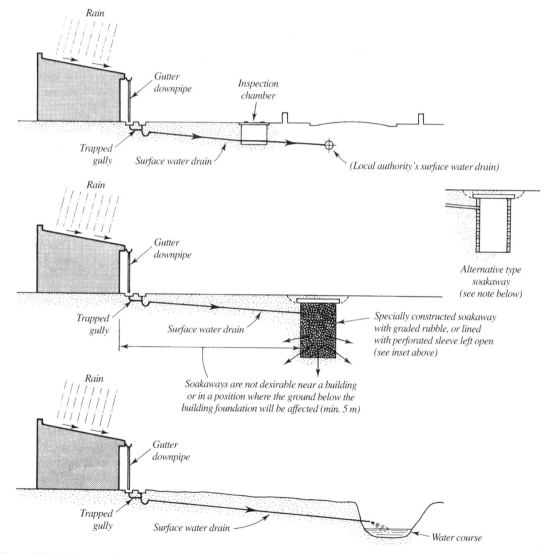

Figure 11.5 Rainwater collection and methods of disposal.

Water falling on a building at roof level is collected and discharged into a rainwater pipe system by means of strategically located rainwater outlets. If a building is lower than about five storeys, easier means of access and maintenance generally make a system of collection involving guttering more acceptable. The water is conveyed to the vertical rainwater pipes, which discharge either directly into the drain below ground or over trap-seal gulleys at ground level. Pipe sizes for the design of rainwater drainage systems in the United Kingdom depend upon the expected risk involved: 50 mm/h of rainfall for flat roofs and other open paved areas; 75 mm/h for sloping roofs; or 150 mm/h when very occasional overflowing of rainwater outlets or gutters cannot be tolerated. (This amount is likely to fall only during short periods of heavy storms.)

When considered together, there are three basic drainage systems approved by local authorities for the disposal of soil, waste *and* rainwater (Fig. 11.6).

Combined system   Both drains discharge into a common sewer. This is a simple and economic system because it involves less pipework and is easy to maintain. However, it has the disadvantage that vast amounts of liquids must pass through the sewage treatment works, particularly after heavy rainfall.

Separate system   Two drains are provided. One of them receives the collected rainwater (or surface water) and conveys it directly to a suitable outfall without treatment, e.g. nearby water course or river. The second drain takes the soil and waste discharge and conveys it to the sewage treatment installation. This obviously involves more drainage pipes, but avoids the risk of overcharging the sewage treatment plant during periods following heavy rainfall.

Partially separate system   A combined drain is used for soil, waste and rainfall. However, a second drain is also available to regulate the amount of rain or surface water discharging into the combined drain, according to the capacity of the sewage treatment installation.

11.5 Means of access to drains

Access to drains is necessary for inspection, testing and maintenance. The following means are acceptable:

- rodding eye
- access fitting
- inspection chamber
- manhole

Rodding eyes may be used at the head of a drain. They are in effect an extension of the drain up to surface level, terminated with an adaptor to a screw-sealed surface access plate.

Access fittings are produced in plastic or clayware to suit the drainage system material. They too are limited to rodding, but in both directions. Fittings may be cut or adapted for depths up to 600 mm to *invert* (invert represents the lowest level that water will flow in a drainage channel).

Inspection chambers are produced from a range of materials including brickwork, preformed plastic units and precast concrete sections. Each provides the option of a few branch channels in addition to through-flow. Limited bodily access is possible in depths of up to 1 m to invert.

Manholes are sufficiently spacious at drain level for a person to work in. In depths over 1 m they require step irons or an attached ladder to aid accessibility. They are generally built of concrete (*in situ* or precast) or dense masonry to provide adequate strengths at depths of several metres.

Siting of access points should occur:

- at or near to the head of a drain
- at a change of direction
- at a change of gradient
- at a change of drain pipe diameter
- at junctions (unless each drain can be cleared from another access point – maximum distances as in Table 11.1)
- on long straight runs

11.6 Integration of systems

Whatever internal drainage arrangements are to be incorporated within a building, it is always important to consider carefully their implication on the precise system to be adopted. Not only can considerable economy be achieved by adopting simple systems, but also the visual impact on a building can be quite considerable; pipework can be exposed, or concealed in service walls, floors or ducts, and false ceilings, etc. When requirements for buildings were much simpler, drainage specialists could arrive on site and, after initial competition for space with other trades, install their services. The new approach to the design of installation leading from a more sophisticated knowledge, together with the standardisation of components for greater economy, now requires designers to give much greater thought to the incorporation of even simple drainage systems in a building. Indeed, the integration and coordination of services in relation to structures as a whole now becomes a paramount design

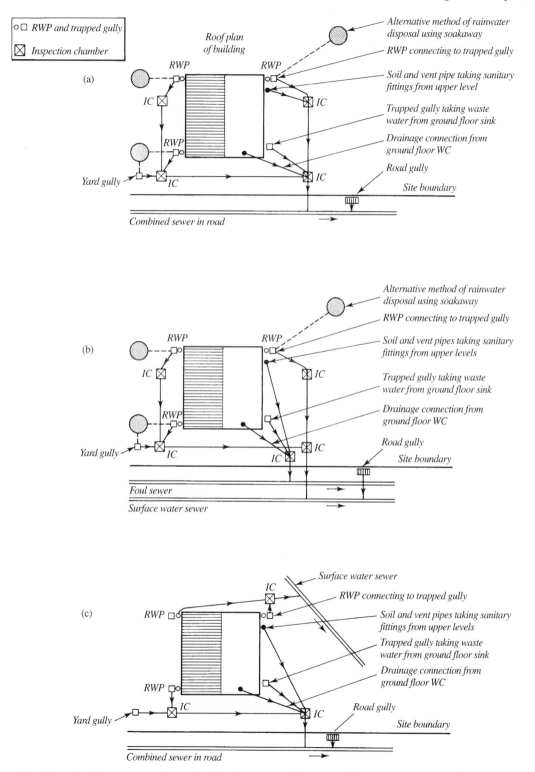

Figure 11.6   Domestic drainage system: (a) combined drainage system; (b) separate drainage system; (c) partially separate drainage system. IC = inspection chamber, RWP = rain water pipe.

Table 11.1 Maximum spacing (m) of access facilities in drains up to 300 mm in diameter

	Small access fitting	Large access fitting*	Junction	IC	Manhole
Start of drain	12	12	–	22	45
Rodding eye	22	22	22	45	45
Access fitting:					
150 mm ø	–	–	12	22	22
150 × 100 mm	–	–	12	22	22
225 × 100 mm*	–	–	22	45	45
IC < 1 m	22	45	22	45	45
Manhole > 1 m	–	–	–	45	90

criterion for such buildings as blocks of flats, offices, hospitals and schools, where all the engineering services (heating, lighting, plumbing, drainage, etc.) could account for 25–50 per cent of the overall capital costs.

Problems relating to the installation of gas, electricity and telecommunications services are discussed in section 17.10.

11.7 Refuse collection and disposal

It is usually necessary for a building to incorporate adequate arrangements in its design for the collection and subsequent disposal of refuse. Methods which can be adopted for this will only operate efficiently if the precise type, form and amount of waste produce has been successfully identified (or anticipated) during the early stages of design investigation. Consideration may then be accurately given to efficient movement patterns of refuse about a proposed building and their influence on required standards of hygiene and safety. Generally, storage for more than a few hours within a building is undesirable, and the small receptacles in which refuse can conveniently be placed temporarily need frequent emptying.

Unobstructed and direct circulation routes to facilitate refuse collection from a building should also be thoroughly planned, thereby furthering the desire for protection from the pollution caused by unpleasant smells, visual horrors and noise. Refuse which is not destroyed at source is taken to local sites where crude selection may take place to permit incineration, consolidation and/or transportation to centralised tips. Some non-toxic refuse may be used to backfill areas subsequently needed for building sites, or used for other forms of land reclamation.

The design criteria for a building will vary when considering refuse collection and disposal methods ap-

plicable to medical, commercial, industrial or domestic activities, for example. A building designed to accommodate complex or multi-purpose activities (hospitals and certain factories) can generate many forms of waste, sometimes toxic, sometimes individually bulky, or sometimes accumulating in vast quantities over relatively short periods. Then different disposal systems, some incorporating incinerators, may have to be adopted within a building, requiring great skill from the designer to ensure maximum operational efficiency. But when convenient to the size and form of refuse, disposal can be satisfactorily accomplished by an independent *water-borne pipe system* installed within a building. This is very similar to the soil and waste *drainage system* already described, except that the refuse is conveyed to an external pit, from which it is collected by specialists at convenient time intervals.

Less costly methods are available, one of which involves the use of a *dry chute*, suitable for medium-rise multi-storey flats or maisonettes. The method consists of a vertical arrangement of jointed impervious pipes to provide a tube into which refuse is placed via a chute. The outlet of the tube deposits the refuse into bins conveniently located to facilitate mechanical emptying by the vehicles of the local authority cleansing department (Fig. 11.7).

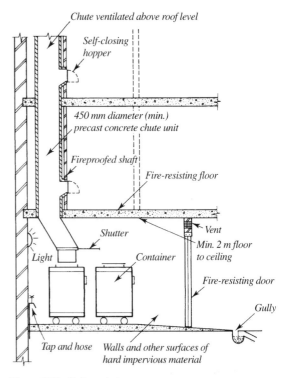

Figure 11.7 Refuse chute.

Small amounts of refuse, such as are generated in houses, can be conveniently stored in dustbins or plastic bags for eventual collection by the local authority. Some domestic refuse – generally waste matter that putrefies – can be forced through a grinding unit below a sink unit, which then allows it to pass through a normal waste pipe and into the drainage system beyond.

Further specific reading

Mitchell's Building Series

Environment and Services	Chapter 9 Water supply
	Chapter 10 Sanitary appliances
	Chapter 11 Pipes
	Chapter 12 Drainage installations
	Chapter 13 Sewage disposal
	Chapter 14 Refuse collection and disposal
Internal Components	Section 2.3.7 Integration of services within demountable partitions
	Chapter 3 Suspended ceilings
	Chapter 4 Raised floors
External Components	Section 8.2.1 Weather exclusion of roofings
Structure and Fabric Part 2	Section 7.3 Ducts for services

Building Research Establishment

BRE Digest 81   *Hospital sanitary services: some design and maintenance problems*
BRE Digest 248   *Sanitary pipework* Part 1: Design basis
BRE Digest 249   *Sanitary pipework* Part 2: Design of pipework
BRE Digest 292   *Access to domestic underground drainage systems*
BRE Digest 308   *Unvented domestic hot water systems*
BRE Digest 365   *Soakaway design*

Building Regulations

H1   Foul water drainage
H2   Cesspools, septic tanks and settlement tanks
H3   Rainwater drainage
H4   Solid waste storage

12 Security

Like many other performance requirements, the *security* aspects of a building involve the immediate physiological and psychological well-being of the occupants. The main areas of concern relate to *unauthorised entry* into a building, *vandalism*, protection against *disasters* and the reduction of *accidents*. The degree of risk associated with a particular building must be established during initial design stages, so that appropriate security measures can be incorporated without hindering the occupants from carrying out their activities. The primary and often most economical defence usually lies in the fabric of a building and therefore involves the use of compatible construction methods (materials and techniques of assembly).

12.1 Unauthorised entry

Unauthorised entry can be achieved either *visually* or *physically*. Both involve invasion of the private zones of a building. But whereas the former may be no more than inconvenient, the latter often involves violence towards people, and/or damage and theft of furniture, fabrics, machines, personal belongings and livestock. The defence systems involved for the two areas must therefore be considered separately, although they can be resolved together to provide combined security.

Visual privacy can be achieved through external planning arrangements of a building which provide an acceptable degree of *remoteness* between observer and observed. When adequate distances for this are not available (see section 17.2), or it is required to augment the remoteness for even greater privacy, the design of a building and its immediate surroundings plays a more important role. The location of the more private zones of a building can be away from the general fields of vision. Alternatively, high perimeter walls, trees and plants, landscaping, projecting wings and outbuildings can be employed as visual barriers.

Although less used today owing to energy conservation requirements, large windows which let in more daylight make room interiors more visible from outside. Cross-lit rooms also reduce privacy levels by silhouetting figures. Although net curtains or slatted blinds may be considered a satisfactory solution by some people, it should still be possible to obtain the levels of daylight the window was designed to provide and for the occupants to enjoy uncurtained views out of a building without loss of visual privacy.

Figure 12.1 illustrates some of the principles involved when trying to provide reasonable visual privacy between dwellings based on the *Design Guide for Residential Areas* published by the County Council of Essex. They involve the interrelationship between acceptable 'eye-to-eye' remoteness and the use of screening devices above eye level.

As an initial means of defence against unauthorised *physical* entry, the approach routes to a building should be carefully organised to prevent, or at least limit, the possibility of uncontrolled usage. This principle is always considered during the initial design stages of a building accommodating strongrooms, security stores, prisons, etc. However, there has been a certain amount of complacency about the planning of a building having less dramatic risks. This has led to the development of the use of numerous elaborate security measures or costly security patrols, often added to a building as an afterthought in an attempt to justify incomplete initial design investigations. Nevertheless, although entrances can be positioned to provide good, well-lit visibility

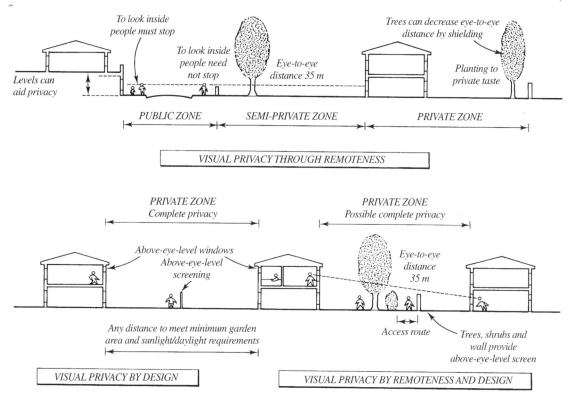

Figure 12.1 Visual privacy provided by building location and design. (Adapted from material in *Design Guide for Residential Areas* published by Essex County Council)

and control, problems may still arise in a building requiring a high degree of security but where easy access by the public is necessary (banks, hotels, offices, police stations, museums, exhibition halls, etc.). And the final defence against unauthorised entry into a building must rely on some form of locking mechanism.

However, a conflict often arises between providing doors and windows that are secure but which allow the legal occupants of the building to escape unhindered during a fire. A compromise is sometimes required when selecting locks (ironmongery), particularly for doors, although locks are now available which appear to satisfy both criteria. A similar form of compromise is often required in the design of fitments, which must freely exhibit goods while protecting them from shoplifters.

Methods of construction should augment the initial design considerations which prevent *illegal* entry. For example, as certain lightweight constructions can be easily dismantled – boards removed from wall-cladding systems or tiles lifted from roofs – care must be taken in their detail design to ensure their use is compatible with the security risk of a building. Windows and doors, normally regarded as a means of giving

access, or light and ventilation, should also be considered as a means of admitting a criminal; they must be carefully designed and protected accordingly.

Once the design of a building fabric has provided the maximum amount of security possible, further protection can be given, if necessary, by electronic devices, alarm bells and surveillance by contracted security patrols. At the planning stage it is advisable to consult the crime prevention officer (CPO) of the local police force for advice about the degrees of risk involved. The building's insurers will also establish their minimum standards for acceptance of risk, and the architect will need to ensure such minimum standards are observed during design and construction.

The Avon and Somerset Constabulary have produced a booklet entitled *Security in Design* which gives an excellent checklist of important considerations under the following headings: assessment of crime risk; layout of site; building design (access from exterior); building design (access – windows and glass); building design (interior); ironmongery; intruder alarm systems; strongrooms and safes; special security planning; building site security (contractor); existing standards on security; references.

The building contractor will also be concerned with security during the construction of a building. Fences and hoardings are necessary to prevent unauthorised entry into the site, which could result in theft of materials, equipment, tools, etc., as well as vandalism. This form of protection will also assist in protecting passers-by, who may otherwise be accidentally hurt as a result of certain construction activities. Permanent boundary fences and walls can provide an initial defence against illegal entry into the grounds of a completed building.

12.2 Vandalism

Vandalism is a continuing social problem which defies complete resolution, although as for unauthorised entry the design of a building can greatly lessen the likelihood of its occurrence. A building and the adjoining spaces need to provide a means of positive identity to owners and normal users as well as the community in general.

The problem ranges from graffiti on accessible surface finishes to physical damage of a building fabric involving defacement, breakage, or complete destruction by demolition or fire. Graffiti can be avoided if not eliminated by the judicious selection of surface finishes; but physical damage is a serious problem which may result in a building employing fortress-like construction methods. Everything must be robust and secure from attack where wilful damage is likely to occur, so the building's appearance and other performance requirements may suffer. The ordinary use of materials will have to be avoided: glazed areas should be of limited size and reinforced; doors of solid construction; walls of dense robust materials which are not easily ignitable, etc. Under certain conditions, suitably selected materials may require further protection by barriers or screens. Accidental damage can also be avoided by these devices.

To some extent the effects of vandalism can be lessened by a building owner through a design which provides conspicuous observation points, allows damage to be promptly repaired, and litter to be minimised and efficiently cleared.

12.3 Disasters

As far as the effects on a building are concerned, disasters can take the form of those which happen accidentally and usually through previous ignorance of a phenomenon, or those which could happen as a direct or indirect result of planned actions. An example of an accidental disaster would be a gas explosion causing collapse of a building. This occurred in a block of flats called Ronan Point and prompted legislative measures to avoid the progressive collapse of a construction as illustrated in Figure 12.2. Other examples of learning from disasters with previously unknown causes are, unfortunately, numerous. The use of high alumina cement (HAC) for the load-bearing reinforced concrete members in a building where warm humid atmospheric conditions exist (swimming pools) produces deterioration of the structure and subsequent collapse. Inadequate bearing for prestressed reinforced concrete beams causes their collapse. A lack of fire barriers in cavity constructions allows the rapid spread of fire from one part of a building to another. Failures such as these and the Ronan Point disaster can cause serious loss of life and cost enormous sums of money to rectify.

Experience of disasters, together with the continuous research by investigative authorities (Building Research Establishment, Princes Risborough Laboratory, TRADA Technology Ltd, etc.) and testing organisations (British Board of Agrément) help to reduce the likelihood of their recurrence. However, there is always need for a designer to proceed cautiously by seeking out maximum information regarding the performance criteria with new products, materials and untried methods of construction.

Specific forms of construction are also available to lessen the effects of natural disasters resulting from earthquakes, floods, hurricanes, etc.

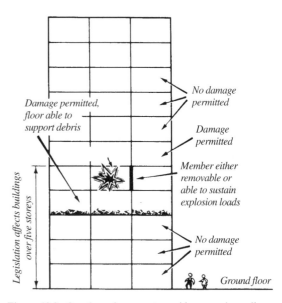

Figure 12.2 Legal requirements to avoid progressive collapse in tall buildings.

12.4 Lightning

The risk of lightning striking a particular building is very low. Therefore, provision for lightning protection is a risk assessment by the building owner or, more realistically, the building insurer. Houses are rarely protected, but larger commercial and industrial premises will be assessed on the basis of their size (height and plan area), contents, purpose, construction materials (exposed metalwork), degree of isolation, likelihood of thunderstorms in the locality and general topography.

A lightning protection system is designed to attract a lightning discharge and direct it to earth through a path of low impedance, thereby limiting the amount of damage that would otherwise occur to the more vulnerable parts of a building. BS 6651: *Code of practice for protection of structures against lightning* provides guidance on air termination conductor locations for various applications. Generally, air terminations comprise a series of conductor strips interconnected to form a grid, with no part of the roof further than 5 m from the grid. Variations are made to suit roof profile, with prominent features such as apexes and spires suitably protected.

Vertical or down conductors are spaced at one per 20 m of building periphery for buildings up to 20 m height and one per 10 m periphery for buildings in excess of 20 m height. Structural steel and metal pipes are bonded to the down conductor. Conductor metals include aluminium, copper and alloys, phosphor-bronze, galvanised steel and stainless steel in 10 mm diameter rods or 20×4 mm strips. Earth terminations are rods driven into the ground to sufficient depth to provide a low electrical resistance. Maximum test resistance is 10 ohms.

12.5 Terrorism

Acts of terrorism or war often involve the use of explosives to cause damage to buildings. For economic reasons there can be few defensive measures taken to avoid complete or even partial destruction of a building constructed using normal techniques. However, when necessary, a building can be specially designed to withstand a certain degree of anticipated damage. Today it seems that the ultimate form of this type of construction is one which must resist the light and heat, blast wave, tremors and fallout from a nuclear explosion. Construction methods can take the form of relatively simple 'do-it-yourself' techniques or those which involve housing large structures almost entirely below ground and which incorporate complicated life-support systems.

12.6 Accidents

A building must be designed to ensure that the human activities it accommodates are carried out with the maximum amount of comfort, safety and efficient enjoyment. It is also important that the construction of a building is carried out with reasonable comfort, high degree of safety and, hopefully, enjoyment (see section 15.7).

Mention has been made under *dimensional suitability* (appropriate size) regarding the need to manufacture components to sizes which are sympathetic with building operations and use (see Chapter 4). However, even appropriately-sized components must be used wisely because it is not uncommon for people either to cause damage or to be damaged, quite accidentally, as a result of ordinary daily activities (Fig. 12.3). Careful consideration must be given to work surface heights in kitchens, offices or workshops, juxtaposition of conflicting activities within confined spaces, suitable clearances around specific activities, etc. The study of the relationship between people and their environment is known as *ergonomics* and applies physiological and psychological reasoning to anthropometric data. Figures 12.4 and 12.5 show typical examples.

The design of spaces in and around a building should take into account the appropriately safe features and dimensions for stairflights, treads, risers and landings; slopes of ramps; heights and profiles for handrails and guard-rails, etc. Circulation areas must be planned to give efficient movement patterns which avoid mixing of opposing activities. In this respect, special consideration must be given to the problems associated with the circulation of disabled people through a building, and into the various spaces it accommodates (see section 12.6.1). Gas and water services should be separated from electrical services to avoid risks of explosion, fire and electrocution. If the opening parts of any windows above ground floor level are lower than 1020 mm from the inside floor, additional protection will be required against people falling out. Beyond third-storey height, the minimum internal height of the window board should be increased to 1120 mm. Whenever large areas of clear glass are incorporated in the construction of walls, precautions must be taken to ensure that people are made aware of its presence. Large uninterrupted areas of clear glazing, such as that found between two parts of the same building at the same level, must incorporate markings at a height of 1500 mm to prevent unaware people colliding with them.

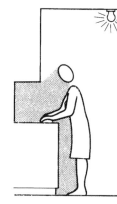

Dangerous designs of kitchen fitment which cause accidents

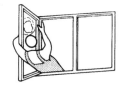

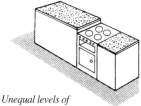

Dangerous practices in cleaning the outside of windows from inside a building

Unequal levels of working planes liable to cause accidents

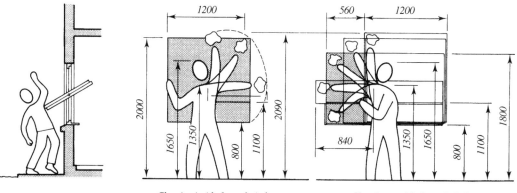

Dangerous window design

Cleaning inside face of window from inside building without steps and/or extension aids

Cleaning outside face of window from inside building. Maximum reach determines size and shape of fixed windows (shaded)

Figure 12.3   Design factors influencing safety in a building.

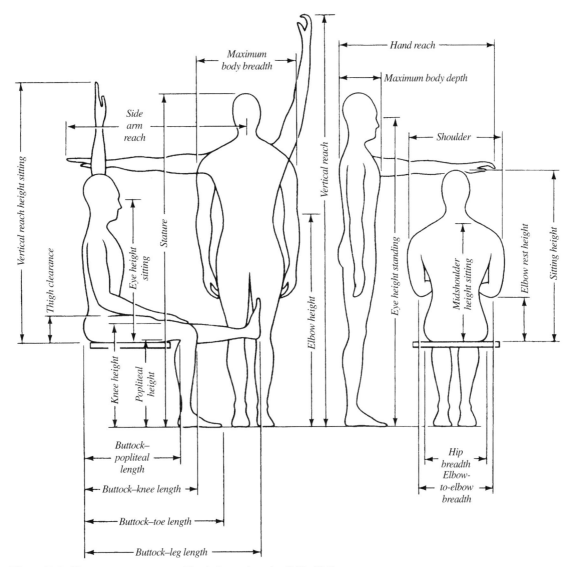

Figure 12.4 Human measurements used by designers (see also Table 12.1).

Table 12.1 Principal average human dimensions

Position from Figure 12.4	Male (mm)	Female (mm)
Maximum body breadth	450	400
Hip breadth	370	400
Hand reach	850	780
Eye height, standing	1630	1550
Stature	1740	1650
Popliteal height	440	440
Knee height	590	590
Elbow rest height	220	210
Midshoulder height, standing	580	540
Eye height, sitting	780	730
Sitting height	900	840

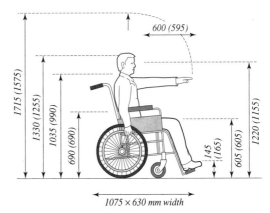

Figure 12.5 Typical wheelchair and occupant dimensions (female dimensions in brackets).

12.6.1 Accessibility for disabled people

Estimates of the number of disabled people in the UK vary. Figures between 8 per cent of population and 8 million persons provide some scale of the need for building designers, constructors, owners and occupiers to facilitate for the disabled. Legislative measures to ensure that disabled people can access buildings include the Disability Discrimination Act introduced in 1995 and then in subsequent parts up to 2004, the Disability Discrimination (Employment) Regulations 1996 and Part M of the Building Regulations endorsed by BS 8300: *Design of buildings and their approaches to meet the needs of disabled people.*

The term 'disabled' covers a wide range of incapacities, but it is the wheelchair-dependent person that is of principle concern to the building designer. A person in a wheelchair can occupy about five times the space needed by an ambulant person. Therefore the following should be incorporated into the construction of new dwellings:

- Building entrance minimum 900 mm wide.
- Firm level access to the building – maximum slope 1 in 20.
- Level (or close to level) principal entrance threshold.
- Entrance door minimum 775 mm clear width.
- Corridors/passageways minimum 750 mm width.
- Stair minimum 900 mm width. Handrail both sides.
- Light switches, power, telephone and aerial sockets at 450–1200 mm above finished floor level.
- WC provision on the entrance or first habitable storey. Door opens outwards. Clear wheelchair space of 750 mm in front of WC with preferably 500 mm each side of the WC measured from its centre.
- Flats to have lifts and stairs which enable disabled residents to access other floors (see Building Regulations Approved Document M2: Section 9).

Buildings other than dwellings should have the following provisions:

- Ramped and easy access to buildings. Minimum width 1200 mm, maximum gradient 1 in 20.
- Tactile pavings (profiled).
- Dropped kerbs.
- Handrails at changes in level.
- Guarding around projections and obstructions.
- Wheelchair manoeuvrability in entrances.
- Entrance width minimum 800 mm clear space.
- Internal door openings, minimum 750 mm clear space.
- Corridors/passageways minimum 1200 mm wide.
- Lift facilities, see Building Regulations Approved Document M2 and BS 5776: *Specification for powered stair lifts*, or BS 6440: *Powered lifting platforms for use by disabled people.*
- Stairs minimum 1000 mm wide, rise between landings 1800 mm maximum, step rise maximum 170 mm, step going minimum 250 mm and a handrail each side.
- Wheelchair-access WC provision on each floor.

12.6.2 Safety glazing regulations

The Building Regulations Approved Document N details acceptable standards for glazing materials and protection, including the critical areas shown in Figure 12.6 which require safety glazing.

The critical locations are defined as follows:

- glass within 800 mm of the finished floor
- glass within 1500 mm of the finished floor and contained in a door or adjacent side panel.

Glazing is accepted as safe if it satisfies one of the following definitions:

- It has 'safe break' characteristics determined in BS 6206: *Specification for impact performance requirements for flat safety glass and safety plastics for use in buildings.*
- It is robust or in small panes (see below).
- It is permanently protected by a screen (Fig. 12.7).

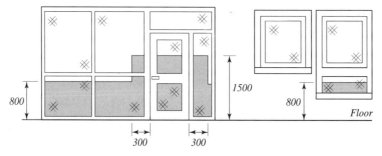

Figure 12.6 Location of safety glass: shaded areas are critical locations for safety glazing; all dimensions are in millimetres.

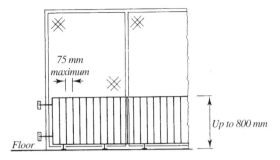

Figure 12.7 Screen protection of glazed areas must prevent the passage of a 75 mm sphere, it must be robust, and it must be difficult to climb (vertical rails).

Table 12.2 Maximum dimensions of annealed glass panels

Annealed glass thickness (mm)	Maximum width (m)	Maximum height (m)
8	1.10	1.10
10	2.25	2.25
12	4.50	3.00
15	any	any

The terms *robust* and *small panes* are defined as follows:

Robust　This term applies to inherently strong materials such as polycarbonates. Annealed glass may also be acceptable as defined in Table 12.2.

Small panes　Small panes can be isolated or in small groups separated by glazing bars. The maximum pane width is 250 mm and the maximum pane area is 0.5 m². The nominal thickness should not be less than 6 mm.

Ideas of design safety must be extended to appropriate surface finishes for a specific activity: non-slip floor tiles in swimming pools; non-combustible wall surfaces along fire-escape routes; antistatic finishes in operating theatres.

The *use of colour* can play a very important role in helping to prevent accidents, although it should never be used to justify the safety of a badly designed feature. Apart from adding further dimensions to the shape and form of objects and planes, the addition of colour can act as a form of safety language. For example, because red is universally regarded as a warm and arousing colour, it can be used effectively to highlight equipment required to be used urgently in the event of danger, e.g. outbreak of fire. According to the Young–Helmholtz theory, human beings use a minimum amount of energy when reacting to red, so their response is relatively rapid. Green requires slightly more energy and blue even more. When the safety precautions for certain areas of a building must be emphasised, the use of varying tones of grey in the colour scheme should be avoided; they produce slow reactions to danger because they tend to camouflage real conditions due to lack of contrast. The reflectance value of certain colours should similarly be investigated to avoid problems from glare, and also the effects which tungsten or fluorescent lights have on certain colours used to indicate safety precautions.

The use of colour as a *safety coding device* is employed to identify electrical wiring and circuits, hot and cold water services, and various gas supplies, as well as in industrial environments involving machines, steel plants, refineries, cranes and boilers. Some notable designers have used the resulting aesthetic qualities in the production of a building which represents modern hi-tech attitudes towards performance requirements (see Chapter 2).

Section 3.3 mentioned the need for specialised maintenance equipment when the design of a building makes conventional methods difficult. Whatever form is necessary, every effort must be made to ensure that maintenance personnel, such as window-cleaners, electricians, plumbers, can carry out their work in safety (see also Health and Safety at Work, Etc., Act 1974). For example, above the third-storey height of a building (and also below where access for ladders is not convenient), windows should be designed so that they can be cleaned and reglazed from inside, unless there are balconies and special devices incorporated for external maintenance. The maximum human reach is 550 mm for cleaning windows through or across an adjacent opening; side-hung opening windows should have easy-clean hinges which produce a clear gap of 95 mm between frames when the window is open.

Further specific reading

Mitchell's Building Series

Internal Components　Section 6.2.6 Strength and stability of doors
Chapter 7 Ironmongery

External Components　Section 3.2.8 Security of external glazing
Section 7.2.6 Strength and security of external doors

Building Research Establishment

BRE Digest 289   *Building management systems*
BRE Digest 428   *Protecting buildings against lightning*
BRE Digest 448   *Cleaning buildings: legislation and good practice*

Building Regulations

M   Access to and use of building
N   Glazing – safety in relation to impact, opening and cleaning (All parts)

13 Cost

CI/SfB (Y)

The design of a building must be judged not only by its appearance and the way it performs, but also by how much it costs. However, a true financial value is often difficult to assess because of the complexities of the building industry. Figure 13.1 shows the approximate apportionment of expenditure attributed to UK main contract awards, accounting for about £17.25 billion per year (2000) (see also Figure 15.2). Once most of the other industries know what is to be produced, they can reasonably accurately formulate the precise processes involved in manufacture, and stipulate under what conditions the work will be carried out. The production of a building often does not rely on such entirely rational decisions. The form of a building can be influenced by many interrelated factors (location, planning constraints, soil conditions, availability of resources, etc.), and the balance between appearance, comfort and, inevitably, convenience is finally resolved through subjective criteria. Methods of construction depend upon the knowledge and skill of the designer as well as the availability of materials and specific technical ability. The functional performance can be expressed in terms of the cost of renewing or replacing fittings and fabric, and methods required for adaptation to meet changing needs, including those of heating, lighting and other services; all these factors will be affected by the day-to-day use of a building.

13.1 Acquisition costs

Acquisition costs cover the financial aspects involved in the creation of a building – investment negotiations, professional fees, cost of land, building design and construction, etc. Further comments regarding most of these are made in Part B and only a few aspects are elaborated here.

The *cost of land* varies widely according to location, quality, size and its value in terms of suitability for a particular building type, e.g. factory, office, hospital, housing. The consequences of these factors need to be carefully considered so that cost comparisons can be made with alternative sites. The value of the site after a building has been erected on it usually remains in proportion to the value of a particular development unless subsequent political or physical phenomena are likely to result in a change in its original characteristics, e.g. motorway construction, tunnel formation, aircraft

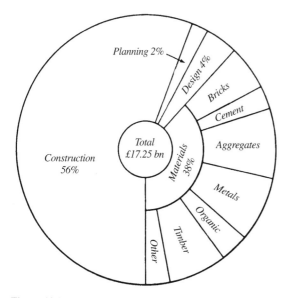

Figure 13.1 A guide to annual expenditure on UK construction projects.

flight paths, compulsory purchase, change in planning zone designation, etc. The cost of land is high and growing faster than the other factors associated with the requirements for a building, and it is therefore essential that the land is used to its maximum efficiency (see section 17.2). A more intensive use of land can be obtained by using a greater number of floors. Usually, however, the cost of suitable construction methods modifies the potential savings, and the necessity to incorporate lifts or mechanical ventilation systems will also add considerably to the subsequent running cost of a building.

There are many financial considerations associated with the *design processes* of a building. A prime consideration is that government grants and subsidies are available towards the acquisition costs of certain types of building, e.g. buildings accommodating manufacturing activities. Available funds are usually given for initial construction rather than towards subsequent running costs and are usually fixed in relation to the size of the proposed project. Unfortunately, this rarely encourages economic design because large, rather than efficient, forms of building provide better initial investment. Having partly financed such projects, unscrupulous developers or clients often leave their tenant users to run a building as efficiently as they may – any 'improvement' being at their own expense.

The cheapest form for a building is based on a compact cube, with no changes in wall, floor or roof planes. This form is difficult to achieve in truly optimum functional design terms, and may not be practicable or architecturally desirable (Fig. 13.2). Ideally, the spaces which a building provides should be the optimum usage (floor area and room height), and each activity accommodated planned to minimise the areas devoted to circulation as this inevitably contributes towards the time and cost necessary for communication. The siting of a building must give the best relationship with regard to access and public zones, and should also provide the best position for the most pleasing composition. Planning for maximum daylight penetration into a building and minimum exposure to the effects of adverse climatic conditions can effectively reduce the initial and subsequent running costs for sources of artificial lighting and heating. The use of deep or interior spaces will also require mechanically-produced environments. Noise, particularly in urban areas, is another important environmental factor related to cost. Tolerable noise levels in city centres can only be obtained by using sealed double glazing for windows, which again necessitates the use of artificial ventilation and perhaps air-conditioning.

The cost consequences of the *construction methods* used for each part of a building must be examined in a similar way to the cost consequences associated with the planning. Mention has already been made of the theoretically unlimited availability of material and technical resources for construction, and the limitations imposed by current economic constraints (section 1.3) and other practical considerations (section 3.3).

During very early development stages of a building, the designer must be aware of the constraints that may be imposed by resources upon otherwise economic planning arrangements. Whenever possible, acquaintance should be made with available materials and technical ability, and a building produced which reflects them, rather than construction methods which are beyond economical means. Certain building arrangements are financially viable because they are based on the mobility of cranes and/or gantries. And the lifting capacity of tower cranes having certain radii will limit the size of components which can be placed at the outer edges of a building under construction (Fig. 13.3).

Single-storey building with identical plan area			
Internal wall			
Foundations	*100*	*105*	*125*
External walls	*100*	*125*	*150*
Floor	*100*	*100*	*100*
Flat roof	*100*	*105*	*120*
Overall	*100*	*108.75*	*123.75*

Figure 13.2 Comparison of elemental and overall cost per unit for buildings having the same plan area but different proportions.

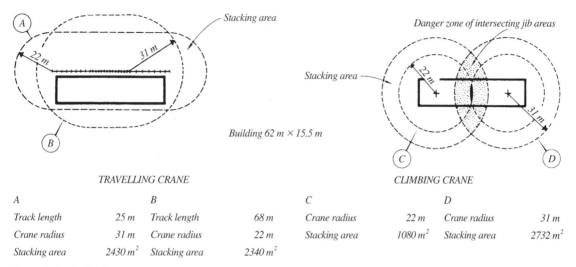

Figure 13.3   Radii of tower cranes and their effects on building shapes.

A		B		C		D	
Track length	25 m	Track length	68 m	Crane radius	22 m	Crane radius	31 m
Crane radius	31 m	Crane radius	22 m	Stacking area	1080 m²	Stacking area	2732 m²
Stacking area	2430 m²	Stacking area	2340 m²				

One aspect that needs amplification concerns the relationship between *traditional* construction methods which involve the shaping of materials on site by skilled and semi-skilled labour, and *industrialised* construction involving the erection of large, virtually finished factory-made components on site, usually by semi-skilled labour (see Chapter 4). Both forms of construction use the two basic resources, materials and labour, but the interrelated cost implications associated with each form affect their economic suitability.

In theory the use of factory-made components for a building should generate considerable savings due to the possibility of large production runs. But these savings may not always be available. The cost of basic materials accounts for about 50 per cent of the total cost of materials used in traditional construction methods. After manufacture, the factory-assembled component must be moved for storage, then eventually transported to site, hoisted and fixed in position. These processes often lead to the incorporation of additional construction features (and materials) for handling during transportation and protection against damage during erection. The made component also needs storage or shelter facilities on site, as well as special plant (cranes, etc.) for its manipulation into position in a building. For example, a precast reinforced concrete wall component will need additional reinforcement to withstand the pressures of general handling and transportation; shuttering for *in situ* concrete connections to the structures of a building, or some form of bolted plate connection device; and a fairly sophisticated jointing system (gasket and

mastics) for connection with adjoining components. If the component is delivered in a finished condition, particular care must be taken when it is being positioned, and afterwards while other trades are in close proximity. This adds considerably to the costs of handling and to the cost of site organisation and supervision.

The ultimate cost of a factory-made component, therefore, could be considerably higher than a similar form of traditionally constructed 'on site' component. On the other hand, traditional methods are subject to weather conditions unless special, and costly, precautions are taken; working conditions are far less congenial than in the factory, amenities poor and safety precautions sometimes primitive. A brief comparison is given in Figure 13.4.

13.1.1 Types of building system and components

- Linear
- Planar
- Box

Linear (skeletal)   Although strictly not a prefabricated system, linear components consist of structural units composed of columns, beams, frames and trusses of steel or concrete, produced or manufactured in a factory. Site assembly is usually by bolting the components together to create repeat modular cells. The infilling construction can be with traditional materials (see section 5.4).

Planar (panel)   These systems include prefabricated floor and wall panels. They are often used in

	Traditional construction	Industrialised construction
Advantages	Well-tried construction principles Greater possibility for design experimentation Users of building generally familiar with construction methods Generally easily capable of adaptation Flexibility in tolerances during construction Flexibility in design by varying assembly of relatively small components Quality control required but not to factory standards	Standardised specifications/drawings Social conditions in a factory better than on site Elements and/or components fabricated under controlled conditions High degree of accuracy possible Site assembly period likely to be shorter Materials storage on site reduced Large percentage of work performed at one permanent location Repetition and standardisation, therefore less opportunity for error Economies of scale Comparatively stable workforce Coordination of component design, production processes and marketing
Disadvantages	Specifications/drawings vary for each project Work may take longer due to sequential nature of trades Work on site dependent upon material availability and storage Materials and components on site could be damaged Site work often undertaken in poor and harsh conditions Social/welfare facilities relatively poor Work locations dispersed and temporary Several tasks (skilled and unskilled) required to complete a project Relatively high turnover of operatives Varied and dispersed authorities, i.e. local authority, client, designer, main contractor, etc.	System requires production in large quantities and therefore subject to market forces, including government economic policies Designs largely dependent on manufacturing processes; aesthetic appeal often of secondary importance Factory manufacturing processes monotonous to workers Extra cost for transportation of large elements/components Usually 'specialist' site contractor required Accuracy of elements/components requires careful setting out, assembly and site control May not be cheaper than traditional construction method although less labour force employed Building usually costly to adapt Limitations of 'open' and 'closed' systems (see section 15.6)

Figure 13.4   Comparison between traditional construction and industrialised construction.

conjunction with a linear support frame or traditional masonry cross walls. Components may be structural, i.e. load bearing, or for simple division of interior space. Prefabrication may include the first fix of plumbing and electrical services, as well as incorporation of thermal and sound insulation. Popular applications include residential buildings, schools and small hotels (see section 5.5).

Box (three dimensional)   Units are prefabricated as a box or module, sometimes referred to as a 'pod' or 'mould'. Typical applications are single-storey-height plastic or concrete sanitary units containing WC, bath/shower and basin, or office units all pre-wired, pre-plumbed and pre-decorated for installation and placement by crane. Units can bolt together and stack, but they are usually applied in conjunction with a conventional steel or concrete support frame.

13.2 Running costs

Running costs include expenses incurred for general maintenance, cleaning and servicing of a building, and for renewing or repairing the fabric and fittings, as well as payments for heating, lighting, ventilation and services. Allowances for taxation rates and insurances should also be made to establish total running costs.

Section 1.3 has already commented on how the acquisition costs affect the running costs, and further points are made in each chapter concerned with performance requirements. As mentioned in Chapter 3, it is generally wisest to select more costly components, construction methods and servicing systems because they tend to require low maintenance. The alternative use of low cost components, etc., requiring high maintenance, takes no account of the inconvenience, disruption and loss of earning power which arises with the need for repair or replacement. Techniques and materials which are only of value in minimising construction costs therefore fall short of what is really required if the designer is to be in a position to provide a building which offers the best value for money. It is also very important that the designer informs prospective users about matters concerning running costs of a building he or she has designed; see section 16.8.

13.3 Operational costs

Operational costs are generally (although not exclusively) associated with non-domestic buildings and include the cost of salaries for employees; provision of amenities which give congenial working conditions (bar, canteen, gymnasium, etc.); machinery, power, and materials used by the processes accommodated in a building; and the costs of adapting a building to meet changing user requirements. These costs must therefore be supplied by, or worked out in close conjunction with, the client and the client's advisers.

Sometimes the layout and design of a building dominates the way certain operations are carried out, the number and type of staff employed, and hence the overall operational costs. The types of activity most affected in this way involve transport systems or product manufacture and handling, but even their efficiency will finally be determined by the calibre (and salaries) of supervision and management staff.

When possible, the prospective building owner must be made aware of the effects which other cost factors of a design have on operational costs. For example, the choice of temperature control in offices can have an effect on the efficiency of employees. During sudden cold spells, a slow room-heating response may produce errors or temporary cessation of clerical performance, giving rise to subsequent financial losses. To ensure continuous internal heating levels in a building, it is possible to increase the acquisition cost of construction (say by double glazing or extra thermal insulation) and/or to increase running costs for fuel (say by using relatively expensive forms of quickly-responding temperature controls).

13.4 Cost evaluation

The true cost evaluation of a building should involve a close examination of viable alternative acquisition, running and operational costs. The depth of investigation possible, and therefore the accuracy in achieving the 'true value' of a building, will depend on the information available and skill of interpretation during the inception, preliminary and detail design stages. Although *cost analyses* (Fig. 13.5) are available through technical publications for numerous schemes which may be similar to one in hand, normally they are not able to supply information regarding operational costs or the effects of solely financial design decisions on building users. When pertinent, this information allows the historical cost analysis data to be used for the future creation of a building specifically oriented towards definable long-term financial goals.

Various techniques exist for the cost analysis of a proposed building. They include a *cost-in-use* method which analyses the relationship between acquisition and running costs in order to achieve optimum performance requirement criteria; a *cost-effectiveness* method which analyses the relationship between acquisition, running and operational costs so as to permit adaptation of certain performance requirements to allow for *specific* user attitudes; and a *cost-benefit* method (the most difficult) which analyses economic building performance criteria in terms of *variable* user-benefits.

The evaluation of capital cost of construction and the financial reality of ongoing maintenance is part of terotechnology. This multi-disciplined awareness for the aftercare of new buildings is defined in BS 3811: 1993 *Glossary of terms used in terotechnology* as 'concern with the specification and design for reliability and maintenance of plant, machinery, equipment, buildings and structures; with their installation, commissioning maintenance, modification and replacement; and with feed back of information on design, performance and costs'. In other words, the central idea involves the pursuance of economic *life-cycle* costs for a building through combined design, technical, financial and managerial skills.

Historically, very few building owners and users calculated the true costs of their building (Fig. 13.6). Now, with the benefits of considerable research into value engineering and life-cycle costing, an analytical approach to the various means of achieving a client's objective, i.e. a building, and its longer-term running costs is not only possible but common practice.

Value engineering is an exercise which identifies the optimum means of construction, eliminating unnecessary costs while retaining a high regard for specification, quality and product usefulness. It is a team approach including the building designer, client representative or adviser in construction practice and a quantity surveyor for financial evaluation. Such disparate professional backgrounds should combine to realise the optimum construction process in the client's initial and longer-term interests. Clients (and designers) need to

Cost summary

	Cost per m² (£)	Per cent of total
SUBSTRUCTURE	**37.31**	**3.48**
SUPERSTRUCTURE		
Frame/upper floors	104.04	9.96
Roof	46.47	4.33
Rooflights	2.82	0.26
Staircases	31.48	2.93
External walls	67.96	6.33
Windows	71.70	6.68
External doors	4.03	0.38
Internal walls and partitions	78.44	7.31
Internal doors	28.22	2.63
Group element total	*435.16*	*40.54*
INTERNAL FINISHES		
Wall finishes	29.87	2.78
Floor finishes	43.47	4.05
Ceiling finishes	14.17	1.32
Group element total	*87.51*	*8.15*
SERVICES		
Sanitary appliances	2.92	0.27
Services equipment	22.84	2.13
Disposal installations	4.41	0.41
Water installations	12.02	1.12
Space heating/air treatment	139.71	13.02
Electrical services	127.89	11.92
Lift & conveyor installations	14.48	1.35
Protective installations	34.03	3.17
Communication installations	13.37	1.24
Builders' work in connection	24.36	2.27
Group element total	*396.03*	*36.90*
PRELIMINARIES & INSURANCE	**86.37**	**8.07**
Total	*1073.24*	*100.00*

Cost analysis

SUBSTRUCTURE

FOUNDATIONS/SLABS £37.31/m²
Bored in-situ piles, pile caps and RC slab with service trenches

SUPERSTRUCTURE

FRAME AND UPPER FLOORS £104.04/m²
RC frame comprising exposed columns integrated with RC slabs, with a feature exposed soffit finish. Steel column and beam framework at fifth floor.

ROOF £46.47/m²
A system of flat roofs with a roof garden and paved terraces at fifth floor, in specialist felt roof system. Flat roofs at high level, and to plant areas, in sheet metal with exposed felt systems.

ROOFLIGHTS £2.82/m²
Large circular double-glazed aluminium mono-pitch rooflight to north atria with individual small lights to top of escape staircases. Individual polycarbonate rooflights to south atria

STAIRCASES £31.48/m²
Steel feature main staircase with mild steel and hardwood balustrade, Jura limestone finish. Five RC secondary access/escape stairs with steel balustrades and handrails. Steel feature balustrades to all atria with glass infill panels and hardwood handrails

EXTERNAL WALLS £67.96/m²
Building clad with Jura limestone and Cornish granite. Aluminium rainscreen panelling to stair towers and plantrooms. PPC louvres to plant rooms

WINDOWS £71.70/m²
High-quality aluminium flat-panelled/double-glazed window/walling system. External solar shading to main elevations

EXTERNAL DOORS £4.03/m²
Aluminium-framed glazed doors in window walling system generally on ground and fifth floors. Two pairs aluminium-framed electric sliding doors to main entrance; revolving door to south elevation

INTERNAL WALLS AND PARTITIONS £78.44/m²
Blockwork walls to cores and plantrooms. Solid and glazed office partitioning with doors. Hardwood single-glazed atria screens and doors

INTERNAL DOORS £28.22/m²
Solid-core hardwood-veneered, or painted, flush doors; hardwood frames; stainless-steel ironmongery

INTERNAL FINISHES

WALL FINISHES £29.87/m²
Plasterboard dry linings; ceramic tiles to WC areas and kitchen/café; painted blockwork to plantrooms

FLOOR FINISHES £43.47/m²
Raised floor, sealed as a plenum; ceramic tiles to WC areas; carpet tiles to office and circulation areas; Jura limestone to atria; hardwood flooring to restaurant, café and boardroom areas

CEILING FINISHES £14.17/m²
Painted fairface concrete soffits; plasterboard ceilings to atria and storage areas; metal-pan suspended ceilings to kitchen, gym, print and post rooms

FITTINGS AND FURNISHINGS

FURNITURE £30.66/m²
Reception desk, blinds, signage, mirrors, shelving and fittings, café servery and shelving, gym mirrors and cupboards, boardroom cupboards, coffee points, atrium access equipment and safety wires

SERVICES

SANITARY APPLIANCES £2.92/m²
Individual WCs with basins; disabled WCs and associated fittings; cleaners' sinks

SERVICES EQUIPMENT £22.84/m²
Contractor-designed kitchen and servery installation; specialist-designed audio-visual suite

DISPOSAL INSTALLATIONS £4.41/m²
General soil, vent, rainwater and waste installations; copper and cast iron

WATER INSTALLATIONS £12.02/m²
Hot and cold water to kitchen, café, WCs, coffee points and sundry basins

SPACE HEATING/AIR TREATMENT £139.89/m²
Displacement-air ventilation system. Perimeter LPHW heating to all areas except atria. Fan-coil units to meeting areas, with ice thermal storage. VAV a/c to boardroom and kitchen/restaurant at fifth floor.

ELECTRICAL SERVICES £127.89/m²
Floor-box power arrangement to offices. Lighting control system. Feature lighting and power outlets. UPS computer backup system

LIFT AND CONVEYOR INSTALLATIONS £14.48/m²
Two banks of two 13-person passenger lifts to core areas. One goods lift to north core

PROTECTIVE INSTALLATIONS £34.03/m²
Internal/external CCTV system, intruder-detection system and an access-control system. Internal fan and window-actuated smoke-control system

COMMUNICATION INSTALLATIONS £13.37/m²
Telephone and data-wiring system to floor boxes

BUILDERS' WORK IN CONNECTION £24.36/m²

PRELIMINARIES AND INSURANCES

PRELIMINARIES, OVERHEADS & PROFIT £86.57/m²
Excludes dayworks

EXTERNAL WORKS

LANDSCAPING, ANCILLARY BUILDINGS £1,078,387
Access road with surface parking. Paving, steps and walls. Block of 12 garages, cycle shed and substation. Landscaping and planting. Two copper-roofed, steel-framed, covered walkways. Below-ground drainage, petrol interceptor and holding tank

Figure 13.5 Example cost analysis. (Reproduced courtesy of the *Architects' Journal*, from 'A new home for Longmans', 30 May 1996)

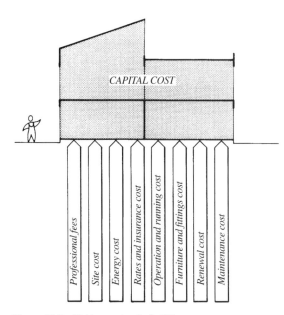

Figure 13.6   Hidden costs of a building.

be made aware of life-cycle costs so that they apply their minds to the best available data in a structured and systematic way in order to obtain a clearer understanding of the future cost implications for a proposed building. Today the high costs of materials, labour and energy imply that the running costs of buildings are of greater importance relative to acquisition costs.

Despite possible short-term reductions in prices, the long-term trend is likely to continue upwards, and operational costs must be used more fully to buffer the effects of overcostly capital investments. A true understanding of this will facilitate the preparation of estimates to give the realistic cost consequences of building designs against which the more subjective judgements of appearance, comfort and convenience can be set. For this purpose, comparative tables can be prepared which set out the costs and values arising from each design alternative: preferred appearance, comfort and convenience factors can be compared against the monetary figures which sum up the cost consequences of the design.

Further specific reading

Mitchell's Building Series

External Components	Section 2.2 Economic and social factors of prefabrication Section 2.6 Fast-track construction with prefabricated components
Materials	Introduction and within each chapter on materials
Structure and Fabric Part 1	Section 2.6 Economic aspects of building construction
Structure and Fabric Part 2	Chapter 1 Contract planning and site organization

Building Research Establishment

BRE Digest 247	*Waste of building materials*
BRE Digest 259	*Materials control to avoid waste*
BRE Digest 374	*Relocatable buildings: structural design, construction and maintenance*
BRE Digest 397	*Standardisation in support of European legislation*
BRE Digest 423	*The structural use of wood-based panels*
BRE Digest 447	*Waste minimisation on a construction site*
BRE Digest 450	*Better building: Integrating the supply chain; a guide for clients and their consultants*
BRE Digest 452	*Sustainable building design: using whole life costing and life cycle assessment*

Part B

An analysis of a building in terms of the processes required, the Building Team which implement them, and the methods used in communicating information

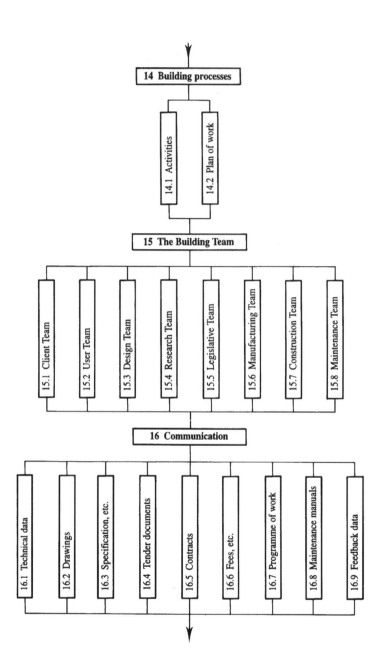

14 Building processes

14.1 Activities

14.2 Plan of work

15 The Building Team

15.1 Client Team

15.2 User Team

15.3 Design Team

15.4 Research Team

15.5 Legislative Team

15.6 Manufacturing Team

15.7 Construction Team

15.8 Maintenance Team

16 Communication

16.1 Technical data

16.2 Drawings

16.3 Specification, etc.

16.4 Tender documents

16.5 Contracts

16.6 Fees, etc.

16.7 Programme of work

16.8 Maintenance manuals

16.9 Feedback data

14 Building processes

CI/SfB (A1)/(A9)

14.1 Activities

The processes currently involved in the erection of a building are very complex. This is primarily because they concern many different categories of personnel, and forms of communication, resources and operational skills. Erection of a building includes the following activities:

- The initial decision to build
- Securing of financial resources
- Selection of appropriate location
- Appointment and briefing of suitable members to be involved with design and construction operations
- Definition of precise functional requirements
- The design process and decisions on how to build
- Implementation of erection
- Operations necessary to maintain the building in the state of continuous performance for which it was intended
- Operations necessary to adapt building to new functions

All these activities may be affected by an elaborate system of approvals, controls, checks and cross-checks which not only involve nearly all the members of the entire building team, but also a variety of outside bodies in varying administrative, technical, financial and fiscal capacities. Furthermore, whereas the creation of a building was formerly a leisurely occupation ultimately dependent upon craft-based skills, the whole process is now greatly influenced by the desire to achieve profit on financial investments as soon as possible and the exploitation of machine technology. Therefore, although most buildings of today are far more complex and sometimes much larger, the timescale from the realisation of the need for a building to its use after erection is generally far shorter than at any previous period in the development of building.

In view of these complexities, a great deal of attention has been devoted to methods of achieving clarity of communication between the participants. The difficulty has been bridging the gap between professional activities. Consultants, designers and builders have stood divided as a result of social and educational background, and contractual relationships which have encouraged apportionment of blame when deficiencies occur, rather than encouraging teamwork.

The need for increased effectiveness – better coordination of effort and control of all the complex operations involved in a building – has been the subject of public and private research for many years. The Tavistock Institute reports, *Communications in the Building Industry* and *Interdependence and Uncertainty*, published in the mid-1960s, indicate that this is not a recent area for concern. The Wood Report of 1975 acknowledged some improvement:

> *The traditional separation between design and construction was found to have diminished with consequent advantages all round . . . Contractors have much to offer at the design stage, especially by way of advice on constructional implications of design solutions and decisions . . . yet, methods of procurement are still such that they are brought in too late for their advice and experience to be of practical use . . . the original problem still exists.*

An awareness of the poor correlation between design and construction led to the concept of 'buildability'. In 1979 the Construction Industry Research and Information Association (CIRIA) embarked on several years' research into this subject, endorsing the need for builders' expertise at the design stage. In 1983 they defined *buildability* as 'the extent to which the design of the building facilitates ease of construction, subject to the overall requirements for the completed building'.

Emphasis is clearly that design must be with regard to the practicalities of construction, therefore traditional adversarial relationships between architects and builders should be replaced with a good working partnership for the client to obtain value for money. In recent years this has been endorsed with the establishment of many design-and-build practices.

In 1983 the National Economic Development Office (NEDO) produced its findings into construction delays, *Faster Building for Industry*. Since then the topic has been the subject of many research papers, presentations and professional journal articles. These compounded and eventually encouraged the government to engage Sir Michael Latham to chair a major initiative into improving productivity in the industry. The outcome in 1994 was a 130-page document, *Constructing the Team*, aimed specifically at providing better value for the client, with the objective of a 30 per cent real cost reduction in productivity targets. In principle the recommendations include formation of a client forum, government commitment to education, training and guidance on practice, consideration of liability and compulsory building user insurance, a research initiative funded by levy and adjudication to be promoted as a means of professional dispute resolution. Overall it places greater emphasis on the professions sharing responsibility for building construction, rather than the traditional practice of ill-defined and segregated interests.

A few years later in 1998, Sir John Egan was appointed to lead another government inquiry into the performance of the construction industry. This resulted in the publication, *Rethinking Construction*, more generally known as the Egan Report. It was highly critical of the industry's efficiency, emphasising the needs for clients to receive value for money rather than price. The report calls for cuts of 10 per cent in construction time and costs, with reciprocal gains in profit and turnover. This is to be achieved through cooperation of parties to the project (partnering), further integration of design with construction, increased standardisation, leaner construction techniques to minimise material waste, phasing out of inflexible contracts and less emphasis on competitive tendering for contractor selection.

The Housing Grants, Construction and Regeneration Act of 1996, known as the *Construction Act*, came into effect in 1998. This aims to reduce the number of legal claims and acrimony between parties to the building contract, by effecting rules on fair payment and a right to adjudication. It is all-embracing, covering construction contracts between contractors and their subcontractors, and contracts between clients and consultants including architects, engineers and quantity surveyors. In addition to encouraging better contractual relationships, it should reduce the number of long-running disputes, legal bills, personal losses, stress and business failures in the construction industry.

It is often stated that the only common factor underlying the whole process is the economic one, and this should therefore provide the basis of organisational procedures. Economics are significant in assessing the nature of the individual contributions of the members of the building team and often explain the reasons for the elaborate system of mutual controls, checks and doubts usually blamed for delays and 'inefficiency' within the building process. Perhaps the future will provide greater professional cooperation, and mutual distrust will no longer interfere with progress.

14.2 Plan of work

Figure 14.1 indicates a plan of work which was originally published in the first edition of the *Royal Institute of British Architects' Handbook* in 1964. Its intention was to provide a model procedure for methodical working of the design team employed on projects which have sufficient common factors to make it widely relevant. The plan of work has subsequently become widely known among other professions concerned with the design and construction of buildings as it is capable of being used in a variety of ways. It can assist the planning of projects and be adapted to form the basis for control of organisational procedures. In developing the model, certain assumptions were made:

- It relates to a building costing £300 000 (1964 price) which uses a full design team.
- An architect is the principal designer and leader of the design team.
- The architect is appointed at an early stage of the building process.
- Each stage follows sequentially.
- The cycle of work in *each stage* involves the following items:
  - stating objective and assimilation of relevant facts
  - assessment of resources required and setting up of appropriate organisations
  - planning the work and setting timetables

Stage	Purpose of work and decisions to be reached	Tasks to be done	People directly involved	Commonly used terminology
A Inception	To prepare general outline of requirements and plan future action.	Set up client organisation for briefing. Consider requirements, appoint architect.	All client interests, architect.	Briefing
B Feasibility	To provide the client with an appraisal and recommendation in order that he may determine the form in which the project is to proceed, ensuring that it is feasible, functionally, technically and finacially.	Carry out studies of user requirements, site conditions, planning, design, and cost, etc., as necessary to reach decisions.	Clients' representatives, architects, engineers and QS according to nature of project.	
Stage C begins when the architect's brief has been determined in sufficient detail.				
C Outline Proposals	To determine general approach to layout, design and construction in order to obtain authoritative approval of the client on the outline proposals and accompanying report.	Develop the brief further. Carry out studies on user requirements, technical problems, planning, design and costs, as necessary to reach decisions.	All client interests, architects, engineers, QS and specialists as required.	Sketch plans
D Scheme Design	To complete the brief and decide on particular proposals, including planning arrangement appearance, constructional method, outline specification, and cost, and to obtain all approvals.	Final development of the brief, full design of the project by architect, preliminary design by engineers, preparation of cost plan and full explanatory report. Submission of proposals for all approvals.	All client interests, architects, engineers, QS and specialists and all statutory and other approving authorities.	
Brief should not be modified after this point.				
E Detail Design	To obtain final decision on every matter related to design, specification, construction and cost.	Full design of every part and component of the building by collaboration of all concerned. Complete cost checking of designs.	Architects, QS, engineers and specialists, contractor (if appointed).	Working drawings
Any further change in location, size, shape, or cost after this time will result in abortive work.				
F Production Information	To prepare production information and make final detailed decisions to carry out work.	Preparation of final production information i.e. drawings, schedules, and specifications.	Architects, engineers and specialists, contractor (if appointed).	
G Bills of Quantities	To prepare and complete all information and arrangements for obtaining tender.	Preparation of Bills of Quantities and tender documents.	Architects, QS, contractor (if appointed).	
H Tender Action	Action as recommended in relevant NJCC *Code of Procedure for Selective Tendering.*	Action as recommended in relevant NJCC *Code of Procedure for Selective Tendering.*	Architects, QS, engineers, contractor, client.	
J Project Planning	To enable the contractor to programme the work in accordance with contract conditions; brief site inspectorate, and make arrangements to commence work on site.	Action in accordance with RIBA Plan of Work.	Contractor, sub-contractors.	Site operations
K Operations on Site	To follow plans through to practical completion of the building.	Action in accordance with RIBA Plan of Work.	Architects, engineers, contractors, sub-contractors, QS, client.	
L Completion	To hand over the building to the client for occupation, remedy any defects, settle the final account, and complete all work in accordance with the contract.	Action in accordance with RIBA Plan of Work.	Architects, engineers, contractor, QS, client.	
M Feedback	To analyse the management, construction and performance of the project.	Analysis of job records. Inspections of completed building. Studies of building in use.	Architect, engineers, QS contractor, client.	Feedback

Figure 14.1   RIBA plan of work. (Reproduced courtesy of RIBA Publications)

- carrying out work
- making proposals
- making decisions
- setting out objectives for the next stage.

This plan of work will be used as a basis for the text in Chapters 15 and 16. Figure 14.2 indicates the relationship of each subject to the plan of work. Further reference should be made to the *Architect's Job Book*, 6th edition.

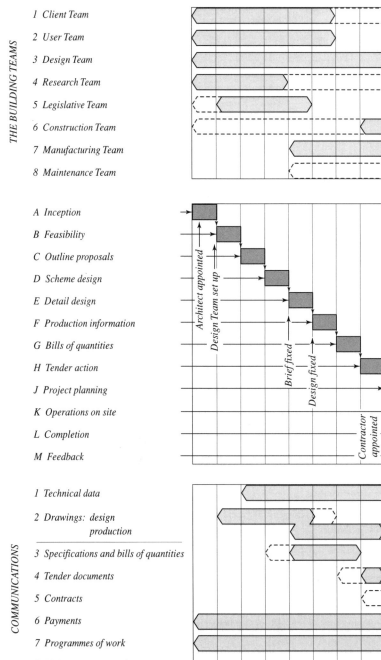

THE BUILDING TEAMS

1 *Client Team*
2 *User Team*
3 *Design Team*
4 *Research Team*
5 *Legislative Team*
6 *Construction Team*
7 *Manufacturing Team*
8 *Maintenance Team*

A *Inception*
B *Feasibility*
C *Outline proposals*
D *Scheme design*
E *Detail design*
F *Production information*
G *Bills of quantities*
H *Tender action*
J *Project planning*
K *Operations on site*
L *Completion*
M *Feedback*

COMMUNICATIONS

1 *Technical data*
2 *Drawings: design production*
3 *Specifications and bills of quantities*
4 *Tender documents*
5 *Contracts*
6 *Payments*
7 *Programmes of work*
8 *Maintenance manual*

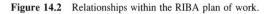

Figure 14.2   Relationships within the RIBA plan of work.

Further specific reading

Mitchell's Building Series

Internal Components	Chapter 1 Component design
External Components	Section 2.2 Prefabricated components: economic and social factors
	Section 2.6 Fast-track construction
	Section 2.8 Prefabricated component design: accuracy and tolerances
Structure and Fabric Part 1	Chapter 2 The production of buildings
Structure and Fabric Part 2	Chapter 1 Contract planning and site organization

Government Reports

Latham, M. *Constructing the Team* (The Latham Report), The Stationery Office (1994).
Egan, J. *Rethinking Construction* (The Egan Report), The Stationery Office (1998).

15 The Building Team

CI/SfB (A1m)

Building is a group activity and its success depends on a good understanding and cooperation between a large number of people. The participants involved can be conveniently arranged into groups or teams according to their particular interest and/or involvement as follows:

- Client Team
- User Team
- Design Team
- Research Team
- Legislative Team
- Manufacturing Team
- Construction Team
- Maintenance Team

Figure 14.2 associates each team to the RIBA plan of work for a medium-size building project as described in Chapter 14. In this way the activities of each can be approximately related to the various stages of the design and building process. However, this represents only one method of sequential timetabling and also indicates the principal designer of a building to be the leader of the building process. Various permutations can be adopted involving different 'leaders', or no 'leaders' at all, as well as employment of teams at different stages according to the precise circumstances. Figure 15.1 indicates the interrelationship of the various teams in the sequence to be considered in this chapter.

15.1 Client Team

The client, or prospective building owner, has the responsibility for defining the building to suit needs, establishing and providing the necessary finances, agreeing design and construction phases, timetabling and, of course, fulfilling the management and running of the completed project. This implies that, besides producing a clear and accurate *brief* (or list of requirements) for a building and maintaining a strategic knowledge throughout the building process, the client should be able to make prompt decisions when requested. But the client usually resolves many of them helped by professional advisers.

Under the Construction (Design and Management) Regulations 1994, which form part of the Health and Safety at Work, Etc., Act 1974, the client is required to appoint a planning supervisor and to nominate the principal contractor as soon as practicable. One person can perform both roles. The client may decide to appoint his or her architect as the planning supervisor; the client does this using RIBA's special form of appointment. If the architect is appointed as planning supervisor, this should be seen as an additional, separate function to the design service. The following three documents are required from the planning supervisor:

- notification of the project to the Health and Safety Executive (HSE);
- a health and safety plan with contributions from the main/principal contractor and designer;
- a health and safety file, completed after construction for future reference by the client.

The planning supervisor is also required to ensure that all contractors are adequately competent, particularly with regard to health and safety procedures. The planning supervisor will advise the client and principal contractor accordingly. The principal contractor contributes to the health and safety plan, maintains a health and safety file as work proceeds, prepares risk assessments and communicates all findings and filed material to the planning supervisor and design team.

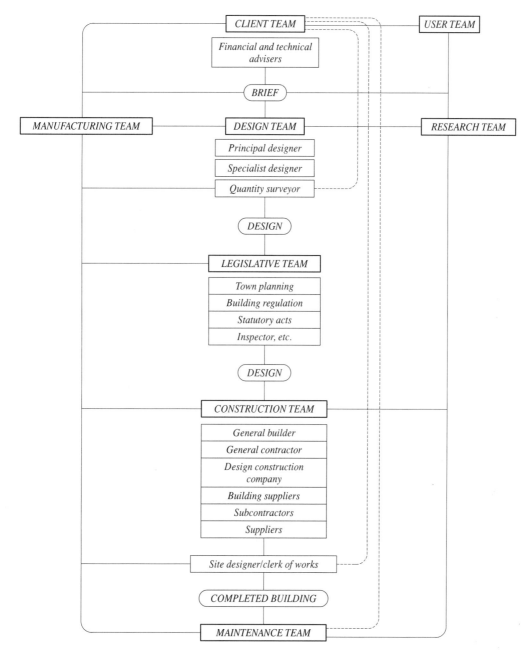

Figure 15.1 Sequential relationships within the Building Team.

The type of client varies but can include a domestic couple, a director of a small company, or a carefully constituted project team of a large industrial complex consisting of departmental heads controlling finances, building skills and other specialisms. Large organisations very often incorporate a committee of laypersons backed by consultants, such as exist in government administrative departments. Nevertheless, whatever the combination or form of the client organisation, its responsibility is always to provide clear and concise instructions to the building professionals.

The British Property Federation is an association devoted to the interests of property owners and others with a major concern for property. Their main objective is to promote a better understanding between property owners (ranging from large development companies to owner-occupiers of single dwellings) and the public, government and local authorities. The Federation provides general advice on all matters relating to the law, management and administration of property, and on taxation, housing and rating problems.

15.1.1 The necessity to build

A potential client must establish whether to build or not to build, and it is first necessary to carry out a comprehensive appraisal of needs. Having decided that a new building is necessary to provide additional or alternative space, it is important that consideration is then given to when the space will be needed. For people who are not connected with the building industry, the timescale for building often seems surprisingly (and unnecessarily) long. A minimum of six months will be required between making an appraisal and the start of even a quite modest building. Larger buildings can take several years in resolving the problems associated with land acquisition, establishment of rights, development permits, planning permission, building approval, contractor selection and subsequent erection. Initial delays may also be encountered while seeking out and appointing a suitable team of professionals (lawyers, financiers, designers, etc.) to advise on these activities.

There are various options open to a person or organisation requiring more space other than commissioning a new building. For example, there may be the possibility of the purchase or lease of existing property which can be altered, adapted, extended or renovated in a suitable manner. Alternatively, there is the possibility of the purchase or leasing of a precisely suitable building. Developers will often build office blocks and shopping areas speculatively, and many town councils will build factory units for sale or lease. These buildings usually, however, require fitting out to greater or lesser amounts by the eventual owner or lessee. Clients wishing to investigate the potential use of an existing building, or one which is under construction having been commissioned by others, should seek the advice of the *estate agents* dealing with the particular interest. These are professionally qualified people involved with the buying, selling and managing of property generally. They may belong to the Royal Institution of Chartered Surveyors (RICS), to the Auctioneers and Estate Agents Institute (AEAI) or to the National Association of Estate Agents (NAEA).

15.1.2 Financial resources

Most building is undertaken from money made available in the form of a loan, so interest rates (loan charges) are important. In this respect, the government has direct influence and can use the building industry as a regulator for the economy of the country. Taxation and/or adjustment to the bank rate on loans can cause boom or slump periods. Funds for a public building are made available from local or national taxation, and there is concern to spend as little as possible with due regard to the value for money. Financial control is implemented by the use of the 'cost yardstick' in housing; cost per place in schools, or cost per bed in hospitals, etc. In the private sector there is a little more flexibility in that the client will spend in proportion to stated aims, which may be as diverse as a temporary-use building or a prestige building. Private funds arise from the profits of industrial undertakings or the savings from individuals which again are closely linked with investment and borrowing potential, mortgage rates and grants to charities, etc.

Privately financed initiatives (PFI) are a combination of public and private interests for public sector building such as prisons, schools, hospitals and roads. This has become known as *public–private partnership* (PPP). The financing objectives can take on a variety of forms. At its simplest, a business may seek to invest in a public sector project for publicity and/or a profit share of its success. More complex is the idea of bonding and sharing, where all team members including designer and main contractor participate financially in their client's project. Some of their costs are deferred until the business is established, with the possibility of a profit share thereafter.

For the year 2000, main contract awards totalled approximately £17.25 billion. Figure 15.2 indicates how that money was variously apportioned.

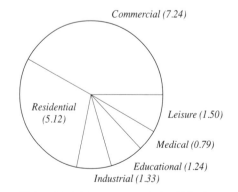

Figure 15.2   Main UK contract awards for 2000: figures in parentheses are expenditures in £billion. (Based on data provided by Emap Glenigan, Bournemouth)

Some building types which the government in power considers should be encouraged for the benefit of a particular region or the country in general may be eligible for financial support from national taxation. Formerly, industrial premises and hotels have been supported in this manner. Indeed, buildings are often only implemented as a result of this financial aid of *grant*, but designers must always confirm the opportunities which may exist for the client. For smaller-scale work, involving the domestic sector, the Housing Acts specify grants which are available for the provision of adequate sanitary appliances where they do not exist, and for essential repairs or building extensions which bring privately-owned homes to a satisfactory standard of habitation. Currently, there are even government grants towards the costs of providing thermal insulation to roof spaces of domestic buildings.

Once money becomes available for a building, the client will require speedy action for its design, construction and subsequent use, so that the lost interest, which would have been gained through alternative financial investments, may be speedily recouped. Indeed, today, speed of use is even more essential to ensure that the value of the original sum is not extensively reduced by the effects of monetary inflation in the cost of labour and materials. The total cost of a building must include the professional fees of the Design Team which the client appoints. In conventional terms the contractor's 'fees' and those associated with initial local authority services (planning, byelaws, water for works, etc.) are included in the costs for the building (see Chapter 13).

15.2 User Team

As the majority of building is for the direct employment of people or involves people for its continued function, i.e. maintenance, a building User Team forms a vital link between design concepts and built reality. Formulating user requirements may be simplified when client and occupants are the same. However, this is not always the case, and it is the initial responsibility of a non-user Client Team to establish user requirements when formulating a brief with their professional advisers.

Typical User Teams which can supply information are tenants' associations, medical associations, consumer associations, tourist boards, unions, etc., and their membership may include professional advisers as well as laypersons. The Design Team often has to draw independently upon the expertise, experience and research of these bodies when advising on a client's own interpretation of user requirements for a proposed

building. In this way the designer can modify certain views and provide for the psychological and physiological well-being of future occupants.

The performance requirements for a building are sometimes derived from feedback information supplied by the users of similar existing designs. This applies particularly to experience gained from designs which resulted in loss of security or accidents (see Chapter 12). Where a novel project is being contemplated, detailed research must be carried out to ascertain the effects which certain designs or construction methods are likely to have on potential occupants; if necessary, a life-size prototype of the proposed enclosure must be made for users to experience and test.

15.3 Design Team

The design of a building can no longer be the total responsibility of one person. There are a great many people concerned with supplying the design expertise which will make a building possible, and these people are known collectively as the Design Team.

Within this team there are two types of building designers: *principal designers* with the responsibility for the overall design of the project, and *specialist designers* who provide expertise concerning certain aspects of a building and whose requirements are often coordinated by the principal designer. The principal designer may also be appointed as planning supervisor – see section 15.1. Cost control and financial advice to client, principal and specialist designers is generally provided by a *quantity surveyor*. The fees for the professional services of principal designers, specialist designers and the quantity surveyor are paid by the client (see section 16.6).

15.3.1 Principal designers

Professional principal designers generally include architects, interior designers and building surveyors. Each is capable of fulfilling the function of a building designer, but their training means they may provide different emphases in the approach to the problems associated with the production of a building.

Architects design and prepare the production information for most building projects, and on small general-purpose buildings their expertise permits them to be the sole designer. They will also inspect the construction work on site and may function as planning supervisor. The title 'architect' is protected under the Architects' Registration Board (ARB), and only persons appropriately qualified and registered can use it.

Most architects are also members of the Royal Institute of British Architects (RIBA) and are governed by its charter, byelaws, regulations and code of professional conduct under, and in addition to, the general law. The purpose of the RIBA as described in its charter is 'the advancement of Architecture and the promotion of the acquirement of the knowledge of the Arts and Sciences connected therewith'. The object of the code is to promote the standards of professional conduct in the interest of the public and consists of three principles:

1. *A member shall faithfully carry out the duties which he undertakes. He shall also have proper regard for the interests both of those who commission and of those who may be expected to use or enjoy the product of his work.*
2. *A member shall avoid actions and situations inconsistent with his professional obligations or likely to raise doubts about his integrity.*
3. *A member shall rely only on ability and achievement as the basis for his advancement.*

Each of these principles contains a number of rules giving instructions about specific aspects of practice. Under Principle 1 an architect must act impartially when interpreting the clauses of a building contract between client and contractor. Under Principle 2 an architect must not take bribes or allow his or her name to promote any service or product. Under Principle 3 an architect must not offer financial inducements to obtain work.

Interior designers can also prepare design and production information for a building, and provide supervision of work but, as their title implies, they may be specifically concerned with the interior of a building and need additional advisers in order to deal with all the design and construction processes involved in a total building. Interior designers can be members of the Chartered Society of Designers (CSD). This organisation has codes of professional conduct which reflect ideals equivalent to those of the RIBA.

Building surveyors are sometimes responsible for the design and supervision of certain building work, although they more usually carry out surveys of structural soundness, condition of dilapidation or repair, alterations/extensions to existing buildings and market value of existing buildings. However, it must be borne in mind that the emphasis of their training lies in the technical aspects of building rather than the aesthetic aspects. They may be members of the Royal Institution of Chartered Surveyors (RICS) and will therefore be governed by its code of conduct (see section 15.1).

Some architects and surveyors have also been members of the Architects and Surveyors Institute (ASI). This is now a specific faculty within the Chartered Institute of Building (CIOB).

15.3.2 Specialist designers

These include civil and structural engineers, services engineers, and those concerned with specific aspects of architecture, including landscape, interiors, office planning, etc. *Civil and structural engineers* are employed to assist principal designers on building projects which contain appreciable quantities of structural work, such as reinforced concrete, complex steel or timber work, or foundations which are either complex or abnormal.

Civil engineering is defined as the construction of roads, bridges, tunnelling, motorways, etc. There is often a certain amount of building work in civil engineering construction and, likewise, many building projects contain civil engineering construction. *Structural engineering* deals especially with the calculation of the structural parts in a building. When projects consist mainly of structural work, the civil or structural engineer will be the principal designer. They can be members of the Institute of Civil Engineers (ICE) or the Institution of Structural Engineers (IStructE). The Association of Consulting Engineers (ACE) issue codes of conduct defining professional responsibility.

Services engineers work with other designers and are concerned with environmental control: lighting, heating, air-conditioning and sound modulation; electrical installations, plumbing and waste-disposal systems; and mechanical services, such as lift installations and electrical conductors. The Chartered Institution of Building Services Engineers (CIBSE) was formed from an amalgamation of the Institute of Heating and Ventilating Engineers (IHVE) and the Illuminating Engineering Society (IES) with the object of promoting the science and practice of services engineering, as well as advancement of education and research in the field. Electrical engineers are governed by the regulations of the Institution of Electrical Engineers (IEE), and mechanical engineers by the Institution of Mechanical Engineers (IMechE). Senior members of CIBSE, IMechE and IEE can also become members of the Association of Consulting Engineers (ACE). Table 15.1 indicates some of the services engineers involved in the design of buildings.

Other specialist designers such as landscape architects, interior designers, graphic designers, space planners, acoustical and production engineers are employed according to the type and complexity of the building

Table 15.1 Some services engineers as specialist designers for the environmental control aspects of buildings

Services engineer	Environmental control aspect
Acoustic	Modulation and audio
Air-conditioning	Heating and refrigeration
Communications	Lifts, hoists, escalators, paternosters and conveyors
Catering	Food preparation and service
Drainage	Above and below ground, and water-borne refuse
Electrical	Heating, air-conditioning, refrigeration and lighting
Gas supply	Heating, including industrial applications
Heating	Gas, electric, oil, solid fuel, solar and ambient
Fire protection	Escape and fire-fighting equipment
Lighting	Natural and artificial
Plumbing	Hot and cold water services
Refrigeration	Preservation and installations
Refuse disposal	Storage, chutes, water-borne and disposal
Sanitation	Pest control, appliances and drainage
Security systems	Protective, anti-theft and pilfering devices
Telecommunications	Telephone, security, cable and TV
Thermal insulation	Fabric design and installation
Ventilation	Natural and artificial
Water supply	Availability and storage

project. Each discipline is represented by a professional institute or association.

15.3.3 Quantity surveyor

An essential part of any design process involves cost control. The costing services for smaller, less complex building projects are generally provided by the principal designer working in conjunction with the client and specialist designers. However, for larger or more complex projects it is usual that a *quantity surveyor* is employed to give cost advice and sometimes a cost control service (see Chapter 13). Chartered quantity surveyors are governed by the codes of professional conduct issued by the Royal Institution of Chartered Surveyors (RICS) and, like other members of the Design Team, their fees are paid by the client. The RICS is also the professional body for certain other members of the building profession (e.g. building surveyors, estate agents).

Until recently, no country, other than the United Kingdom, regularly employed a quantity surveyor for building projects. The quantity surveyor's primary role is to prepare a *bill of quantities* from the drawings and specifications supplied by the Design Team. This itemises the type, form and amount of materials to be used in the construction project. The bill (see section 16.3) will also define the legal requirements for the project, including the form of contract to be adopted between the client and the contractor. The whole document, together with relevant drawings and specifications supplied by the Design Team, will therefore provide a good basis for obtaining prices for the project from a number of interested contractors (competitive tendering).

Prior to the work on the bills of quantities, the quantity surveyor may be expected to give cost advice on any alternative solutions that may be considered during the various stages of the design of a building. Also, during the actual construction period for a project, the quantity surveyor must measure and value the work carried out at regular (monthly) intervals, and submit details to the overall financial administrator (usually the principal designer) for payments from client to contractor. The quantity surveyor also advises on the use of sums of money listed in the bill of quantities for contingency or provisional items, the cost of making variations in areas originally described in the bills or indicated on the drawings, and settlement of the final account for the finished project.

15.3.4 Student assistants and technicians

Principal and specialist designers, as well as quantity surveyors, often employ students of their individual profession to assist them in their work. The students are mostly post first-degree standard and, as such, their work will require close supervision by suitably qualified staff within an organisation. If a part-time or sandwich mode of professional education is not available, the use of this type of assistant may be limited to university vacation times, or the periods designated as 'time out' for *practical training* such as is required by the RIBA and CSD examining boards.

Greater continuity of assistance may, however, be provided by *technicians* who are specially trained in a number of areas concerned with the design features of a building and the subsequent inspection of construction or installation work. According to qualifications and experience, technicians may be responsible for carrying out surveys, technical feasibility studies, as well as the preparation of designs, working drawings and models. They can also become involved in cost analysis, obtaining legal approvals, contract administration and assisting with site inspections. The British Institute of Architectural Technologists (BIAT) represents the specially-trained architectural, structural engineering and quantity surveying technicians. Nevertheless, as far as architecture is concerned, it is important that technician skills are not abused by allowing them to influence the appearance of a building in areas which the principal designer neglects either through choice or ignorance.

15.3.5 Methods of operation

Principal designers, specialist designers and quantity surveyors can work in private practice either as part of a single-discipline firm, or as part of a larger multi-discipline firm. They can also be employed by public bodies such as central or local government, by major and industrial firms, and by building contractors. The fees of the designers and quantity surveyors are paid by the client at prescribed intervals related to the stage of development in a project. But where they are employed as part of a large organisation, these fees will be absorbed as part of the overall charges and paid to the consultants as a regular salary. Depending on the precise nature of a project, and as a rough guide, the *combined cost* of these professional fees, including cost-planning advice, will vary from 12 per cent to 20 per cent of the final construction costs (see section 16.6).

Table 15.2 briefly summarises the sequence of events for a building project, where the principal designer is the team leader. For small uncomplicated contracts, say less than 500 m^2, principal designers may perform all the functions of the Design Team. They will establish the brief, appraise the building requirements, obtain approvals, provide sketch schemes, develop them in detail, present them for evaluation to the client, provide drawings and provisional assessments, compile the production information and supervise the job on site. A short-list of builders will price from detailed specifications and key drawings, and the successful builder will provide a breakdown of the price and the schedule of rates for the main activities. For medium-sized contracts, say 500–2000 m^2, the principal designer will probably work in conjunction with the quantity surveyor. The principal designer is normally selected first and is responsible for the design and management of the project. The quantity surveyor can be appointed on the recommendation of the designer or chosen separately. The designer and quantity surveyor together undertake a general appraisal of the building requirements. For large projects, say over 2000 m^2, the Design Team will comprise the principal designer(s), quantity surveyor(s) and specialist designers; the skills of the specialist designers are related to the particular complexity of the projects. Where circumstances are such that the principal designer is not the leader of the team insofar as management function is concerned, then various divisions of responsibility should be made to all members of the team at the outset of a project (see pages 144–150).

15.3.6 Professional liability

Principal designers are the conventional leaders of the Design Team because, following the client's instruction, they normally produce the design of a building. In this case it is also their responsibility to supervise and coordinate the production of all drawings, schedules and specifications, thereby ensuring the project is carried out in accordance with their aims for the design. Principal designers should make certain that adequate information is available to other members of the Design Team, so decisions relative to the continuity of work can be made at the right time. They must also try to foresee, as far as is practicable, the problems likely to arise and take action on unplanned eventualities. Being leader of the team, therefore, the principal designer must take the ultimate responsibility for the project. Accordingly, any deficiencies which may occur as a result of any design and construction decision, or mismanagement of associated contracts, often make the designers liable to legal claims for negligence. This may appear to be perfectly reasonable, but must be considered with some caution depending upon current legal interpretations of what constitutes the *reasonable* 'skill and care' which should be exercised by designers in the performances of their duties. Certain claims for negligence are just, but it must be remembered that design and construction processes involve innumerable value judgements when balancing the demands imposed by various performance requirements for a modern building. Each demand can only be satisfied at some expense to others. Also, the liability of principal designers often extends into the work of others, in that there is a responsibility for *inspection* and therefore acceptance.

The Civil Liabilities (Contribution) Act makes provision for a writ to name all and any party who might conceivably have any degree of liability for negligence in connection with a claim. Furthermore, the designers are currently liable *for life* for the cost implications (including inflation costs) arising from the errors in decisions made. In the event of a belated claim, the identifiable personal estate of the principal designer may even be penalised after his or her death. It is therefore necessary to take out *professional indemnity insurance* against possible claims, and when unlimited liability periods are required the premiums are very high. Furthermore, it is recommended that designers retiring from private practice continue this insurance cover, although once they are no longer practising it will be given at a slightly lower premium year by year.

Table 15.2 Sequence of events for small building projects following the RIBA plan of work format

Plan of work stage	Principal designer	Specialist designer	Quantity surveyor (QS)
A Inception	Agree extent of work to be done, financial arrangements and appointment with client	–	–
B Feasibility	Elicit all information by questionnaire, etc., from client Carry out studies on site-user information, e.g. boundaries, rights of way, rights of light, easements, services, etc. Make particular enquiries to local authority regarding planning approvals, preservation orders, site lines, availability of services, etc. Prepare feasibility report and present to and discuss with client	Provide initial guide about possible factors influencing design and cost of proposals Assist in preparation of feasibility report and if required help in presentation to client Agree fee for services	Agree extent of work to be done, financial arrangements and appointments with client Produce cost area guides, i.e. cost per m^2 or m^3 Advise on alternative cost factors Discussions on financial aspects of specialist designer's discipline
C Outline proposals	Study circulation and space problems for site and proposed building Try planning/massing solution and investigate alternative costs Produce diagrammatic analysis and discuss problems Try out various general solutions indicating critical dimensions and mean-space allocation Contribute to, prepare and present report to client	If not already done, agree condition of appointment with client Provide more detailed information on design and cost problems Assist in preparation of outline proposal	Estimate cost of project based on outline plans and brief specification Provide cost evaluation of elements for designs as they develop Provide comparative cost information relative to possible alternative materials and construction techniques Prepare elemental cost at sketch design stage and estimate of building cash flow
D Scheme design	Complete outstanding user studies Consult other members of design team Prepare full scheme design, taking individual and group advice Receive and discuss proposals of specialist designers and QS Prepare presentation drawings and obtain client and other members of Design Team approval Prepare more detail drawings for local/central government approvals	Formulation of accurate information regarding space and dimensional requirements Detailed negotiations with principal designer, other specialist designers and QS Preparation of preliminary drawings, specification and schedules indicating proposals; assist with presentation to client Negotiations for local/central government approvals	Evaluate design within cost plan limits Prepare preliminary measures of quantities Advise on alternative cost factors Discussions on financial aspects with specialist designers Advise on running cost for proposed building
E Detail design	Complete user studies Complete final design Close cooperation with specialist designers and QS	Finalise drawings, specification and schedules Close cooperation with principal designer, other specialist designers and QS	Detail consultation with principal and specialist designer
F Production information	Prepare drawings, specifications and schedules Agree drawings, specifications and schedules of specialist designers Close consultation with QS, agree subcontractors and suppliers	Assist with incorporating requirements into principal designer's documents Advise on suitable subcontractor, etc.	Assist with cost information for principal and specialist designer's drawings, specification and schedules Checks on cost factors, agree subcontractors and suppliers

Continued

Table 15.2 (cont'd)

Plan of work stage	Principal designer	Specialist designer	Quantity surveyor (QS)
G Bills of quantities	Complete all information for QS including contractual arrangements Assist in completion of bills of quantities	Assist with preparation of bills of quantities Advise on form of contract, methods of payment, etc.	Preparation of specification and bills of quantities Advise on forms of contract, methods of payments, etc.
H Tender action	Agree list of suitable contractors with client and other interested parties In conjunction with QS issue tender documents and answer queries arising Receive tenders and consider with QS If necessary, interview short-listed contractors Report to client with recommendation of appointment of contractor, adjustment or retendering Draw up and arrange for client and contractor to sign contract documents	When appropriate, advise on selection of contractor and/or subcontractors Assist with recommendation for appointment of contractor	Assistance with list of suitable contractors and issue of tender documents Check arithmetic and content of tenders from contractor Prepare detailed report on tenders for principal designer and give recommendations regarding acceptance, adjustment or retendering Assist in drawing up contract documents
J Project planning	Prepare list of critical dates for contractor's programming and pass on sets of contract and construction information documents to contractor for use on site Arrange start of work and inform site inspectorate	Assist in preparation of critical dates for contractor's programming and supply all necessary information for contractor through principal designer Brief contractor when necessary	Supply cost information relative to site organisation Agree programme of certification with contractor Advise on contract interpretation
K Operations on site	Hold regular site meetings using agenda and minutes Give general supervision and negotiate with subcontractors Issue instructions and certificates for payment in consultation with QS Keep client informed of progress and financial reports Obtain approvals for increased costs and time for completion	Attend regular site meetings Provide general supervision and negotiate with subcontractors Issue instruction and certificates for payment through QS and principal designer Prepare progress reports Obtain approvals for increased costs and delays in contract	Attend regular site meetings Measure and value work in progress for interim payments Provide cost advice on and measure and value instructions and variations Agree costs involved in delays in building programme
L Completion	Agree completion of work with QS and contractor Agree final costs and prepare statement Arrange for any defects to be rectified Arrange for final payments	Agree completion of work with contractor, principal designer and QS Agree final costs for their work and prepare statement Arrange for any defects to be rectified Notify principal designer and agree final payments	Agree completion of work with contractor, principal and specialist designer Prepare final account in agreement with contractor Agree final payment following rectification of any defects Prepare elemental cost of finalised building based on final account
M Feedback	Prepare building owner's manual Arrange for storage of contract documents	Assist in preparation of building owner's manual Arrange for storage of contract documents	Assist in preparation of building owner's manual Advise on running costs of building

15.4 Research Team

Any synopsis of the Building Team, however brief, must include reference to those members who make understanding and development of current construction methods (materials and technical ability) possible, i.e. the *researchers*. The process of building has moved from a craft-based art towards a science-based technology over a period of about 100 years. Latterly this is particularly true in the environmental science areas of energy usage, thermal comfort, sound control and lighting, where quantifiable criteria have found application in the design of the external envelope of a building.

The function of building research through government agency was begun in Britain in 1921 when the Building Research Station was founded, now known as the Building Research Establishment (BRE). The aim of the research is to discover facts by means of scientific study and, in matters concerning building, it covers a very wide area of knowledge requiring controlled programming of critical investigation of chosen subjects. The BRE regularly publishes information, current papers and monthly digests (*Building Research Establishment Digest*) about its work.

The BRE has numerous fields of study and provides an advisory service to the Building Team providing information which includes the following:

Building materials   Research into old materials and new materials to produce an advanced scientific knowledge of their structure and behavioural characteristics so as to produce economic advantage in use.

Structural engineering   Concern with the problems of design, serviceability and safety related to cost.

Geotechnics   Provision of a rational set of principles and analysis to help structural engineers to design a wide range of structures in and on the ground with a calculated risk and with economy.

Mechanical engineering   Investigations into the mechanisation of the building process and the development of new production equipment, and methods and study of new building plant.

Environmental design   Study of the physical environment in and around a building, including research on human needs, the physics of building, and design and development of engineering services, bearing in mind the relevant cost implications.

Building production   Concern with the use of resources of capital, materials and labour for new building, the maintenance of old buildings, and life-cycle costs.

Urban growth   Study of problems associated with urban growth, either in new and expanded towns or in sectors of existing towns.

Separate bodies associated within the BRE also make special studies:

Fire   Studies concerning the behaviour of people subjected to the effects of a building fire and smoke; evaluation of fire statistics; undertaking fire tests; identifying and quantifying fire and explosion hazards, and methods of reducing them; evaluating the performance of structures in fires in relation to future design guidance and the compilation of legislation.

Timber technology   Investigation into the properties and performance of wood, wood-based products and similar materials used in lightweight components; problems relating to the processing of home-grown timbers and jointing components; understanding of the causes of deterioration of wood and wood products with the aim of reducing loss by the evaluation of preventative, remedial and preservative treatments; timber grading by quantifiable stress criteria and the effective design of timber structures.

There are many other advisory bodies allied to the construction industry which are prepared to supply informed opinions on specific problems. Like the BRE Advisory Service, they charge a fee and include the Construction Industry Research and Information Association (CIRIA) which often helps to finance research into aspects of construction by allocating funds to universities, other research bodies and industrial organisations. The Building Cost Information Service (BCIS) provides a cost analysis service and issues reviews of building costs; the RIBA Information Services provide subscribers with computer access to a wide range of technical and practice documentation; and the Society for the Protection of Ancient Buildings (SPAB) advise (without charge) on possible listing of old buildings, as well as give recommendations concerning technical aspects of conservation.

Research and development organisations also exist which are sponsored by industries to promote the scientific and aesthetic use of particular materials. These organisations include the Brick Development Association (BDA); Clay Roofing Tile Council (CRTC); Timber Research and Development Association (TRADA Technology Ltd); British Woodworking Federation (BWF); Constructional Steel Research and Development Organization (CONSTRADO – Steel Construction Institute); British Cement Association (BCA); and Gypsum Products Development Association (GPDA). Other organisations promote the use of data concerning the use and specification of suitable and tried construction methods as well as many other aspects, including documentation methods, testing procedures and the legal requirements associated with the building process. These include the National House Building Council; National Building Specification; Concrete Society; Fire Protection Association; Chartered Institution of Building Services Engineers; Chartered Institute of Building and Tarmac, Black & Veach (TBV Consult). These and many other organisations, too numerous to mention here, can supply information which assists in making critical decisions during design and construction processes.

More detailed reference must be made to the importance of the British Standards Institution (BSI) and the British Board of Agrément. The basic function of the BSI lies in the preparation and promulgation of *British Standards* covering nearly all aspects of productive industry. These standards represent the recommendations made by specialist committees of interested parties in a particular subject and, for the building industry, cover dimensions, quality, performance, safety, testing, analysis, etc., of materials, components and methods of assembly or construction. Standards relating to materials and components are known as specifications, standards for methods of assembly as codes of practice. Most designers, quantity surveyors and builders are conversant with the existence of standards to meet specific requirements, although they are not always very clear as to just what particular recommendations are given. More and more products carry the now familiar Kitemark, the registered BSI symbol, which implies that the product concerned has been tested and found to meet the appropriate requirements. Furthermore, British Standards recommendations regarding appropriate methods of construction frequently form the basis of the mandatory minimum requirements contained in the Building Regulations.

The United Kingdom is now part of the European Union, and many British Standards have been harmonised with the Comité Européen de Normalisation (CEN) requirements for European standardisation, e.g. BS 65 on vitrified clay pipes for drains and sewers has become BS EN 295. Many European products carry a CE mark, which signifies safety, durability and energy efficiency. This should not be treated as a mark of performance and is not intended to substitute for the BS Kitemark. International standards are also available. They are prepared by the International Organisation for Standardisation and are prefixed ISO followed by a reference number, e.g. ISO 2808: *Paints and varnishes*. It is also possible that British, European and International Standards have comparable objectives, e.g. BS EN ISO 9288: *Thermal insulation*.

The principal objective of the British Board of Agrément, which employs the resources of the BRE and other organisations, is to help to bring into general use in the building industry new materials, products, components and processes. Accordingly, the Board offers an assessment service based on examination, testing and other forms of investigation. The reports include details of the methods of assessment and testing, and are known by the acronym MOATs. The aim is to provide the best technical opinion possible within the knowledge available, and those innovations satisfying critical analysis relative to their intended performance are issued with an *Agrément certificate*. Products or processes already covered by a British Standard do not normally fall within the scope of the Agrément Board; items to be covered by a British Standard will be issued with a certificate renewable after three years.

Agrément certificates, therefore, not only supply a critical analysis of new materials, products, components and processes for members of the Design Team, but also further the production of new British Standards and often provide proof that certain innovations suit the requirements of the Building Regulations.

15.5 Legislative Team

15.5.1 The planning process

Modern planning for development originates from 1947 legislation which required all local planning authorities to determine development plans for their specific areas. It also required formal applications for permission to build. Procedures are now effected by the Town and Country Planning Act of 1990, through central government, county planning authority and district planning authority (the local authority).

- Central government is represented by the Secretary of State for the Department of Transport, Local Government and the Regions (DTLR). Administration of government town and country planning policy is through departmental circulars or planning policy guidance documents known as PPGs. These refer to such issues as housing policy, green belts, telecommunications, recreation facilities, highways, transportation and industrial, retail and commercial development.
- The county planning authority's main planning responsibility is to prepare and adopt a strategic plan for its county. This is known as the *Structure Plan*. A structure plan is wide-ranging, establishing a general framework for development in accordance with guidance from the DTLR. The plan is the subject of public consultations and usually remains in place for 15 years. It will provide target figures for housing development within each district and overall policies for land development and use.

● The district (local) planning authority's main responsibilities are to produce a *Local Plan* for its area and to process all but the most complex of local planning applications. The local plan is prepared within the framework of the structure plan and includes details for development and use of land. Draft plans are prepared for public consultation before modifications are submitted and approved by the district council. Prior to final acceptance, a public inquiry is held in the presence of an inspector appointed by the DTLR. Subject to the inspector's report and modifications, the plan becomes statutory.

Structure plans, local plans and planning applications may be referred to the Secretary of State where matters remain unresolved at a lower level. Where a planning authority is found to have acted unreasonably, the Secretary of State can award costs against the authority.

15.5.2 Application to build

After the client's brief for a building has been established, it will be necessary for the Design Team to start negotiation with the local authority in which it is to be situated in order to clarify certain legal requirements. The location and design of a building are controlled by the planning authority, and the construction by the building control department. These are supported by many other legal requirements concerning fire precautions, clean air, highways, factories, offices, shops, railways, etc.

Although in theory a designer must be conversant with all the legal requirements affecting a building, this can be an impossible task. There are probably more than a thousand Acts of Parliament which make reference to the building process and at any time during the course of construction (and even after occupation) a building is liable to be inspected by such diverse officials as factory inspectors, health and safety inspectors, water inspectors, petroleum officers, planning and building control officers, fire-fighters and police officers. Although not necessarily guaranteeing the avoidance of demands being made during a late stage of construction, demands which cause additional expense and loss of building time, it is the responsibility of a designer to be aware of the relevant legislation influencing the design of a particular building. At least this awareness means that contact can be made with the appropriate controlling person or organisation for specific advice during the various design stages of a building.

Town and Country Planning Acts are administered through the local authority by professional planners who are members of the Royal Town Planning Institute (RTPI) and perhaps the Royal Institute of British Architects. In dealing with the creative, environmental, social and administrative aspects of planning, the planning officers frequently employ specialist advisers, e.g. sociologists, ecologists, statisticians, economists, planning technicians, geographers, architects, graphic designers and landscape architects.

Initial advice will be given on the suitable location of a building in a selected area, *zoning*; amount of building permitted on a particular site (plot ratios and densities); the influence of previous planning decisions affecting the site; tree preservation; car parking; building lines; general public circulation; road widths and proposed widenings; and changes of use from one activity to another. Advice will also be given when work is to be carried out on existing buildings of special architectural or historic interest. Such buildings will have become *listed* by the local planning authority in an effort to safeguard heritage, and any proposed demolition, extensions, alterations or modifications require special consent through the *Listed Building Consent* procedure. Listing places a responsibility on owners to ensure that their buildings are maintained properly and not altered or demolished. Selection may be for architectural or historic reasons, constructional style or as a valued part of an attractive grouping. Most listed buildings are Grade II, but where they have exceptional features may be II*. Grade I listings include buildings of exceptional interest which could be town halls, churches and country houses.

A building preservation notice can also be served on any individual when a building of special interest is proposed for demolition. Indeed, whole areas of countryside and urban development can be designated or zoned as a *conservation area*, and either no development permitted or only that permitted which is necessary for overall maintenance of features required to be preserved. Local authorities have a statutory responsibility to designate areas of conservation. Within these, even the smallest scale of works will require planning permission, i.e. *Conservation Area Consent*. This includes building a porch, replacing windows or minor alterations which outside of conservation areas can be undertaken without planning permission (permitted development). Some planning authorities also issue guidelines about the appearance of a proposed building in an attempt to control the general character of a neighbourhood. (See DOE Circular 8/87 *Historic Buildings in Conservation Areas: Policy & Procedures*.)

Having incorporated the requirements of the planning officer (or negotiated compromises), a formal application can be made, which consists of a set of drawings indicating proposals and documentation giving details of ownership (certificate of ownership, under Article 7 of the Town and Country Planning Act), land usage, densities, etc. If the application is successful, permission will be granted for the proposals to commence construction *as far as the provisions of the Town and Country Planning Act are concerned*. Planning applications can be made in two consecutive stages, a fee being paid to the local authority for each according to the size of the proposed projects. *Outline planning consent* gives permission to the principle that a building, as yet not designed in detail, may be erected in a certain locality, e.g. a shop or office in an area or zone of an urban area designated for commercial usage; *full planning consent* gives permission for a specific building to be erected when precise details are finalised and proposals are known.

Before making a planning application for an industrial building, it may be necessary to obtain an *industrial development certificate* from the Department of Trade and Industry. This procedure attempts to control the effects of indiscriminate development and also to focus industrial activities on areas of high unemployment. For similar reasons, in highly congested urban areas, such as London, request for planning permission to erect an office building may have to be accompanied by an *office development permit* which is issued by the Department of Transport, Local Government and the Regions.

The Town and Country Planning Act stipulates that a decision of approval or rejection of a project should be given within 8 weeks of receipt of application, but most authorities currently request an extension of this period rather than formally reject a proposal owing to the lack of sufficient time for detailed consideration. Appeals against rejections, or any conditions imposed on a consent, can be made to the Secretary of State.

15.5.3 Building control

Building control legislation in England and Wales (Scotland and Northern Ireland have separate systems) is covered by the Building Act 1984 and the Building Regulations 1991 and 2000. The Act is a consolidating statute drawing together requirements previously found in such documents as the Public Health Acts 1936 to 1961, the Health and Safety at Work, Etc., Act 1974, the Fire Precautions Act 1971 and the Housing and Building Control Act. The Building Act calls for a series of *approved documents* (see Table 15.3) to give practical guidance on some of the ways of meeting building control requirements and these combine to form the basis of the Building Regulations. A *manual*, giving general guidance in the form of explanatory

Table 15.3 Approved Documents to the Building Regulations

Approved Document	Application
A	Structure
B	Fire safety
C	Site preparation and resistance to contaminants and moisture
D	Toxic substances
E	Resistance to the passage of sound
F	Ventilation
G	Hygiene
H	Drainage and waste disposal
J	Combustion appliances and fuel storage systems
K	Protection from falling, collision and impact
L	Conservation of fuel and power
M	Access to and use of buildings
N	Glazing – safety in relation to impact, opening and cleaning

Also:
Approved Document to support Regulation 7 – Materials and workmanship.
Thermal insulation: avoiding risks, 2nd edition, BRE Report 262.

notes, accompanies the approved documents. The main purpose of the regulations is to ensure the health and safety of people in or about a building, although the regulations are also concerned with energy conservation and access to buildings for the disabled.

Building work controlled by the regulations includes new buildings, extensions and material alterations to existing buildings, as well as the provision, extension or material alteration of controlled services or fittings and the work required for a change of use. Certain small buildings and extensions (normally under 30 m^2) and buildings used for special purposes, are exempt from the regulations. These can include conservatories, sheds, greenhouses, car-ports and porches.

To ensure compliance with the provision of the regulations, the Client Team (acting on the advice of the Design Team) can have the work approved and supervised under construction by the local authority *or* by a privately-employed approved inspector acting with the local authority. The two methods are independent and affect the way in which initial building control approval is sought.

Under full local authority control, two main options are available. One is to deposit working drawings for approval, to be given conditionally or in stages within 5 weeks, or by agreement, within 8 weeks. Although having the drawings passed gives some protection in the event of subsequent problems, there is no need to wait this long before starting work. The alternative option involves giving a *building notice* to the local authority who may then ask for drawings to help them when inspecting the work. This option is not permitted for the construction of shops or offices. For both options, 48 hours' notice must be given before work commences and, if the local authority considers that the regulations are contravened at any stage, they must serve a notice requiring rectification, unless they have approved the drawings. There are procedures for challenging the views of the local authority and there is no need for certification upon completion of the work.

When the local authority is fully involved in this way, building control officers, inspectors or district surveyors will inspect the work during the various stages of construction to ensure compliance with the regulations. These inspectors are usually members of the Royal Institution of Chartered Surveyors (RICS) or have qualifications of other professional bodies concerned with the construction of buildings. Some building designers may wish to discuss proposals for use of materials or certain constructional strategies with these officials prior to the submission of drawings. A fee is payable to the local authority for this whole approval process as prescribed in the *Building (Prescribed Fees) Regulations*. Instead of seeking local authority approval, which involves checking calculations, etc., certificates of compliance can be supplied instead. These are prepared by approved persons, who are professionals in those areas of design (members of CIBSE or IStructE).

The alternative method of obtaining approval under the Building Regulations, using the privately-employed *approved inspector*, requires the submission of an *initial notice* to the local authority. This notice must be accompanied by certain drawings and evidence of insurance. The local authority must accept or reject the initial notice within 10 working days, and once accepted, their powers to enforce the regulations are suspended. It becomes the duty of the inspector to notify the Client Team if the work contravenes the regulations. If the defective work is not remedied within 3 months, the initial notice must be cancelled. On satisfactory completion of work, the inspector issues a certificate to the local authority and the Client Team.

The fee payable to an approved inspector is a matter of negotiation with the Client Team. Currently most approved inspectors come from the National House Building Council (NHBC). However, there is increased interest from many professional bodies, including the Institute of Building Control (IBC), the Chartered Institute of Building (CIOB) as well as RIBA, ICE, IStructE and RICS.

In addition to the Building Act and Building Regulations, a designer may need to consult the inspectors who administer the many other acts directly affecting the building process. These include the Office, Shops and Railway Premises Act; Factories Act; Clean Air Act; Chronically Sick and Disabled Act; Civil Amenities Act; Noise Abatement Act; Licensing Act; Housing Acts; and the Health and Welfare Act, etc. There are also peripheral 'laws' in respect of cost, dimensional coordination of certain building types and costs allowances (yardsticks). Account must also be taken of the recommendation and requirements of the service authorities: gas, water, electricity, telecommunications, drainage, fire brigade and police. The rights of the building owners or landowners adjoining the site of a proposed building are also protected by legislation and require attention from a designer: rights of light, rights of way, etc.

A building site can be a place of danger if care is not taken or there is a lack of experience at the management level. There is a vast amount of legislation related to this problem, but currently the principal acts are the Factories Act 1961; the Office, Shops and Railway Premises Act 1963; and the Health and Safety at Work, Etc., Act 1974. These statutes incorporate many other significant regulations, known as Statutory Instruments. These include:

- Management of Health and Safety at Work Regulations.
- The Construction (Design and Management) Regulations.
- The Construction (Head Protection) Regulations.
- Lifting Operations and Lifting Equipment Regulations.
- The Construction (Health, Safety and Welfare) Regulations.

These acts and their subsidiary regulations require the builder and client to ensure that the building site maintains safe and healthy conditions for employees. Emphasis is on an assessment of risks by all involved in the construction process. Specific work situations are defined in the regulations, in addition to provision of adequate air quality, safe use of transport and traffic facilities and site accommodation sufficient to include first aid and other welfare installations. Further emphasis is on good site management, documentation of proceedings and organised planning. The general public should be adequately protected from dangers resulting from site operation. The regulations should be on display for employees to read, and adequate instruction must be given to ensure safety consciousness. HSE inspectors have the power to enter, inspect and examine a building site at all reasonable hours and to make examinations and enquiries to ensure compliance with the regulations. Inspectors must be informed when serious accidents occur, or when disasters happen such as when cranes or lifting appliances collapse, or if there is any fire or explosion.

The builder is required to employ a competent person to inspect and supervise certain construction activities. Most of the larger building organisations have a safety officer, either employed or under contract from a group safety scheme. This officer will visit site offices, workshops and site works, liaise with site personnel and, where necessary, report to the organisation's safety committee and to the local authority factory inspector.

The Design Team also has legal responsibilities under the Health and Safety at Work, Etc., Act 1974 because their projects must not create hazards for building operatives during construction. Although this may influence a change between initial concepts and the final design, the safety requirements are entirely reasonable. A building organisation has the right to refuse to allow its employees to become involved in its construction methods which are liable to be dangerous.

15.6 Manufacturing Team

The Manufacturing Team supply the materials, components and equipment which are used during the construction processes of a building, and therefore incorporate many organisations and interests. Traditionally, construction processes relied on the supply of readily available materials easily converted into manoeuvrable forms or sizes which could be adapted further by skilled workers on site to suit a particular design. With the need to economise in labour and reduce costs, building procedures became more rationalised. Materials were formed into readily usable components during manufacturing processes, and were assembled with few adaptations after delivery to site. This *rationalised traditional* construction procedure reduced the number of separate operations and saved time on site.

However, the continual advancement of technology and increases in complexity and size of buildings has generated even more complex construction processes. Manufacturers must extend their services from the supply of single components to the supply of much larger parts of a building (elements), and indeed whole buildings. Site operations are reduced to a minimum using mechanical plant, and methods of building become largely concerned with the organisation of the systematic supply and assembly of prefabricated items, i.e. *system building* (see also section 13.1).

Some manufacturers produce items which will not normally fit with the components of other manufacturers, and the resulting method of building is commonly known as *closed-system building*. When component design is coordinated between the manufacturers of different products so that they can be used together without alterations, or become interchangeable, the building method is called *open-system building*.

The purpose of this brief account of an aspect of the evolution of building processes is to indicate the vital role which the Manufacturing Team have on the design and development of a building (see Fig. 15.1). In many respects, they should be members of the Design Team. Generally, however, manufacturers are often only concerned with the entire suitability of their particular product as it leaves the factory, and it is up to the Design Team to assess its performance relative to other criteria. (Mention has already been made of the influence of the Research Team and construction specialist in assisting the work of the Design Team in this area.) Whether producing materials, equipment, components or building systems, Manufacturing Teams often incorporate their own research and development organisation to test products. Manufacturing Teams will also employ public relations organisations to produce information about their products for circulation to members of Client, Design, Research, Legislative and Construction Teams.

15.7 Construction Team

A study of building workers carried out by the Building Research Establishment lists over 50 separate occupations associated with construction. Therefore, the erection of a building depends on an industry where total reliance is placed on the diverse attitudes, abilities and adaptability of its workers. Conventionally, these workers were grouped under 'trade' headings according to their skills (Table 15.4), and 30–40 years ago most were employed by a *main contractor* managing and directing all works on a site using a general supervisor to coordinate the work of each trade or *subcontractor* supervisor. Today most specialist trades are employed as *nominated subcontractors* by the client or principal designer on behalf of the client; relatively few key trades are employed directly by the main contractor as *ordinary subcontractors* (known also as nonnominated subcontractors or domestic subcontractors).

Nominated subcontractors may be required to design and provide specialist elements within a building from a statement of performance requirements, but the main contractor is still entirely responsible for the satisfactory completion of the work involved. It is also quite common for the client or principal designer on behalf of the client to employ *nominated suppliers* for certain specialist materials, components or equipment which are to be used or fixed into position by the main contractor.

Table 15.4 Basic list of trades required for erection of a simple building

Trade	Job
Asphalter	Roof, floor and wall (basement) finishes
Bricklayer	Laying brickwork
Carpenter	Structural and carcassing timber work
Concretor	Placing concrete
Drainlayer	Providing below-ground drainage
Electrician	Electrical installation
Excavator	Levelling site and digging drain/ foundation trenches
Floor tiler	Internal floor finishes
Gas-fitter	Gas installation
Glazier	Fixing glass
Joiner	Timber work to finished components
Metalworker	Sheet metal applications (roofing)
Painter and decorator	Finishing components
Paver	External paths/road finishes
Plasterer	Plastering walls/ceilings, screeding and rendering
Plumber	Plumbing installation, flashing and gas pipes (interior)
Scaffolder	Erecting scaffolding and working platforms
Steel erector	Erecting steel columns and beams
Steel fixer	Cutting, shaping and positioning steel reinforcement
Tiler and slater	Roof finishes

Although this system of site organisation remains normal with most small and many medium-size building firms, there is an increasing tendency for the larger main contractors to become *building managers*, responsible for the coordination of the erection of a building using *only* nominated subcontractors or suppliers. Perhaps the main reason for this is the fact that the continuous employment of their own trade operatives cannot be guaranteed during periods of economic recession. Enforced redundancies are sometimes contractually difficult and likely to prove expensive.

15.7.1 Available skills

The training of skilled building workers has traditionally been very much a matter for employers and unions. The City & Guilds of London Institute has a long history of examining for skilled building workers. It now joins others to provide syllabuses and training courses for National Vocational Qualifications (NVQs), which form the basis for college and employment assessment. In order to compete with the other employment areas in terms of time-related earning power, the basic apprenticeship period for skilled building workers was reduced and has resulted in a gradual but continuous 'deskilling' of traditional crafts. Furthermore, the greater emphasis

now being placed on academic subjects in secondary schools, together with the unattractiveness of adverse climatic conditions on exposed UK construction sites, has resulted in far fewer recruits than previously. Government attempts through the Construction Industry Training Board (CITB) and the Training Services Agency (TSA) have endeavoured to boost training of skilled staff. So has the trend towards providing an artificial climate around a building site by transparent protective sheeting, or moving most processes into a factory producing preformed units for final and speedy erection on site. Nevertheless, the reduction of practical tests in favour of the theoretical assessments of craft skills and knowledge has produced building managers rather than craft operatives.

The importance of the move away from the traditional skilled building workers on site lies in the need for building designers to concentrate earnestly upon the selection (or implementation) of suitable construction methods which are known to be realistic, relative to the depth of *practical expertise* likely to be available at the time when a project is to be erected. There is therefore an even greater need than before for close consultation between Design Team (including quantity surveyor) and Construction Team, especially if hitherto untried materials and/or techniques are being contemplated. If this is not possible, the Design Team must include sufficient expertise (perhaps through specialist designers) to be able to supply a potential main contractor with more detailed specifications and drawings of the chosen construction method than perhaps was necessary for corresponding innovations in the past.

Notwithstanding the above comments concerning the role of trade skills, it is important to realise that mechanisation based on the development of petrol, diesel and electric engines, pneumatic and hydraulic engineering, has influenced building methods considerably. In recent years mechanisation has become universal on site: excavators have replaced hand-digging of foundation trenches; mechanical hoists have largely replaced the necessity to carry materials up ladders by hand; pumping techniques make concrete more manoeuvrable without the need for wheelbarrows; cranes transport large building components; hand-held power tools have replaced hammer and chisel; mobile heaters have been introduced for drying out, etc. Indeed, certain building techniques rely on the use of equipment specially designed to carry them out. The increased use of plant or machines means that semi-skilled operatives are often in the majority on a building site. However, completely unskilled activities still remain, mainly to service skilled and semi-skilled workers; they are fulfilled by labourers.

The Construction Confederation, incorporating the National Federation of Builders, is one of the most influential employment representatives within the building industry and has affiliated organisations representing specific trades. The Confederation of British Industry (CBI) represents the interests of most building firms in the United Kingdom. Professional qualification for builders is through membership of the Chartered Institute of Building (CIOB). This institute concerns itself with promoting education and management training for a diversity of disciplines within the construction industry.

15.7.2 Types of building organisation

The term 'main contractor' used earlier will now be investigated in greater detail since, for the efficient and economic construction of a building, it is imperative that the right type of organisation is selected. Criteria generally relate to size and complexity of the project in hand, although the speed with which it can be erected and the special resources which may be provided by a particular builder play an important role in selection.

The sometimes loose separation between a simple and complex project relates more to the intended purpose (design) of the building rather than actual size. Single or predominantly single-storey projects are not necessarily simple buildings, and multi-storey or large-span projects not necessarily complex. The general configuration of a building, the amount and degree of complication of the services it accommodates, the characteristics of the site, and the complexity or otherwise of the construction method are more accurate divisions between simple and complex. The selection of a suitable builder may additionally be influenced by members of either the Client Team or the Design Team who may have preference for a builder with whom they are familiar. Generally, however, the available range of suitable builders can be narrowed by their ability to build a particular project to suit the intended financial outlay allocated to it.

Although it may be more complicated, as just explained, main contractors can be divided into three basic groups: *general builders*, *general contractors* and *design and construction companies*. Currently, each could be coordinated by building managers, but this role is mostly associated with the larger organisations of the latter two groups. All could therefore employ trade operatives as part of their regular staff, and use subcontractors to deal with the specialist construction areas necessary for a particular project.

General builders   General builders will take on a wide range of work, but most concentrate on particular types and sizes of projects which are seldom in excess of £500 000 (Fig. 15.3) The *smallest* firms will be the 'jobbing builders' concerned with minor repairs of existing buildings. Proprietors are usually trade operatives (bricklayers or carpenters) who may employ subcontractors for other skills such as plastering, electrical and plumbing installations, etc. The larger firms will tend to be management organisations, employing administrative staff, including site managers, quantity surveyors, site engineers and safety personnel, perhaps with a few skilled and other operatives. They

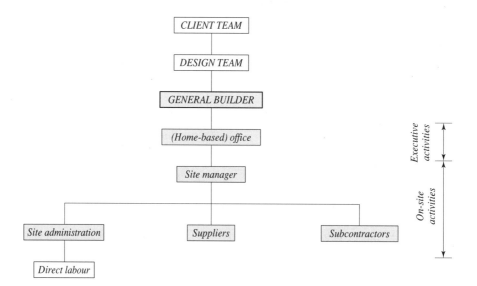

Figure 15.3 General builders.

normally operate in a particular area and therefore avoid excessive commuting distances. When their own building staff, plant and/or material resources are insufficient to carry out a particular trade or specialism, subcontractors will be employed.

Some organisations with continuous building programmes undertake part or even all of the construction processes themselves through the agency of capital works departments which directly employ all the personnel concerned. These capital works departments are usually associated with maintenance and the building of small general-purpose works, or sometimes housing under the control of local government.

General contractors   General contractors will range in size and include multinational organisations having international, national and regional headquarters (Fig. 15.4). These firms will probably carry out both building and civil engineering projects of a very large size, e.g. multimillion pounds, but will also carry out much smaller projects requiring a special expertise they may have developed. For example, many general

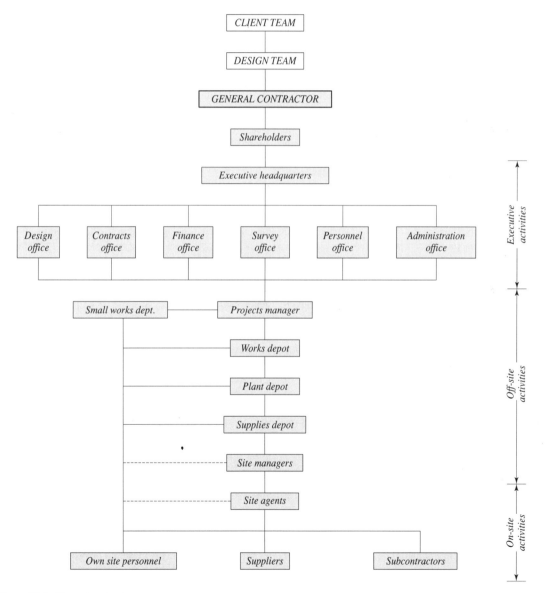

Figure 15.4   General contractors.

contractors specialise in particular types of work based on local traditions or availability of workers: shopfitting; specialist joinery work; special expertise in bricklaying, or concrete work resulting from particular skills in shuttering and form work. Some include research and development divisions, and have evolved individualistic construction methods based on the exploitation of certain materials such as precast concrete units, steel frame components or load-bearing timber panels.

The general contractor's organisation must necessarily be divided between office and site activities. Offices will often be concerned with estimating, tendering, site planning, construction process and planning, quantity surveying, cost control, and the bulk purchasing of materials and hire of plant. In large firms there will be separate transport, personnel and employee welfare sections.

The work on site will be under the control of a *site* or *project manager*, who may coordinate many other projects on different sites. The resident contact on a particular site will be the *general supervisor* or *agent* who will be responsible for the contractor's own employees and hired subcontractors.

A site office will employ staff to keep careful records of the work in progress, and assist in effective control and costing. This staff will also be concerned with time sheets, delivery and store records, weather records and details of work progress.

Drawing-office staff may also be employed, either at the main headquarters or on a particular site. Their function is to make larger and more detailed working drawings when it is felt special information is required, such as the temporary work for shoring up excavations or methods of constructing shuttering for *in situ* concrete work required by a particular building design. It may also be necessary for drawing-office staff to confer with a designer regarding alternative methods of construction when greater efficiency can be achieved as a result of a general contractor's special resources.

A great deal of contracting work is carried out for and by government agencies, and runs into several millions of pounds of work each year. This is true of both local government and the privatised industries, such as transport, gas and electricity, which have their own building works departments known as *direct labour organisations*.

Design and construction companies These provide a service which is more elaborate than the previous two, as they will undertake the responsibility for *both* design and construction of a building project (Fig. 15.5). This type of contractor is usually a specialist in one

form of building, such as housing, factories or offices, using a particular form of construction.

Design and construction companies therefore combine the services of the Design Team with those of general contracting, and are employed directly by a client for a particular project. In effect, the company provides a 'package deal' in which a contractor is responsible for all the major decisions on design and technical matters, prepares plans and specifications, obtains approvals and carries out the construction. They either employ salaried designers to prepare the design for projects, or pay a fee to independent principal designers and specialist designers. Arrangements are also made with building specialists and subcontract staff. There is an area of free enterprise building in which contracting firms will seek out and acquire land, negotiate planning permission and develop a project to their own specification before setting out to find a customer for the project (Fig. 15.6). Conversely, with many overseas projects, fee-paid consultant designers may work as a team with contractors to make an all-in offer for design and construction supervision.

15.7.3 Other personnel on site

During the construction period of a building site, other personnel who may not be directly involved with the practical work of a building organisation will also be present, and the coordinating role of the site supervisor, agent or manager must also include responsibility for making sure their needs are recognised. This includes consideration for their safety and welfare while they are on site. Appropriate and adequate insurance must be taken out, to compensate for accidents and the possible financial loss caused by delays resulting from any accidental damage they may cause while on site. Among those covered are the following personnel.

Client Team Visits may be made, but it is important that they only observe and not, for example, issue instructions. This will only confuse channels of communication and eventually cause the Design Team to lose control of the running of the project.

Nevertheless, when a client requires a regular check on the quality of the work, a *clerk of works* can be employed with responsibility for ensuring that the building organisation strictly adheres to the documentation and instructions supplied by the Design Team (and agreed by the client). On small projects the clerk of works would visit the site at regular intervals to inspect critical work, but on larger projects the clerk would probably be resident on site and have an office there.

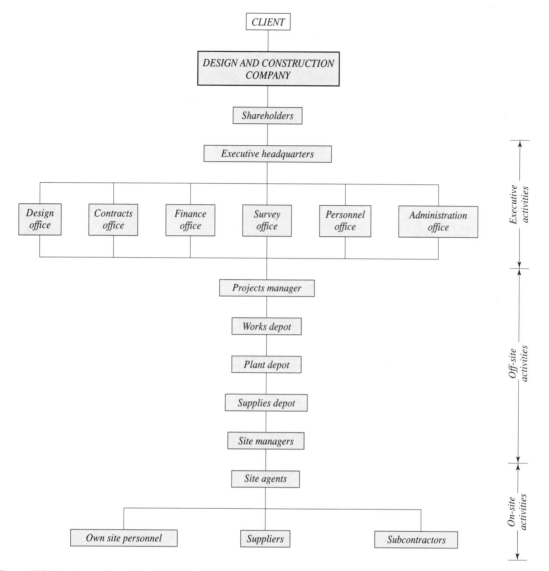

Figure 15.5 Design and construction companies.

Although paid for by the client, the employment of a clerk of works usually follows the recommendation of the principal designer, especially when it is considered that a particular building project requires a fairly high degree of supervision. For similar reasons, specialist designers may also recommend that a clerk of works specialising in their discipline should be employed when a building organisation is lacking the necessary coordinating expertise.

Records will be kept by a clerk of works of all events, such as delay periods caused by inclement weather, strikes or unavailability of materials, components and labour. This information is invaluable in assisting the Client and Design Teams to establish the need for extensions of construction time for a project which is requested by a builder or contractor.

Clerks of works will have been trained in all aspects of construction and contract management, and are often those of craft background who have decided to turn their acquired experience of building into an overall supervisory skill. The professional body representing clerks of works, examining and issuing their code of professional conduct, is the Institute of Clerks of Works of Great Britain Incorporated (ICW).

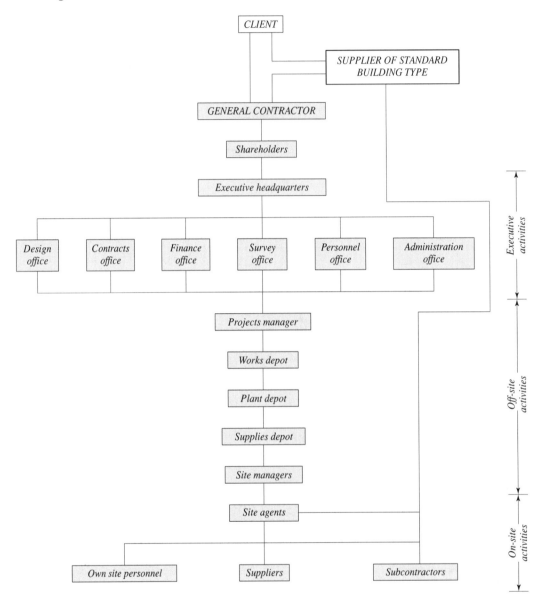

Figure 15.6 Package-deal contractors.

Design Team This includes the quantity surveyor. Visits will be made to deal with queries, supply information, issue additional instructions, establish financial criteria and monitor progress. For this purpose the principal designer usually organises site meetings at regular and frequent intervals, the first one or two taking place before any work commences. All members of the Design and Construction Teams may not necessarily have to attend every meeting, but it is essential that individual members are present when issues are to be raised concerning their expertise. The agendas for site meetings must therefore be carefully planned in advance. Other meetings, probably less formal, will take place as necessary to ensure the smooth running of a project.

For large or complicated projects it may be desirable to have a member of the Design Team permanently resident on site during construction activities. Depending on the nature of the project involved, he or she may be a representative of the principal designer, quantity surveyor or any of the specialist designers. In certain cases a member from each may be necessary. The function of this resident designer or quantity surveyor is to answer and authorise action on any day-to-day queries a contractor may have, and to supply detailed information about areas of the project when the drawings, schedules and specifications issued by his or her office require further clarification or amendment. In fulfilling these duties, the resident designer or quantity surveyor liaises with his or her office, who will provide the overall communication links between other members of the Building Team.

Research Team   Visits may be made to monitor any work they have recommended, or to give advice concerning special problems which have arisen during the course of construction. Information gathered during these visits may form the basis of future useful research.

Legislative Team   Members are likely to visit the site by direct invitation or often to carry out spot checks on parts of the work relevant to their delegated power. The building control officer or district surveyor will regularly visit the site, usually by invitation of the builder, to inspect aspects of construction required by the provision of the Building Regulations (Fig. 15.7).

When approved inspectors have been commissioned to ensure compliance with regulations, they will also be visiting the site. Highway engineers, public health inspectors and town planning officers will check compliance with approvals and investigate complaints made by the public concerning such matters as the builder's access into a site, mud on roads, noise obstruction, and other nuisances or infringements.

Although not strictly members of the Legislative Team, the public utility services organisations (gas, water, electricity, telecommunications, etc.) will periodically check for possible damage to existing installations and/or organised work to new installations. Police may wish to inspect the site during the day or night to discuss security measures or to resolve problems concerning road obstruction caused by unloading lorries, etc.; the fire officer will check for fire hazards and to establish that recommendations have been incorporated as the work proceeds; the safety officer and a representative from the Health and Safety Executive will examine the way that safety, health and welfare facilities are maintained on site.

Construction Team   Members, either permanent or visiting, are not all involved in practical work. They include contract administration staff, health officials,

---

LONDON BOROUGH OF ENFIELD	**FOR OFFICIAL USE ONLY**
For the attention of the Chief Building Surveyor Date .... **BUILDING CONTROL STATUTORY NOTICE** (Building Regulation 14)	**FEE PAID**
Plan No. BC .... Nature of Works	**£...........TO PAY**
Address .... I hereby give notice that:– (a) the above works will commence on .... at .... a.m/p.m (b) the under-mentioned work or building will be ready for inspection on	
Name and Address of Builder	

Tel. No. .... Signature of Builder ....
1.   Excavations.   6.   Drains (after haunching or
2.   Foundations.   surrounding and backfilling)
3.   Damp Proof Course   7.   Occupation of Building.
4.   Hardcore/Concrete Oversite. 8.   Completion.
5.   Drains (before haunching or covering)
 	
Note:– (a)   delete items not applicable.
        (b)   48 hours notice required under Building Regulations for item (a)
        (c)   24 hours notice required under Building Regulations for items
                   (b) 1,2,3,4 and 5.
        (d)   not less than 7 days notice required for item 7.
        (e)   not more than 7 days notice required for items 6 and 8.
 	
BC/15
	AREA

Figure 15.7 Typical notification card used to inform a local authority of progress in building work on site and to request an official inspection for compliance with the Building Regulations.

draughting technicians, secretaries, canteen staff and union officials. On larger projects, subcontractors may have a similar range of staff on site and will have regular visits from their management staff. Suppliers will deliver materials and components to designated storage areas.

Apart from a building site being inspected by members of the above teams, special visitors often make applications or are permitted by a building organisation to observe the construction work during progress. These visitors include interested professional institutions, user teams, councillors, foreign visitors and students of the building design and construction professions. With this type of visitor the site agent or manager must be quite certain that proper insurance cover exists, in case the visitors injure themselves or cause damage to the site, perhaps delaying a project's completion.

15.8 Maintenance Team

The chosen design and construction method of a building must take into account the effects which time will have on their performance (see Chapters 3 and 13). Because of the complicated requirements, their interrelationships and the multiplicity of stylistic conventions which influence the selection of design and construction methods used today, it is sometimes necessary for the Design Team to consult certain specialists who, during the investigatory phases of design, can offer advice which goes towards the assurance of satisfactory performance standards for the intended life span of a building. These specialists collectively form a consultative *Maintenance Team* who, although they may not necessarily become involved with the physical processes, can use their acquired experience and research to advise a designer on a suitable solution to a particular problem, especially where the more normal procedures of maintenance are impracticable. For example, very tall tower blocks often present external cleaning problems which could be solved by incorporation of one of several special items of equipment: cantilevered gantry devices and specially profiled curtain wall mullions to allow safety clips to be inserted for external manual cleaning; or strategically positioned sparge pipes which allow water to be sprayed on façades, thereby eliminating the need for external manual cleaning altogether. Inevitably, the appearance of the building will be affected by such devices and it is therefore very important that the Design Team and consultative Maintenance Team work in close harmony (Fig. 15.8). The precise methods adopted for subsequent maintenance and cleaning will also be influenced by the attitude of the Client Team towards the running costs of a building.

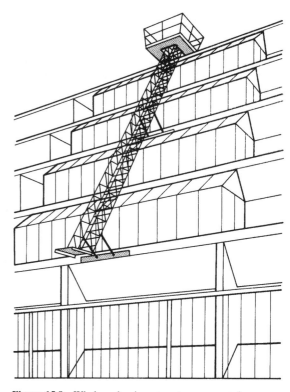

Figure 15.8   Window-cleaning apparatus necessary for complicated façades may become a permanent feature, affecting the appearance of the whole building.

The Maintenance Team will therefore give valuable advice which affects the ultimate cost evaluation of a particular project. Three-dimensionally profiled glass façades often present particular financial problems associated with both maintenance and cleaning. Additional costs can be incurred in the design, manufacture and installation of special climbing gantries to facilitate accessibility, and additional costs are sometimes involved in special training programmes for maintenance and cleaning personnel. Nevertheless, for certain prestige building types the aesthetic desirability of three-dimensional glass façades may more than offset additional costs.

Normal maintenance procedures can usually be formulated using the expertise of the Design Team without the need for special consultation, e.g. methods of access to services; dimensions of ductwork and crawlways within ducts to permit space for repair or alterations; the routines necessary for continued or adjustable service supply (heating, lighting, ventilating, drainage, etc.); and the care needed to maintain finishes, furniture and fitments. On completion of a project, the Client Team must be presented with a maintenance manual compiled by the Design Team which incorporates the advice of the consultative Maintenance Team. This manual describes how a building can be expected to perform, what measures have been taken to ensure that it does, and what action must be taken in the future (see section 16.8).

The aftercare or continued maintenance of a building following its completion will often be contracted out to a facilities management company. This simplifies the building owner's responsibility for engaging caretaking staff. For an agreed fee, many facilities management contractors provide a full package of building maintenance, repairs, replacements, security, car parking attendance, catering, and so on. The British Institute of Facilities Management (BIFM) represents the professional interests of facilities management organisations.

Further specific reading

Mitchell's Building Series

External Components   Chapter 2 Prefabricated building components

Structure and Fabric Part 1   Chapter 2 The production of buildings

Structure and Fabric Part 2   Chapter 1 Contract planning and site organization
Chapter 2 Contractors' mechanical plant

Building Research Establishment

BRE Digest 289   *Building management systems*
BRE Digest 397   *Standardisation in support of European legislation*
BRE Digest 408   *A guide to attestation of conformity*
BRE Digest 448   *Cleaning buildings: Legislation and good practice*
BRE Digest 450   *Better building: Integrating the supply chain: a guide for clients and their consultants*

16 Communication

CI/SfB (A3/A8)

The operational procedures and other management activities associated with the design, construction and subsequent performance of a building rely a great deal upon quite complex information being transferred between the various participants of the Building Team. Ideas and technical data must be dispersed to a wide range of people at both professional and non-professional, skilled and unskilled levels. For this reason, methods used for communication should not only clarify issues, but also attempt to bring harmony to the work processes involved in the creation of a building, and foster the cooperation which ensures maximum contributions from all those with tasks to perform.

Communication methods have developed in type and sophistication to meet the needs of the disparate parties increasingly becoming involved with the creation of a building. Traditional methods such as memos, letters, reports, minutes, schedules, diagrams and drawings are gradually becoming universally standardised in format to simplify understanding and to speed up production. Computers, telecommunication systems, recording tapes and closed-circuit television make it easier to achieve instant communication.

The introduction of microchip technology has changed attitudes towards efficiency and communication within today's building industry. Design aims, methods of communicating design information and technical data can be instantly dispersed between designer, specialist designer, client and contractor by coding through the universal digits of the computer. This process raises possibilities of absolute understanding and, in the face of ever-increasing building costs, the concept of keeping options open until the last minute. For example, it is now possible to design a school and produce all the drawings, schedules, specifications and other contract documents in less than three days when using *computer aided design* (CAD) technology. This process obviously involves a modification in the way the various teams operate when creating a building and how they deal with design and construction problems. The designer will no longer have to ponder over a drawing board in search of a solution to a problem through reliance on his or her intuitive skills and ability to interpret technical facts through memory or prolonged research. Instead, he or she will receive multifarious solutions from a computer, all with equal technical and economic competency; the problem becomes one of selection. Any selection process must include a realisation of aesthetic ideals, and although the computer can incorporate ergonomic and other data about physical comfort, it cannot (as yet) design a building to give full psychological delight to a human being. A virtual reality visit via the monitor, through and around a visualisation is about as close to actuality as we can get. The techniques applicable to computer technology involve the production and manipulation of information, information storage and information transmission. The extent to which these activities can be carried out depends upon the capacity of the computer: its size and cost, its programming capability and complexity. Information technology computers range from pocket-size multifunction dictaphones, telephones, calculators and e-mailers to larger but fully portable laptop equipment; from desktop equipment to large processors, including 'intelligent' machines networked with similar computers at remote locations. Needless to say, the competence of the operator is of paramount importance.

Figure 14.2 illustrates the RIBA plan of work for a building of moderate cost and which involves the activities of a full team of designers with a principal designer as leader, as described in Chapter 15.

The following areas of communication complement the progress of the work:

- Technical data
- Drawings
- Specification of works and bills of quantities
- Contracts
- Tender documents
- Fee accounts and certificates of payments
- Programmes of work
- Maintenance manual

16.1 Technical data

The time has long passed when members of the Building Team were able to rely only on the technical data contained in a few textbooks for the design and construction of a building. The development of new uses for materials and of new techniques of construction, together with continual research into all aspects of building, has undoubtedly led to an information explosion in the building industry. Technical data is now readily available from research organisations in the form of research bulletins and papers, and from manufacturers of materials, equipment, components and building systems in the form of trade literature and/or actual samples of products, mostly distributed by travelling sales representatives.

Computer websites have been produced by manufacturers, commercial organisations and research institutes. These give simple access to a wealth of information on products, services and new developments. Listings of websites are found in trade and professional directories. Agencies produce, under licence, computer-compatible compact discs containing regulations, legislative documents, standards and a variety of technical data. These are normally available for an annual fee or subscription, to include updates.

But the vast amount of new and important information supplied through these organisations makes accumulation and categorisation of the knowledge difficult, and thorough assimilation virtually impossible. The industry therefore needed to establish suitable methods of presenting technical data to the Building Team, and to recommend a method of cataloguing documents, etc., so they could be readily found when required. The International Council for Building Research (CIB) published a list of headings for the guidance of authors of technical literature which classified a method of giving information based on sequential performance criteria. From 1961 onwards, a method of classification

for technical information began to be adopted which developed into the currently used *CI/SfB system*.

This system was first created in Sweden by what was then known as *Samarbetskommittén for Byggnadsfrågor* (the Coordinating Committee for Building) and related more to the specific requirements of the building industry than the universal decimal classification (UDC) system employed in UK libraries. However, in order to make the SfB system relate precisely to the UK building industry, the RIBA initiated modifications that led to the adoption of the CI/SfB system (CI stands for Construction Index). Figure 16.1 gives the four tables of this modified system (Tables 1–3 constitute the original SfB system), and each table represents a major group of subjects or concepts now available in the building industry. Broadly speaking, the tables represent the design process for a building as it proceeds through levels of increasing detail. The construction work proceeds through these same levels but in reverse order. Therefore, CI/SfB is a common language for communicating information to all members of the Building Team. Figure 16.2 illustrates how the system may be used.

Whether the CI/SfB system is used totally or only partially depends on the size of the organisation requiring to classify information, or the degree of cataloguing necessary for easy retrieval by an individual. For example, members of the Client Team may only require information relating to Table 0 (physical environment); or a manufacturer to Tables 1, 2 and 3 (elements, construction form and materials); whereas Design and Construction Teams may need to make use of all four tables.

A selection of critical technical data can be obtained from a *Building Centre* located in one of many areas in the UK, and which forms the recognised meeting places for manufacturers, users and designers. The Building Centre, London, published a short but useful guide to manufacturers of trade literature for the building industry; it stated:

Trade literature may be defined as information which enables the user to select, specify and utilize a product in service. This information also helps him to compare similar products and services. He may thus elect the ones that best suit his requirements. The fully informative piece of trade literature is a great aid in the preparation of drawings, specifications and bills of quantities.

Building Centres incorporate libraries of building and engineering product literature (including samples) classified under the CI/SfB; they provide an information service which includes a computer link to European

CI/SfB

Table 0
Physical environment

Planning areas

0 Planning areas
01
02 International, national scale planning areas
03 Regional, sub-regional scale planning areas
04
05 Rural, urban planning areas
06 Land use planning areas
07
08 Other planning areas
09 Common areas relevant to planning

Facilities

1 Utilities, civil engineering facilities
11 Rail transport
12 Road transport
13 Water transport
14 Air transport, other transport
15 Communications
16 Power supply, mineral supply
17 Water supply, waste disposal
18 Other

2 Industrial facilities
21–25
26 Agricultural
27 Manufacturing
28 Other

3 Administrative, commercial, protective service facilities
31 Official administration, law courts
32 Offices
33 Commercial
34 Trading, shops
35–36
37 Protective services
38 Other

4 Health, welfare facilities
41 Hospitals
42 Other medical
43
44 Welfare, homes
45
46 Animal welfare
47
48 Other

5 Recreational facilities
51 Refreshment
52 Entertainment
53 Social recreation, clubs
54 Aquatic sports
55
56 Sports
57
58 Other

6 Religious facilities
61 Religious centres
62 Cathedrals
63 Churches, chapels
64 Mission halls, meeting houses
65 Temples, mosques, synagogues
66 Convents
67 Funerary, shrines
68 Other

7 Educational, scientific, information facilities
71 Schools
72 Universities, colleges
73 Scientific
74
75 Exhibition, display
76 Information, libraries
77
78 Other

8 Residential facilities
81 Housing
82 One-off housing units, houses
83
84 Special housing
85 Communal residential
86 Historical residential
87 Temporary, mobile residential
88 Other

9 Common facilities, other facilities
91 Circulation
92 Rest, work
93 Culinary
94 Sanitary, hygiene
95 Cleaning, maintenance
96 Storage
97 Processing, plant, control
98 Other: buildings other than by function
99 Parts of facilities; other aspects of the physical environment; architecture, landscape

Table 1
Elements

(– –) Sites, projects, building systems

Substructure

(1–) Ground, substructure
(10)
(11) Ground
(12)
(13) Floor beds
(14)–(15)
(16 Retaining walls, foundations
(17) Pile foundations
(18) Other substructure elements
(19) Parts of elements (11) to (18) Cost summary

Structure

(2–) Primary elements, carcass
(20)
(21) Walls, external walls
(22) Internal walls, partitions
(23) Floors, galleries
(24) Stairs, ramps
(25)–(26)
(27) Roofs
(28) Building frames, other primary elements
(29) Parts of elements (21) to (28) Cost summary

(3–) Secondary elements, completion if described separately from (2–)
(30)
(31) Secondary elements to external walls; external doors, windows
(32)*Secondary elements to internal walls; internal doors
(33) Secondary elements to floors
(34) Secondary elements to stairs
(35) Suspended ceilings
(36)
(37) Secondary elements to roofs; rooflights, etc.
(38) Other secondary elements
(39) Parts of elements (31) to (38) Cost summary

(4–) Finishes if described separately
(40)
(41) Wall finishes, external
(42) Wall finishes, internal
(43) Floor finishes
(44) Stair finishes
(45) Ceiling finishes
(46)
(47) Roof finishes
(48) Other finishes to structure
(49) Parts of elements (41) to (48) Cost of summary

*Use for doors generally if required

Services

(5–) Services, mainly piped and ducted
(50)–(51)
(52) Waste disposal, drainage
(53) Liquids supply
(54) Gases supply
(55) Space cooling
(56) Space heating
(57) Air conditioning, ventilation
(58) Other piped, ducted services
(59) Parts of elements (51) to (58) Cost summary

(6–) Services, mainly electrical
(60)
(61) Electrical supply
(62) Power
(63) Lighting
(64) Communications
(65)
(66) Transport
(67)
(68) Security, control, other services
(69) Parts of elements (61) to (68) Cost summary

Fittings

(7–) Fittings
(70)
(71) Circulation fittings
(72) Rest, work fittings
(73) Culinary fittings
(74) Sanitary, hygiene fittings
(75) Cleaning, maintenance fittings
(76) Storage, screening fittings
(77) Special activity fittings
(78) Other fittings
(79) Parts of elements (71) to (78) Cost summary

(8–)*Loose furniture, equipment
(80)
(81) Circulation loose equipment
(82) Rest, work loose equipment
(83) Culinary loose equipment
(84) Sanitary, hygiene loose equipment
(85) Cleaning, maintenance loose equipment
(86) Storage, screening loose equipment
(87) Special activity loose equipment
(88) Other loose equipment
(89) Parts of elements (81) to (88) Cost summary

External, other elements

(9–) External, other elements
(90) External works
(98) Other elements
(99) Parts of elements Cost summary

*Use only (7–) if preferred

Figure 16.1   CI/SfB construction classification. (Reproduced courtesy of RIBA Publications)

Construction classification

Table 2
Constructions

A* Constructions, forms

B*

C*

D*

E Cast in situ work

F Block work; blocks

G Large block, panel work;
large blocks, panels

H Section work; sections

I Pipe work; pipes

J Wire work, mesh work; wires,
meshes

K Quilt work; quilts

L Flexible sheet work (proofing);
flexible sheets (proofing)

M Malleable sheet work;
malleable sheets

N Rigid sheet overlap work;
rigid sheets for overlapping

P Thick coating work

Q

R Rigid sheet work; rigid sheets

S Rigid tile work; rigid tiles

T Flexible sheet work;
flexible sheets

U

V Film coating and impregnation
work

W Planting work; plants, seeds

X Work with components;
components

Y Formless work; Products

Z Joints

*Used for special purposes
eg specification*

Table 3
Materials

a* Materials
b*
c*
d*

Formed materials eo

e Natural stone
f Precast with binder
g Clay (dried, fired)
h Metal
i Wood
j Vegetable and animal materials
k
m inorganic fibres
n Rubbers, plastics etc.
o Glass

Formless materials ps

p Aggregates, loose fills
q Lime and cement binders,
mortars, concretes
r Clay, gypsum, magnesia and
plastic binders, mortars
s Bituminous materials

Functional materials tw

t Fixing and jointing materials
u Protective and process/property
modifying materials
v Paints
w Ancillary materials

x
y Composite materials
z Substances

*Used for special purposes
eg resource scheduling by computer*

Table 4
Activities, requirements

Activities, aids

(A)   Administration and
management activities, aids
(Af)   Administration, organization
(Ag)   Communications
(Ah)   Preparation of documentation
(Ai)   Public relations, publicity
(Aj)   Controls, procedures
(Ak)   Organizations
(Am)   Personnel, roles
(An)   Education
(Ao)   Research, development
(Ap)   Standardization, rationalization
(Aq)   Testing, evaluating

(A1)   Organizing offices, projects
(A2)   Financing, accounting
(A3)   Designing, physical planning
(A4)   Cost planning, cost control
tenders, contracts
(A5)   Production planning, progress
control
(A6)   Buying, delivery
(A7)   Inspection, quality control
(A8)   Handing over, feedback
appraisal
(A9)   Other activities; arbitration
insurance

(B)   Construction plant, tools
(B1)   Protection plant
(B2)   Temporary (non-protective)
works
(B3)   Transport plant
(B4)   Manufacture, screening,
storage plant
(B5)   Treatment plant
(B6)   Placing, pavement, compaction
plant
(B7)   Hand tools
(B8)   Ancillary plant
(B9)   Other construction plant, tools

(C)*

(D)   Construction operations
(D1)   Protecting
(D2)   Clearing, preparing
(D3)   Transporting, lifting
(D4)   Forming; cutting, shaping,
fitting
(D5)   Treatment; drilling, boring
(D6)   Placing; laying, applying
(D7)   Making good, repairing
(D8)   Cleaning up
(D9)   Other construction operations

Used for special purposes

Requirements, properties

(E/G) Description [2]
(E)   Composition [2.01/2.03]
(F)   Shape, size [2.04/2.06]
(G)   Appearance [2.07]

(H)   Context, environment [3]

(J/T) Performance factors [4/5]
(J)   Mechanics [4.01]
(K)   Fire, explosion [4.02]
(L)   Matter [4.03/4.06]
(M)   Heat, cold [4.07]
(N)   Light, dark [4.08]
(P)   Sound, quiet [4.09]
(Q)   Electricity, magnetism,
radiation [4.10]
(R)   Energy [4.11]
Side effects, compatibility
[4.12/4.13]
Durability [4.14]

(S)

(T)   Application [5]

(U)   Users, resources

(V)   Working factors [6]
(W)   Operation, maintenance
factors [7]

(X)   Change, movement, stability
factors

(Y)   Economic, commercial factors
[8/10]

(Z)   Peripheral subjects; forms of
presentation; time; space [11]

Note: Codes in square brackets above
are from CIB Master Lists for structuring
documents relating to buildings,
building elements, components,
materials and services (CIB Report
No 18, 1972).

CI/SfB is a flexible system ie free
codes throughout the tables can be
used for special purposes on
information which is not for general
publication; for Tables 0 and 1,
headings and codes in colour will
often provide sufficient breakdown

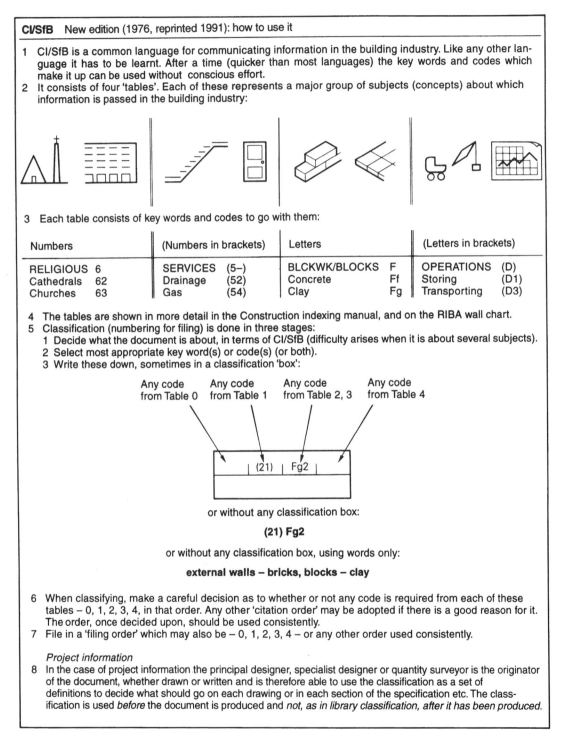

CI/SfB   New edition (1976, reprinted 1991): how to use it

1   CI/SfB is a common language for communicating information in the building industry. Like any other language it has to be learnt. After a time (quicker than most languages) the key words and codes which make it up can be used without conscious effort.

2   It consists of four 'tables'. Each of these represents a major group of subjects (concepts) about which information is passed in the building industry:

3   Each table consists of key words and codes to go with them:

Numbers		(Numbers in brackets)		Letters		(Letters in brackets)	
RELIGIOUS	6	SERVICES	(5–)	BLCKWK/BLOCKS	F	OPERATIONS	(D)
Cathedrals	62	Drainage	(52)	Concrete	Ff	Storing	(D1)
Churches	63	Gas	(54)	Clay	Fg	Transporting	(D3)

4   The tables are shown in more detail in the Construction indexing manual, and on the RIBA wall chart.

5   Classification (numbering for filing) is done in three stages:

   1   Decide what the document is about, in terms of CI/SfB (difficulty arises when it is about several subjects).

   2   Select most appropriate key word(s) or code(s) (or both).

   3   Write these down, sometimes in a classification 'box':

Any code from Table 0   Any code from Table 1   Any code from Table 2, 3   Any code from Table 4

| (21) | Fg2 |

or without any classification box:

(21) Fg2

or without any classification box, using words only:

external walls – bricks, blocks – clay

6   When classifying, make a careful decision as to whether or not any code is required from each of these tables – 0, 1, 2, 3, 4, in that order. Any other 'citation order' may be adopted if there is a good reason for it. The order, once decided upon, should be used consistently.

7   File in a 'filing order' which may also be – 0, 1, 2, 3, 4 – or any other order used consistently.

Project information

8   In the case of project information the principal designer, specialist designer or quantity surveyor is the originator of the document, whether drawn or written and is therefore able to use the classification as a set of definitions to decide what should go on each drawing or in each section of the specification etc. The classification is used *before* the document is produced and *not, as in library classification, after it has been produced.*

Figure 16.2   Using the CI/SfB classification system. (Reproduced courtesy of RIBA Publications)

sources of reference. But specifiers of non-UK products must ascertain that testing data complies with BSI Standards, Comité Européen de Normalisation (CEN) for European Standards, International Organisation for Standardisation (ISO) for international standards or have an appropriate Agrément certificate or its equivalent.

Some Building Centres may provide contact offices for professional and research bodies such as the RIBA, ICE, BRE and TRADA Technology Ltd. Whereas Building Centres primarily provide permanent information resources, manufacturers and trade organisations also frequently sponsor temporary national building exhibitions. Well-designed UK products may be exhibited at a *Design Centre* run by the *Council of Industrial Design* whose chief purpose is the promotion of British design.

Instead of collecting and classifying technical information, or in order to supplement an established technical library, some organisations prefer to employ the services of companies specialising in the dissemination of information. As far as the building industry is concerned, the major distributors of technical information are Barbour Index, Building Products Index, Technical Indexes and RIBA Information Services. These organisations provide a basic library of literature on compact discs (CD-ROM) for computer application which they periodically update. RIBA Information Services currently provide technical resources in the forms of product and practice data. Product data is linked to a product selector, which provides simple computer programs as well as listing firms able to supply products and services; the lists give information on legislation, practice, building failures, design, British Standards and information technology. Less elaborate collective information is available from such sources as the *Architects Standard Catalogue*, *Building Commodity File*, *Specification*, etc. Many manufacturers of commercial products have also supplemented their traditional hard copy trade catalogues with CD-ROMS and video packages.

16.2 Drawings

Drawings of many types provide the main method of communication between all the members of the Building Team. As the information required at any one stage of the implementation of a building will vary, and also be at different levels of complexity according to user requirements, the different types of drawing are divided for convenience into two broad categories.

Design drawings communicate the form of a building in terms of shape, colour and texture; *production* or *working drawings* communicate the technical, physical and economic aspects of a building which are associated with its construction, subsequent use and maintenance. The information conveyed on both types of drawing are generally supplemented by reports, schedules, samples, specifications, models, discussions, etc. In reality, there should be no firm division between design and production or working drawings, just as in reality there should be no division between design ideas and construction. However, certain drawings need to convey more about appearance to less technically-minded parties, whereas other drawings are needed to convey technical information to parties which have priorities besides the overall appearance and function of a project. Though all the Building Team would undoubtedly benefit from studying a comprehensive range of drawings for a particular project, such a need is of most value to the principal designer, who is concerned with the all-embracing quality of the project. It is the responsibility of the overall coordinator of the project (whether principal designer, contractor or some other party) to ensure that the most useful drawings are distributed to the appropriate members of the Building Team at the precise time they are required.

16.2.1 Design drawings

There are two types of design drawings: those concerned with the preliminary investigation processes for a design, and those concerned with the presentation of a design solution. Both are produced during the 'design' stages of a project.

Design drawings for *investigation purposes* communicate information between designers, quantity surveyors, etc., (Design Team) and, if involved during early stages of negotiations, the builder or contractor (Construction Team). In addition, this information about the project will also be passed to the client, and eventually form the basic information necessary to produce a preliminary visualisation or *sketch design*.

The earliest form of investigation drawing is that which provides information about the site (and existing building if a conversion is involved) and the immediate environment, including adjoining structures, roads, services, etc., likely to influence a project. This is known as a *site survey* and will enable the Design Team to begin the design. Ordnance Survey maps are useful initial references (Fig. 16.3).

As the design process continues, the Design Team may require specific information about the cost of different solutions for a potential scheme, and may supply *sketch drawings* to a quantity surveyor for comment.

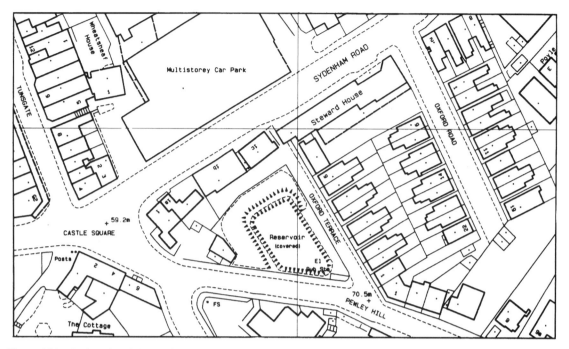

Figure 16.3   Section of an Ordnance Survey 'Superplan' map. (© Crown copyright (399582))

A contractor may also be able to help in establishing economic and efficient construction methods by suggesting the suitable type and availability of labour and mechanical plant for a particular design. Specialist designers usually prepare sketch drawings of their proposals for similar reasons and communicate vital information which enables the principal designer to incorporate them in the design. Drawings for these purposes need not be elaborate when information is being communicated between parties having a common aesthetic and/or technical language.

Computer printouts are increasingly used for the production of technical data (Fig. 16.4) to develop concepts outlined by sketch design decisions. Using this method, speedy and accurate information is made available for assessing various aspects of design, including lighting levels created by different shapes and sizes of window, heat losses for different constructions, and economic spanning methods for alternative structural solutions.

Design drawings for *presentation purposes* are prepared by the designer to illustrate to the client the appearance of a project, the general disposition of the accommodation to be provided, and the effects the overall scheme has on the environment, including details of landscaping and the immediate physical surroundings. Drawings should use techniques which are readily understood, preferably three-dimensional

WALL CONSTRUCTION

MATERIAL	RESISTIVITY	THICK	RESISTANCE
SURF. RES. IN	0	0	0.13
GYP. PLASTER	2	15	0.03
LTWT. CONC. BLK.	5.3	100	0.53
EXP. POLY.	27.2	30	0.83
CAVITY	0	25	0.18
BRICKWORK	1.78	102	0.18
SURF. RES. OUT	0	0	0.05

TOTAL RESISTANCE   =   1.93

'U' VALUE   =   0.52 W/m^2 K

STEADY STATE CONDENSATION ANALYSIS

CONDITIONS:
      OUT. TEMP. = 0°C   OUT. VP = 6 mb
      IN. TEMP. = 20° C   IN. VP = 11.3 mb

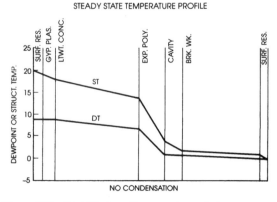

Figure 16.4   Brick/block cavity: computer analysis.

Figure 16.5 Typical perspective drawing.

representations such as perspectives (Figs. 16.5 and 16.6), axonometrics or isometrics. And two-dimensional representations (orthographic – see Fig. 16.12) should be clear and provide easily identifiable features, e.g. people, furniture and trees. The drawings range from simple pencil sketches to highly finished, fully coloured and mounted drawings. Very little technical information is usually given, although this depends on the expertise and requirements of a particular client. Clients having technical expertise influencing the design will probably need information to show how their interests have been incorporated.

Whenever possible, information about the design should be supplemented by scale models and/or sample boards of materials to be employed. Some principal designers prefer to use the skills of *visualisers*, who specialise in artistic techniques so that a design is most effectively presented. Line perspectives can be produced by computers (Fig. 16.6). After being programmed accordingly, computers are also capable of producing a video/virtual reality presentation of a walkabout within a proposed building. This preview can also assist a designer during sketch design stages.

With a few minor additions, design drawings can form a suitable submission to the local authority for planning approval, although generally for this they need not be as elaborately presented.

16.2.2 Production or working drawings

Production drawings are often called working drawings, and are produced by designers in order to communicate technical information throughout the Building Team. According to the nineteenth-century architect Sir Edwin Lutyens, 'a working drawing is merely a letter to a builder telling him precisely what is required of him – and not a picture to charm an idiotic client'. The building process has become much more complex since; the following list describes the uses of production drawings:

- Obtaining official consents and statutory approvals
- Analysing cost factors
- Establishing use of materials
- Informing extent of subcontractors' work
- Providing source of information for other contract documents
- Providing details for tendering
- Indicating contractual commitments
- Providing basis for ordering materials and components
- Establishing type and amount of labour required
- Demonstrating construction detailing
- Forming part of documentation during site meetings
- Indicating degree of supervision
- Providing check for variations from the contract

CELTEL HEADQUARTERS, KAMPALA

NORMAN + DAWBARN LIMITED

Figure 16.6 Computer drawings: three-dimensional representation of different aspects of a building complex. (Reproduced courtesy of the architects, Norman + Dawbarn Limited)

- Assisting the measurement of progress
- Providing guidance on interim financial payments to contractor
- Agreeing completed works and final payment
- Checking on defects in site construction method
- Recording work completed
- Indicating factors to put in maintenance manual
- Analysing factors affecting health and safety
- Providing an information base for structural calculations

Generally, the principal designer of a project prepares information and incorporates the data supplied by others (including specialist designers) to produce a master set of production drawings. For larger projects, certain specialist designers (structural and mechanical services engineers, etc.) prepare independent production drawings, but this must be done in close collaboration with the principal designer in order to ensure unified intentions and maintenance of professional responsibility.

Together with the written descriptions of a project (e.g. bills of quantities, schedules and specifications of works), production drawings are a vital part of the legal documentation upon which contractual arrangements are based. It is important that any alterations which are necessary after the formal issuing of production drawings are communicated to all members of the Building Team as they could result in changes in legal arrangements. Amendments influencing cost and/or uses of materials and labour need particular attention.

A typical basic list of items to be incorporated on production drawings for a small project is indicated in Appendix I. From this it will be realised that they must provide an accurate record of the principal designer's intentions at all stages of the construction process, and the information must be interpreted by many people with different sets of priorities, e.g. building control officers, quantity surveyors, sales staff, supervisors, site operatives and labourers. It is important, therefore, that production drawings are clearly representative and easily understood; are comprehensive and sufficiently detailed for their purpose; and are produced in a format which enables them to be easily collated so that specific drawings can be found by particular users when required.

For this reason, a great deal of work has been carried out, initially by the British Standards Institution, in order to establish a system of coordinating production drawings and the information they communicate so as to avoid errors, inadequate data and omissions. Various standards contain recommendations for the optimum arrangement of production drawings for communication of information, e.g.:

- BS EN ISO 3766: *Construction drawings, simplified representation of concrete reinforcement.*
- BS EN ISO 4157–1: *Building and parts of buildings*; 4157–2: *Room names and numbers*; 4157–3: *Room identifiers.*
- BS EN ISO 6284: *Construction drawings, indication of limit deviations.*
- BS EN ISO 7518: *Construction drawings, simplified representation of demolition and rebuilding.*
- BS EN ISO 7519: *Technical drawings – construction drawings – general principles of presentation for general arrangement and assembly drawings.*
- BS EN ISO 8560: *Construction drawings, representation of modular sizes, lines and grids.*
- BS EN ISO 9431: *Construction drawings, spaces for drawing and for text and title blocks on drawing sheets.*
- BS EN ISO 11091: *Construction drawings, landscape drawing practice.*

The above standards have replaced the withdrawn BS 1192: *Construction drawing practice,* with the exception of Part 5, i.e. BS 1192–5: *Guide for structuring and exchange of CAD data.*

Members of the Building Team first need to know the shape, size and *location* of the building to be constructed and its constituent parts; then about the methods to be adopted for the *assembly* of the parts (type of material and labour required); and finally about details of *components* to be used. Figure 16.7 indicates in greater detail the type and purpose for each category of drawing together with the scales which are recommended for each. Schedules provide more detailed specialised information and are a collection of mostly written information about the repetitive parts of a building, such as doors, windows and finishes (Fig. 16.8).

This method of structuring information greatly assists individuals of the Building Team in identifying the group of production drawings which are able to give the particular information they require. However, consideration must also be given to how this information is presented to assist easy reference and understanding. Drawing sheets should be used which conform to a uniform series of sizes to facilitate handling or storage and the international *A series* of paper sizes (Fig. 16.9) should be adopted for all drawings and written material, including trade literature. Secondly, consideration must be given to one of the standard methods of graphically representing materials, components and dimensions so that a common grammar is established between the members of the Building Team, thus reducing ambiguity and speeding understanding.

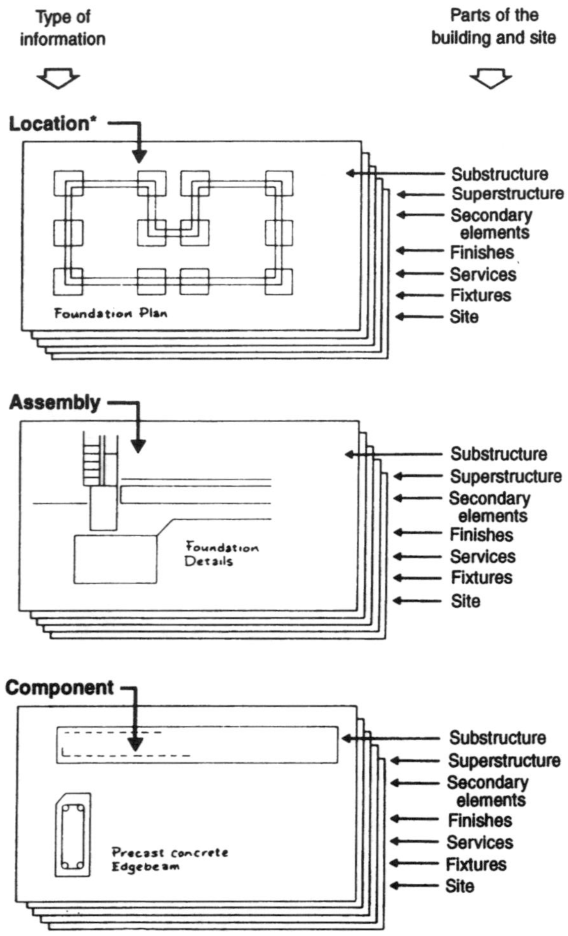

Type of information	Parts of the building and site	Type of drawing and purpose		Approximate scale
Location*	Substructure, Superstructure, Secondary elements, Finishes, Services, Fixtures, Site	**Location**	Site and external works	General site plan	1:2500 / 1:1250
				To identify, locate and dimension the site and external works	1:100 / 1:200 / 1:500
			Building	To identify, locate and dimension parts and spaces within the building and to show overall shapes by plan, elevation and section	1:50
				To locate grids, datums and key reference planes	1:100
				To convey dimensions for setting out	1:200
				To give other information of a general nature for which a small scale is appropriate (e.g. door swings)	
			Element	To give location and setting-out information about one element, or a group of related elements	1:100 / 1:200
			Cross-references	To show cross-references to schedules, assembly and component drawings	1:100 / 1:50
		Schedule	Element	To collect repetitive information about elements or products which occur in variety	
				To collect cross-references to assembly and component drawings	
Assembly	Substructure, Superstructure, Secondary elements, Finishes, Services, Fixtures, Site	**Assembly**	Element	To show assembly of parts of one element, including their shapes and sizes	1:5
				To show an element at its junction with another element	1:10
				To show cross-references to other assembly and component drawings	1:20
Component	Substructure, Superstructure, Secondary elements, Finishes, Services, Fixtures, Site	**Component**	Element or sub-elements	To show shape, dimensions and assembly (and possibly composition) of a component to be made away from the building	1:1 / 1:5 / 1:10
				To show component parts of an *in situ* assembly which cannot be defined adequately on the assembly drawing	1:20 / 1:50

(Location: Foundation Plan; Assembly: Foundation Details; Component: Precast concrete Edgebeam)

* **There will be additional location drawings dealing with the project as a whole**

Figure 16.7 Production of working drawings. (Based on material in *BRE Digest 172*; reproduced by permission of the Controller of The Stationery Office. © Crown copyright)

DOOR GROUP PREFIX LETTERS	REFERENCE NUMBERS	AZA	|1.|2|3|4|5|6|7|8|9|10|11|12|13 INTERNAL DOORS	|1|2|3|4|5|6|7|8|9|10|11|12|13 EXTERNAL DOORS coded XD

(Column layout below: Category · Description · AZA No. · Internal doors 1–13 · External doors 1–13. Double-door columns noted in source: internal 8–9 DOUBLE; external 1, 5, 7, 9 DOUBLE.)

Category	Description	AZA	I1	I2	I3	I4	I5	I6	I7	I8	I9	I10	I11	I12	I13	E1	E2	E3	E4	E5	E6	E7	E8	E9	E10	E11	E12	E13
UPRIGHT LOCKS	WITH ONE KEY	100																										
	LOCKS TO PASS	101	/	/	/		/					/	/	/														
	WITH TWO KEYS	102																										
	REBATED COMPONENTS	103																										
	ROLLER BOLT, ONE KEY	104																										
	ROLLER BOLT, LOCKS TO PASS	105																										
	ROLLER BOLT, TWO KEYS	106																										
	ROLLER BOLT REBATED COMP'S	107																										
UPRIGHT DEADLOCK	WITH ONE KEY	112																										
	LOCKS TO PASS	113								1.						1.		/	/	/		/		/		/	/	/
	WITH TWO KEYS	114																										
	REBATED COMPONENTS	115														/		/	/	/		/		/		/	/	/
LEVER HANDLES	PAIR ON ROSE	130																										
	PAIR ON BACKPLATE, KEYHOLE	131	/	/	/		/					/	/	/														
	PAIR ON BACKPLATE, NO KEYHOLE	132																										
	ESCUTCHEONS	133								2						2		2	2	2		2		2		2	2	2
PULL HANDLES	150 mm CENTRES FIXING	134																										
	225 mm CENTRES	135					/	/	/	/				/		/	/	/	/	/	/	/	/	/	/	/	/	/
	300 mm CENTRES	136																										
FINGER PLATES	300 x 75	140				/		/	/	/				/		/	/	/	/	/	/	/	/	/	/	/	/	/
KICKING PLATES	625 mm WIDE	141										1.																
	725 mm	142				/	/	/	/	2	2		/															
	775 mm	143	/	/	2						2		2								/	/				/	/	
	825 mm	144															/	/	/	/		/	/	/	/			/
	875 mm	145																										
FLUSH AND BARREL BOLTS	PAIR FLUSH BOLTS	150								/						/				/		/		/				
	SOCKET FOR WOOD	151																										
	SOCKET FOR CONCRETE	152																										
	PAIR BARREL BOLTS	153																										
OVERHEAD CLOSERS	FOR EXTERNAL DOORS OPEN OUT	160																										
	FOR INTERNAL DOORS	161				/		/	/				/	/														
	DOOR SELECTOR	162																										
	OVERHEAD LIMITING STAY	163														/	/	/	/	/	/	/	/	/	/	/		
DOOR STOPS	FOR TIMBER	170																										
	FOR CONCRETE	171				/					/	/	/	/														
	POST MOUNTED DOOR HOLDER	172																										
	CABIN HOOK	173																										
LETTER BOX	LETTER BOX	174																										/
HAT AND COAT HOOKS	ALUMINIUM	180																										
	NYLON COATED SECRET FIX	181																										
	PAIR NYLON COATED SECRET FIX	182																										
	NYLON COATED SCREW FIX	183																										

Figure 16.8 Ironmongery schedule.

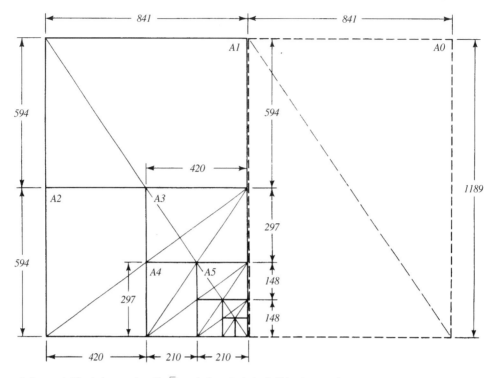

A sizes retain identical proportions (1 : √2), each sheet size being half the size next above

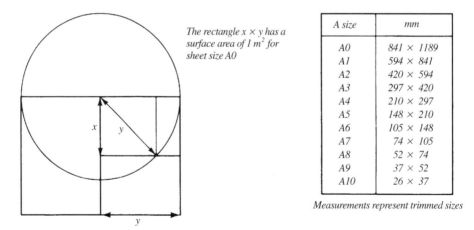

The rectangle x × y has a surface area of 1 m² for sheet size A0

A size	mm
A0	841 × 1189
A1	594 × 841
A2	420 × 594
A3	297 × 420
A4	210 × 297
A5	148 × 210
A6	105 × 148
A7	74 × 105
A8	52 × 74
A9	37 × 52
A10	26 × 37

Measurements represent trimmed sizes

Figure 16.9   Paper sizes: the internationally agreed A series for all written documents, drawings and trade literature.

Figure 16.10 indicates the British Standard recommendations for dimensions on drawings. Some of the current conventions are illustrated in Figure 16.11.

In addition to the above factors it is important that the layout of drawing sheets should be done in a systematic manner; particular attention should be given to the title panel of the drawings, as ambiguity can cause a considerable waste of time when seeking out particular information. Figure 16.10 illustrates a typical title panel,

originally developed from the now withdrawn BS 1192: *Construction drawing practice*.

The messages conveyed by the production drawings are really a translation of the three-dimensional ideas of a designer, so perhaps the best form of graphical presentation should also be three-dimensional, i.e. isometric or axonometric. This form is particularly useful when new construction methods or difficult junctions between several components need amplification (see Fig. 6.8).

DIMENSIONS

Modular and coordinating dims. should be shown with an open arrowhead.

Closed arrows should be used for work sizes and other dimensions. Small dims. are indicated from the outside of their limits. The oblique dash is a common alternative.

Running dimensions

These should only be used in field surveys. They start from a point in a circle and arrowheads face the ongoing dims. only. Written dims. should be near or over that end of the line marked with the arrow. The oblique dash method should not be used as this neither indicates the source nor direction of travel.

Dimensional coordination Structural grid-based axial controlling lines which are shown by a chain-dotted line.

Planning grid (used to locate non-structural elements) based on face controlling lines which are shown by solid lines.

Work and coordination sizes

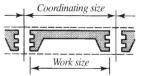

SECTION PLANES

Section planes are usually indicated by arrows pointing from the plane of the section towards the angle of view. If there is more than one section on a drawing or set of drawings they should be identified.

LEVELS

Sectional and elevational levels are indicated by an inverted triangle on the projected line of the chosen plane. State whether it is a structural or a finished level.

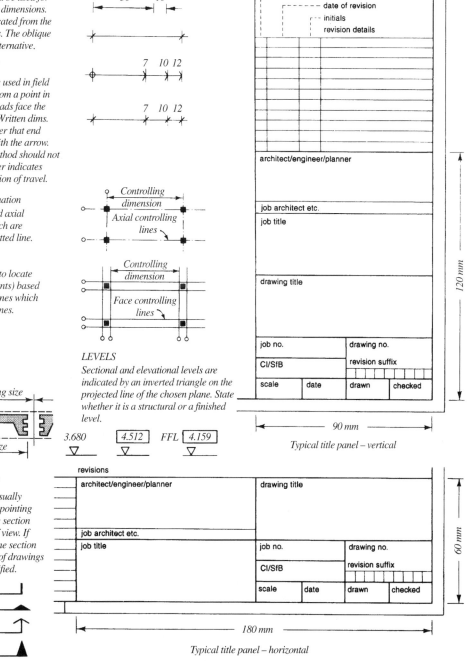

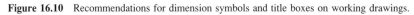

Figure 16.10 Recommendations for dimension symbols and title boxes on working drawings.

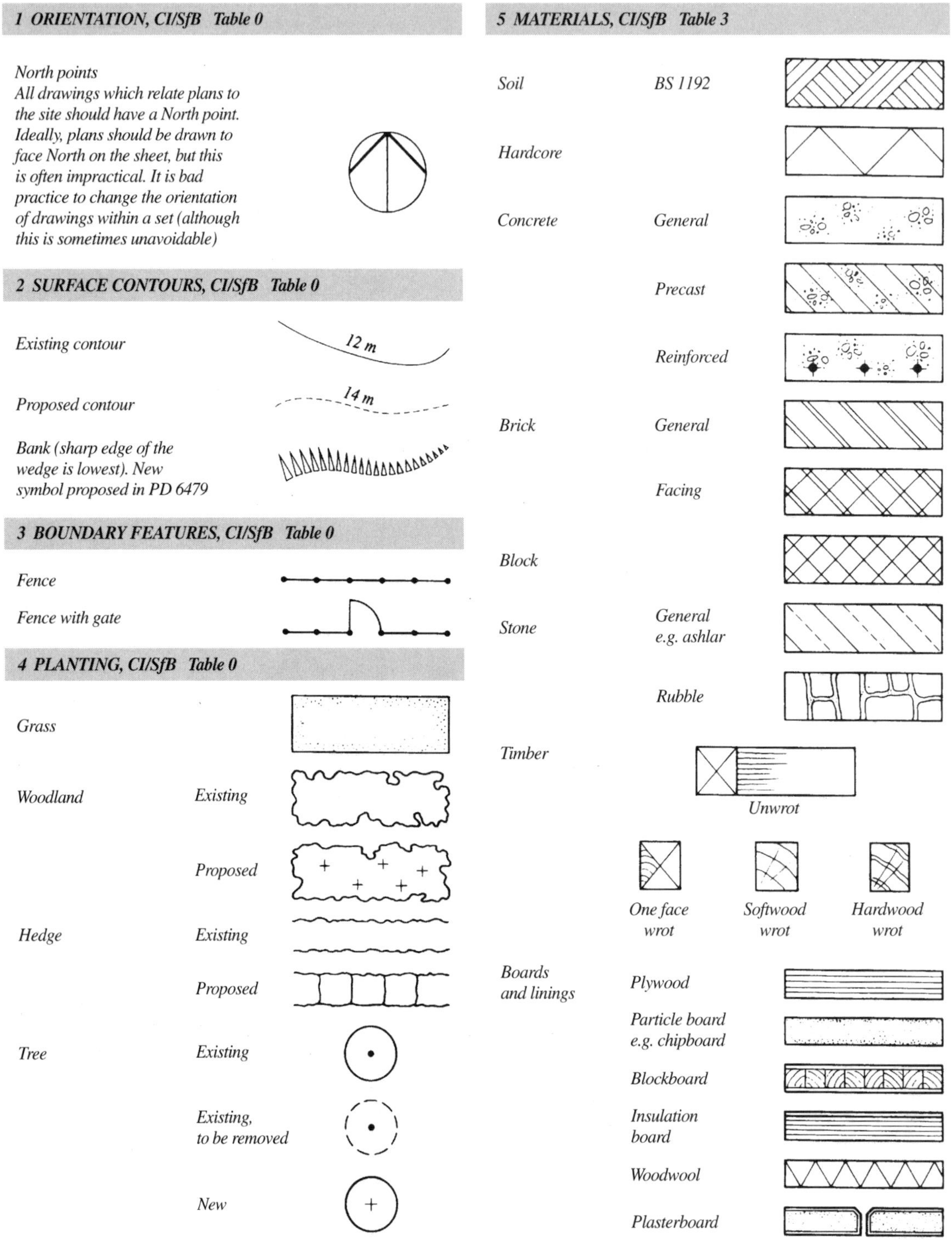

1 ORIENTATION, CI/SfB Table 0

North points
All drawings which relate plans to
the site should have a North point.
Ideally, plans should be drawn to
face North on the sheet, but this
is often impractical. It is bad
practice to change the orientation
of drawings within a set (although
this is sometimes unavoidable)

2 SURFACE CONTOURS, CI/SfB Table 0

Existing contour

12 m

Proposed contour

14 m

Bank (sharp edge of the
wedge is lowest). New
symbol proposed in PD 6479

3 BOUNDARY FEATURES, CI/SfB Table 0

Fence

Fence with gate

4 PLANTING, CI/SfB Table 0

Grass

Woodland — Existing

Proposed

Hedge — Existing

Proposed

Tree — Existing

Existing,
to be removed

New

5 MATERIALS, CI/SfB Table 3

Soil — BS 1192

Hardcore

Concrete — General

Precast

Reinforced

Brick — General

Facing

Block

Stone — General
e.g. ashlar

Rubble

Timber

Unwrot

One face
wrot

Softwood
wrot

Hardwood
wrot

Boards
and linings — Plywood

Particle board
e.g. chipboard

Blockboard

Insulation
board

Woodwool

Plasterboard

Figure 16.11 Recommendations for graphic symbols used on working drawings.

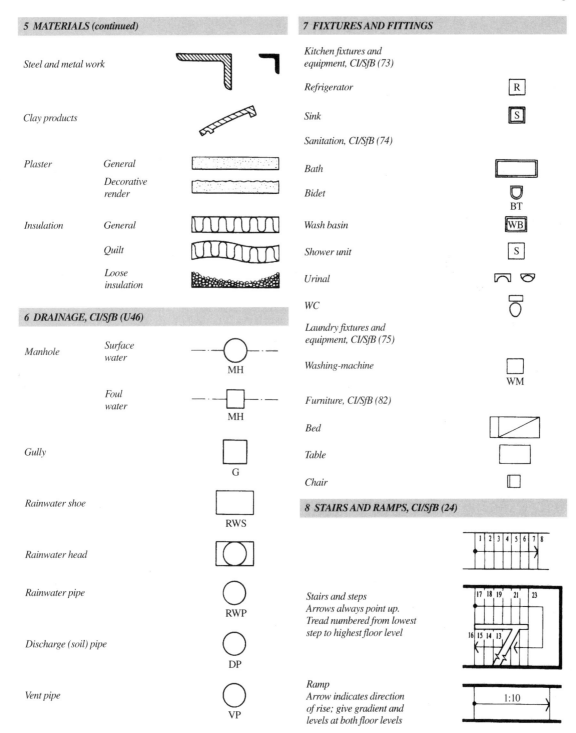

5 MATERIALS (continued)

Steel and metal work

Clay products

Plaster — General

Decorative render

Insulation — General

Quilt

Loose insulation

6 DRAINAGE, CI/SfB (U46)

Manhole — Surface water — MH

Foul water — MH

Gully — G

Rainwater shoe — RWS

Rainwater head

Rainwater pipe — RWP

Discharge (soil) pipe — DP

Vent pipe — VP

7 FIXTURES AND FITTINGS

Kitchen fixtures and equipment, CI/SfB (73)

Refrigerator — R

Sink — S

Sanitation, CI/SfB (74)

Bath

Bidet — BT

Wash basin — WB

Shower unit — S

Urinal

WC

Laundry fixtures and equipment, CI/SfB (75)

Washing-machine — WM

Furniture, CI/SfB (82)

Bed

Table

Chair

8 STAIRS AND RAMPS, CI/SfB (24)

Stairs and steps
Arrows always point up.
Tread numbered from lowest
step to highest floor level

Ramp
Arrow indicates direction
of rise; give gradient and
levels at both floor levels

1:10

Figure 16.11   (Cont'd)

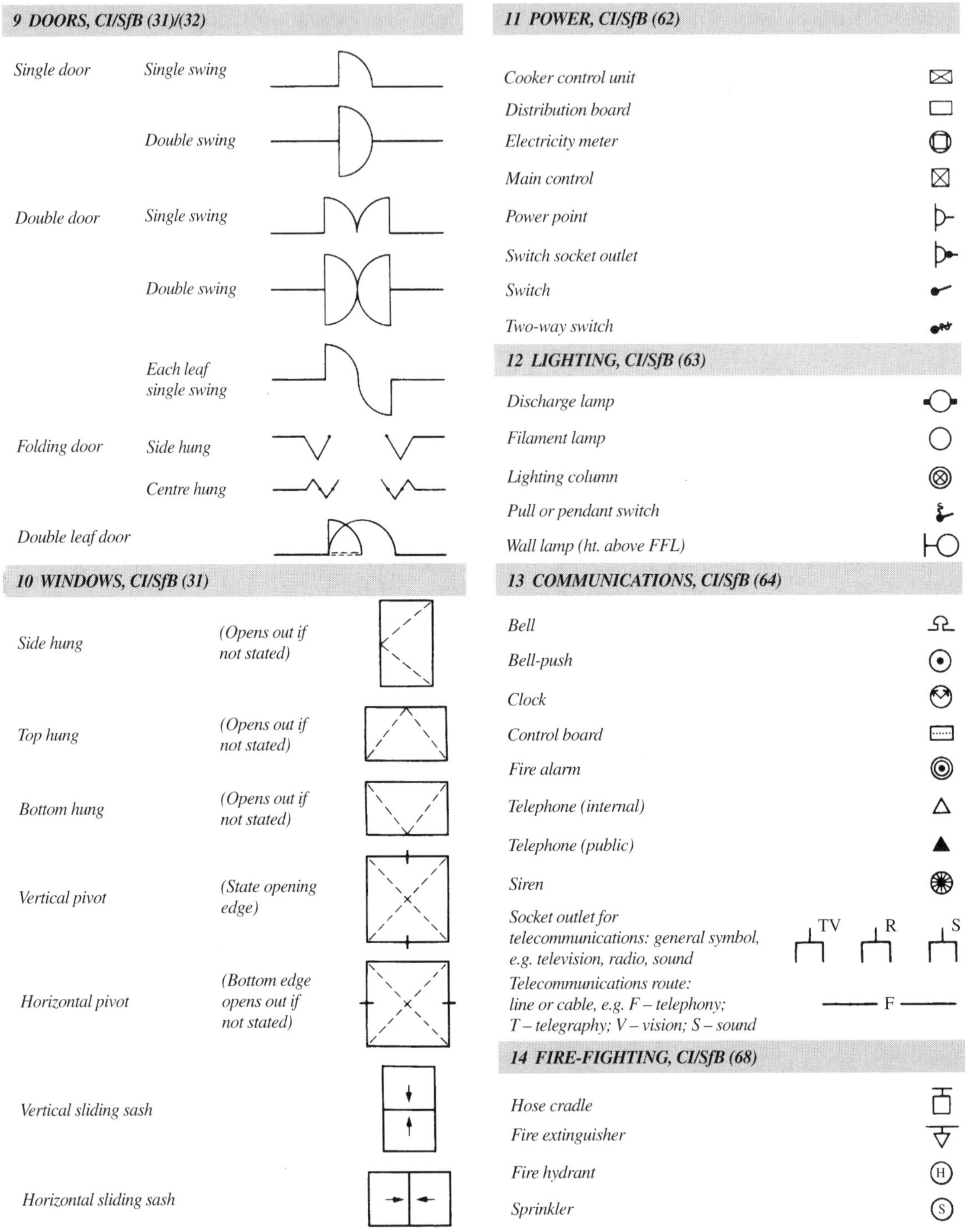

9 DOORS, CI/SfB (31)/(32)

Single door	Single swing	
	Double swing	
Double door	Single swing	
	Double swing	
	Each leaf single swing	
Folding door	Side hung	
	Centre hung	
Double leaf door		

10 WINDOWS, CI/SfB (31)

Side hung	(Opens out if not stated)	
Top hung	(Opens out if not stated)	
Bottom hung	(Opens out if not stated)	
Vertical pivot	(State opening edge)	
Horizontal pivot	(Bottom edge opens out if not stated)	
Vertical sliding sash		
Horizontal sliding sash		

11 POWER, CI/SfB (62)

Cooker control unit	
Distribution board	
Electricity meter	
Main control	
Power point	
Switch socket outlet	
Switch	
Two-way switch	

12 LIGHTING, CI/SfB (63)

Discharge lamp	
Filament lamp	
Lighting column	
Pull or pendant switch	
Wall lamp (ht. above FFL)	

13 COMMUNICATIONS, CI/SfB (64)

Bell	
Bell-push	
Clock	
Control board	
Fire alarm	
Telephone (internal)	
Telephone (public)	
Siren	
Socket outlet for telecommunications: general symbol, e.g. television, radio, sound	TV R S
Telecommunications route: line or cable, e.g. F – telephony; T – telegraphy; V – vision; S – sound	——— F ———

14 FIRE-FIGHTING, CI/SfB (68)

Hose cradle	
Fire extinguisher	
Fire hydrant	
Sprinkler	

Figure 16.11   (Cont'd)

However, three-dimensional representations are very time-consuming to prepare and, for the majority of building work, are not normally necessary. There are two basic methods of communicating information on production drawings. The *conventional method* consists of drawings containing many notes about the construction methods to be employed which are further developed by the clauses of specifications and/or bills of quantities. There is also the *systematic method*, which consists only of outline drawings but makes frequent direct cross-reference (using the CI/SfB system) to supplementary documents such as standard details, schedules, specification clauses and bills of quantities. These forms are illustrated in Figures 16.12 and 16.13.

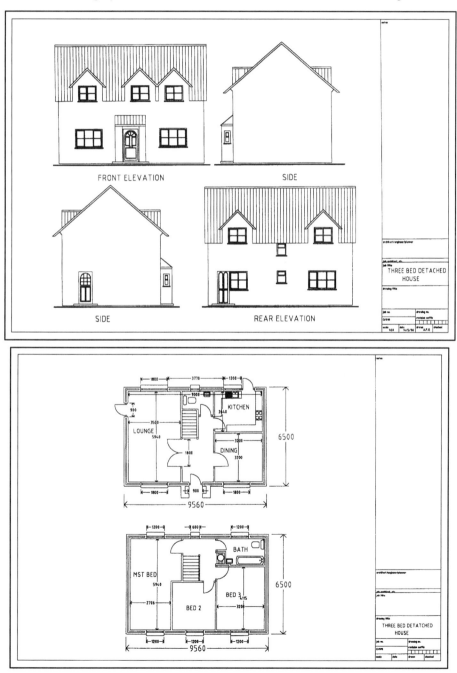

Figure 16.12 Working drawings produced on a computer.

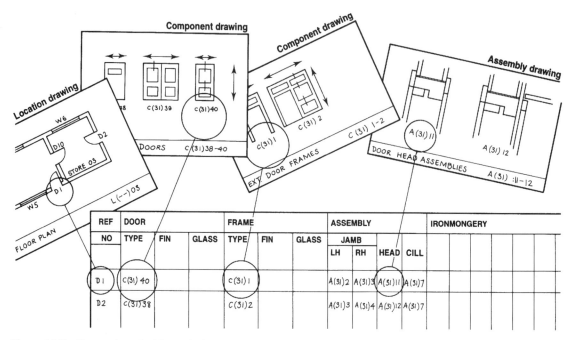

Figure 16.13 Systematic method for producing working drawings. (Based on material in *BRE Digest 172*; reproduced by permission of the Controller of The Stationery Office, © Crown copyright)

A more recent simplification of building information, specification and referencing is a system known as coordinated project information (CPI). Due to a history of confusion, insufficiency and even conflict over conveyance of building information between the professions, the government in conjunction with ACE, BEC, RIBA and the RICS sponsored a new initiative to co-ordinate elements of work into a common arrangement of work sections (CAWS). This method simplifies conventions and applies a letter- and number-coded notation common across the standard method of measurement (SMM), a code of procedure for project specification and drawings code, and the national building specification (NBS) incorporating the national engineering specification (NES). Work sections are coded A to Z (I and O omitted), as listed in Figure 16.14. Each section is subdivided, as shown in Figure 16.15, for masonry, and further subdivided for detailed descriptions. New innovations and products are easily added, e.g. F4. Confusion over section entries is avoided by definition of work sections. So for F10, brick/block walling (Fig. 16.16), inclusions and exclusions are clearly indicated, with exclusions referenced to alternative sections.

The selection of the most suitable form for a particular project depends upon the organisation of the

A	Preliminaries/general conditions
B	Complete buildings
C	Demolition/alteration/renovation
D	Groundwork
E	In situ concrete/large precast concrete
F	Masonry
G	Structural/carcassing metal/timber
H	Cladding/covering
J	Waterproofing
K	Linings/sheathing/dry partitioning
L	Windows/doors/stairs
M	Surface finishes
N	Furniture/equipment
P	Building fabric sundries
Q	Paving/planting/fencing/site furniture
R	Disposal systems
S	Piped supply systems
T	Mechanical heating/cooling/refrigeration systems
U	Ventilation/air-conditioning systems
V	Electrical supply/power/lighting systems
W	Communications/security/control systems
X	Transport systems
Y	Services reference specification
Z	Building fabric reference specification

Figure 16.14 CPI work sections.

F   Masonry

F1	Brick/block walling	F10	Brick/block walling
		F11	Glass block walling
F2	Stone walling	F20	Natural stone rubble walling
		F21	Natural stone/ashlar walling/dressings
		F22	Cast stone walling/dressings
F3	Masonry accessories	F30	Accessories and sundry items for brick/block/stone walling
		F31	Precast concrete sills/lintels/copings/features

Figure 16.15   CPI workgroups.

Building Team, and on the complexity of the building work involved. The conventional form of production or working drawing shown in Figure 16.12 has been produced by a computer; the hand-drawn equivalent, which still forms the major technique of presentation today, is similar to most of the illustrations in this book. Further information concerning this subject can be obtained from BRE Information Paper 3/88 *Production drawings – arrangement and content*.

The use of a 'systematic method' for production drawings has developed with the use of computers, which are able to supply drawings for particular design and construction problems, e.g. repetitive details relating to columns and beams on structural grids.

16.3 Specification of works and bill of quantities

These, together with the production drawings, combine to form the legal documents describing the totality of the construction process. A *specification* is a written document prepared by members of the Design Team and provides fundamental information which, for various reasons, cannot be incorporated on the production drawings for a proposed project. It is an integral part of the design process because it describes the quality of work which is considered necessary during construction. Information for a specification therefore evolves during the preparation of the production drawings, and is subsequently issued in two main parts:

- *Preliminaries* describe the legal contract through which the construction work is controlled, including insurances; the facilities to be provided by members of the Building Team on site; the general conditions of the site and/or existing building to be converted, including details of access routes, roads and local restrictions; and other details concerning the general running of a project, including information about the quality of work required, nominated subcontractor and suppliers, hours of working, generation of noise, dust, etc.

- *Trade clauses* describe the actual construction methods (materials and techniques) to be adopted and may be done in the following ways:
  - precisely describing the materials to be used and the work to be executed under each trade (see page 143);
  - describing the materials and work required for each separate part or element;
  - giving a *performance specification* (statement of requirement) which accurately details the quality of work expected in terms of performance criteria for each part involved, without describing a method of achieving it, e.g. Concrete in foundations, C15 (15 N/mm^2). BS 5328–1: *Guide to specifying concrete*. (See also section 17.4.5).

The purpose of a specification is to provide information to the potential builder or contractor of a proposed project which, together with the drawings, will enable a reasonably accurate price to be submitted or tendered for the work involved. The detailed information it provides could not be adequately indicated on the production drawings because these would become far too complicated, and extremely difficult to read. In order to avoid unnecessary confusion, it is important that notes on a drawing should not duplicate clauses of specification. Drawing notes should only provide a general comment necessary for an overall understanding, and which leads to the more detailed description in the specification. Only for very simple projects involving few trades will drawing notation suffice for describing the work; the legal arrangements are left to the clauses of the formal building contract.

Writing specifications is often thought to be tedious and time-consuming. Over the years attempts have been made to standardise phraseology, rationalise descriptions

---

F10

Brick/block walling

Laying bricks and blocks of clay, concrete and calcium silicate in courses on a·mortar bed to form walls, chimneys, partitions, plinths, boiler seatings, etc.

Included

Brickwork of clay, concrete and calcium silicate
Blockwork of clay and concrete
Special shape bricks and blocks
Specially faced bricks and blocks
Brick facing slips
Brick DPCs
Firebrick work
Brick bands, copings, sills, arches, etc.
Holes, chases, grooves, mortices, cutting, bonding, pointing other than for engineering services
Forming key for asphalt and other applied finishes
Centring
Mortar (Z21)

Excluded

Concrete cavity fill and concrete fill for hollow blocks or reinforced brickwork/blockwork (In situ concrete, E10)
Bar reinforcement for reinforced brickwork/blockwork (Reinforcement for in situ concrete, E30)
Natural stone rubble walling, F20
Natural stone ashlar/dressings, F21
Cast stone walling/dressings, F22
Damp-proof courses, wall ties, forming cavities, etc. (Accessories and sundry items for brick/block/stone walling, F30)
Proprietary metal, concrete, etc. lintels, sills, copings, etc. (Accessories and sundry items for brick/block/stone walling, F30)
Non-proprietary concrete sills, lintels, etc. (Precast concrete sills/lintels/copings/features, F31)
Holes and chases for services (Holes/chases/covers/supports for services, P31)

Figure 16.16   Contents of a CPI workgroup.

and provide a structure to the clauses in order to speed production while still providing an adequate means of communication. The latest version of the *standard method of measurement* (SMM), written to comply

with CPI (section 16.2.2), provides a system advocating the structuring of specification clauses to follow the construction process, starting with excavations and structure, proceeding through all the services to finishing trades and external works. This logic is complemented by the National Building Specification (NBS) in CPI format, although it is also available in CI/SfB coding to suit organisations using either system for collating technical information and production of drawings (Fig. 16.17).

A *bill of quantities* is also a written document providing fundamental information about a proposed project, but varies from a specification in that it arranges the information into a form more suitable for direct pricing by a builder or contractor. Bills of quantities are used for larger and more complicated contracts, and are prepared within the organisation of a Design Team by a quantity surveyor using the information supplied by production drawings and specification notes. A currently acceptable method of presentation for bills of quantities consists of three main parts:

- *Preliminaries*: as described for specifications.
- *Preambles to trades*: a general specification and description of materials and standards of work.
- *The quantities*: a description of the individual items to be priced and also the numbers, amounts or *quantities* of each required for the project.

The Construction Team's or building contractor's quantity surveyors or estimators generally prepare their own list of quantities, when they are not supplied, from information indicated on the production drawings and in the specification. But as the bills of quantities prepared by the Design Team describe the work to be done, the conditions relating to the work *and* measure the materials and components required (and therefore the type and amount of labour), it is easier for a contractor's quantity surveyors or estimators to prepare a more accurate price for a project. (Furthermore, an individual contractor's estimated price is based upon standardised information, and this facilitates the Design Team when comparing it with the prices of any other contractor who may wish to do the work; see the next section.)

16.4 Tender documents

The production drawings, specification and/or bills of quantities together form the initial *contract documents* which are submitted to the Construction Team for pricing. The principal designer may agree with the client to submit these documents to a builder or contractor who is known to be capable of executing the particular work involved. This will produce a *negotiated tender*,

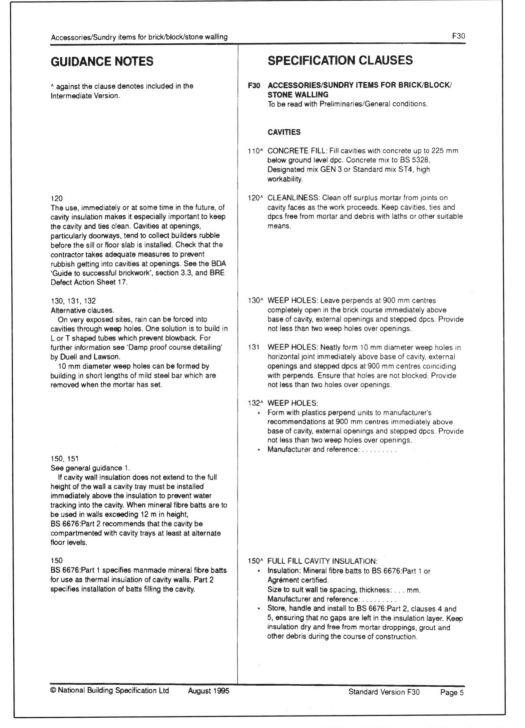

Accessories/Sundry items for brick/block/stone walling   F30

GUIDANCE NOTES

^ against the clause denotes included in the Intermediate Version.

SPECIFICATION CLAUSES

F30   ACCESSORIES/SUNDRY ITEMS FOR BRICK/BLOCK/STONE WALLING
To be read with Preliminaries/General conditions.

CAVITIES

110^   CONCRETE FILL: Fill cavities with concrete up to 225 mm below ground level dpc. Concrete mix to BS 5328, Designated mix GEN 3 or Standard mix ST4, high workability.

120
The use, immediately or at some time in the future, of cavity insulation makes it especially important to keep the cavity and ties clean. Cavities at openings, particularly doorways, tend to collect builders rubble before the sill or floor slab is installed. Check that the contractor takes adequate measures to prevent rubbish getting into cavities at openings. See the BDA 'Guide to successful brickwork', section 3.3, and BRE Defect Action Sheet 17.

120^   CLEANLINESS: Clean off surplus mortar from joints on cavity faces as the work proceeds. Keep cavities, ties and dpcs free from mortar and debris with laths or other suitable means.

130, 131, 132
Alternative clauses.
On very exposed sites, rain can be forced into cavities through weep holes. One solution is to build in L or T shaped tubes which prevent blowback. For further information see 'Damp proof course detailing' by Duell and Lawson.
10 mm diameter weep holes can be formed by building in short lengths of mild steel bar which are removed when the mortar has set.

130^   WEEP HOLES: Leave perpends at 900 mm centres completely open in the brick course immediately above base of cavity, external openings and stepped dpcs. Provide not less than two weep holes over openings.

131   WEEP HOLES: Neatly form 10 mm diameter weep holes in horizontal joint immediately above base of cavity, external openings and stepped dpcs at 900 mm centres coinciding with perpends. Ensure that holes are not blocked. Provide not less than two holes over openings.

132^   WEEP HOLES:
- Form with plastics perpend units to manufacturer's recommendations at 900 mm centres immediately above base of cavity, external openings and stepped dpcs. Provide not less than two weep holes over openings.
- Manufacturer and reference: . . . .

150, 151
See general guidance 1.
If cavity wall insulation does not extend to the full height of the wall a cavity tray must be installed immediately above the insulation to prevent water tracking into the cavity. When mineral fibre batts are to be used in walls exceeding 12 m in height, BS 6676:Part 2 recommends that the cavity be compartmented with cavity trays at least at alternate floor levels.

150
BS 6676:Part 1 specifies manmade mineral fibre batts for use as thermal insulation of cavity walls. Part 2 specifies installation of batts filling the cavity.

150^   FULL FILL CAVITY INSULATION:
- Insulation: Mineral fibre batts to BS 6676:Part 1 or Agrément certified.
  Size to suit wall tie spacing, thickness: . . . mm.
  Manufacturer and reference: . . . .
- Store, handle and install to BS 6676:Part 2, clauses 4 and 5, ensuring that no gaps are left in the insulation layer. Keep insulation dry and free from mortar droppings, grout and other debris during the course of construction.

© National Building Specification Ltd   August 1995   Standard Version F30   Page 5

Figure 16.17   National Building Specification: standardised clauses, based on CI/SfB or SMM format, which can be used to produce a full specification. Product clauses are followed by work standard clauses. The guidance notes in the right-hand column are particularly useful. (Reproduced courtesy of National Building Specification Limited)

where a price for the work is prepared on the basis of discussion around the contents of the contract documents. This method of tendering has the advantage that a considerable amount of time is saved by not involving other builders or contractors, and the selected organisation may well be able to make cost-saving suggestions.

As an alternative, the principal designer may advise the client to obtain *competitive tenders* from a suitable range of Construction Teams (building contractors). Bills of quantities are particularly useful for this purpose as they permit individual Construction Teams/builders to submit tenders on a uniform basis. Equally, the Design Team is able to check and compare the figures of each competitor efficiently before making recommendations to the Client Team. Before this recommendation is given, however, the Design Team must also ascertain the suitability of resources and quality of work available from each competitor. The period of time required for the construction work is also an important consideration. It is not uncommon for the organisation submitting the lowest tender price to be rejected because of doubt in these areas.

Work which is being undertaken by competitive tender is often advertised in professional journals such as *Building*, and with local authority work this is mandatory. On receiving the contract documents, contractors should visit the site of the proposed project in order to base their economic assessments on the actual conditions likely to affect construction work. Each competitor will prepare their estimate for the work and submit a tender, which gives prices for the individual items in the bill of quantities.

Matters to be considered when pricing work include the current cost of materials, machinery and transport, fluctuations in wage rates, and the overhead expenses involved in maintaining an administrative organisation. To this must be added an allowance for waste of materials, depreciation of equipment, labour on non-productive work and a reasonable margin for profit. The profit objective is a corporate decision. It may be determined by the amount of work currently in progress and that anticipated in the near future. Factors such as the possibility of repeat and continuous work from the client/architect will have an influence, as will the anticipation of a continuing aftercare/maintenance contract for the completed building.

A building contractor's profit is not only realised through good organisation of site work, but involves the efficient costing ability of their quantity surveyor or estimator at the time of tendering. Consideration must be given to how much work is to be done within the organisation and how much is to be subcontracted.

Prices must be obtained for materials and labour from subcontractors and suppliers. The management personnel within the organisation must discuss the estimate to establish a figure for *contingency items*, resulting from likely increases in materials and labour, before the contract ends. When the cost of certain items indicated in the contract documents cannot be predicted, such as the degree of excavation, provisional allowances are made on a *daywork basis*. This incorporates the cost of material and equipment required and a *unit rate* for the time taken for the work, which includes the hourly sum paid to the workers, as well as insurance, holiday pay, grants, travel and overtime expenses, and adverse weather payments, etc.

The absence of information caused by poorly prepared, or the non-existence of, specifications or bills of quantities could mean that much of the work indicated on drawings will be priced by using this method, but with additional sums added to cover uncertainties. The lack of information concerning some constructional aspects of a project could result in massive extra costs once the work has proceeded and can therefore lead to serious repercussions.

If the price for the construction work is considered acceptable and the project proceeds, it is quite usual for the quantity surveyor of the builder or contractor to continue with the financial management for the construction work. This entails agreement of interim work, measurement or accurate assessment of completed work and materials on site, and negotiation of payments with the client's quantity surveyor at intervals agreed in the building contract. Details are submitted to the principal designer for confirmation and payment by the client. Similarly, at the completion of the project, the measurement and valuation of all variations leading to the final account must be agreed. These variations will usually be calculated in relation to the itemised prices which have been given in the bills of quantities. These negotiations are considered in section 16.6.

It sometimes happens that the initial contract documents are not fully completed, but it is still desirable to negotiate a price for the intended work or obtain competitive tenders from a builder or contractor. This will enable an early start on site preparation and construction, as well as permitting close cooperation with a builder or contractor during the preparation of production information, of particular advantage when specific construction skills or resources have to be taken into account.

One method of involving specialised construction skills early in the design process is for the principal designer to prepare a 'short-list' of potential contractors, and then interview each to establish their methods of management, construction policy, the possible form and content of their pre-contract input, and their current contractual commitments. The contractor having the best potential suited to the particular project can then be appointed. However, the criteria to be used to calculate the price of the project must be agreed before the contractor's appointment so that a sum can be developed as the design process continues. This method of negotiation requires a high level of trust between the parties involved, specific costing expertise and a thorough knowledge of current building costs.

Alternative methods include a *cost reimbursement scheme*, where the selected contractor is paid for the work executed on the basis of audited accounts to which is added an agreed percentage for overheads and profit. Sometimes an estimated lump-sum price for a project is agreed prior to the commencement of construction work and any discrepancies, savings or extras, are shared equally between client and contractor. This is known as a *target contract*.

When it is necessary to obtain fully competitive tenders for a project without completed tender documents, one of the following techniques may be more appropriate. These methods of tendering can also be used when negotiating a price with a preferred builder or contractor without adopting competitive techniques.

A list of the major items of work is prepared and a selection of suitable builders or contractors are invited to submit a *schedule of rates* for each. Once the work on site starts, the successful competitor will be paid for particular construction work at the agreed rates listed in the schedule, unless there are major changes. Work which the final design requires, and which has not been given a price rate, will be subject to negotiation.

A notional bill of approximate quantities can be prepared by the quantity surveyor, which lists the approximate quantities measured from preliminary design drawings, or the quantities given in a full bill of a similar project. This notional bill will be submitted for tenders and the successful contractor, once appointed, will be paid at the unit price rate agreed for the actual quantity of work measured on site. If there are any items not included in the notional bill, their cost must be negotiated separately.

A two-stage tendering procedure may be adopted. The first stage is based on a notional bill of approximate quantities, as described. Once appointed, the successful contractor helps to develop the design and prepares the second-stage tender based on contract documents which include a full bill of quantities. Rather than appoint

a contractor after the first-stage tender, the client may prefer (or be recommended by the principal designer) to delay appointment until after the second-stage tender, in order to obtain more favourable prices. In this case the pre-contract contractor may be paid a fee for advice and services during the design process stage. Whichever stage of appointment is adopted, construction time can be shortened if the contractor is authorised to prepare the site, or to order long-term delivery of materials during the period between obtaining first and second tenders.

A serial tendering procedure may be appropriate when a number of projects of a similar nature are to be constructed over a considerable period of time. The full bill of quantities for the project first constructed can be progressively updated for subsequent projects.

16.5 Contracts

It is preferable that building operations, like any other activity involving human and material resources, have a formalised set of rules based on a legal contract. This will ensure the rights and responsibilities of each party, and help in the successful production of a building. Various types of agreement can be drawn up, but it is usually advisable to adopt a *standard form of contract* because the individual provisions of non-standard forms will most likely not have been tested in the courts of law, which means the accountability of each party is not always clear. Even though standard forms of contract result from the joint consultations of interested professional bodies and may have been well tested in practice, matters often arise which cause dissatisfaction between various parties, as indicated by the legal sections of most current building trade journals. For this reason, a continuous process of reassessment is necessary; revisions are issued from time to time, particularly in the face of the increasing complexity within the building industry.

16.5.1 Contracts between Client Team and Design Team

Standard forms of contracts are available which stipulate the rights and responsibilities between a client and the principal designer, or specialist designer, or quantity surveyor. They are available from the professional body which represents each consultant, and state the role played by them during the design and construction periods of a project, the fees payable and the liability taken. Although these formal agreements are perhaps the least used (most clients employ consultants through personal recommendation), the increased tendency towards costly litigation to solve disputes makes them almost mandatory in the interest of both parties.

Standard forms of contract between client and building designer have been drawn up by the RIBA. The principal appointing practice documents are:

- Standard Form of Agreement for the Appointment of an Architect.
- Standard Form of Agreement for the Appointment of an Architect for Design and Build.
- Conditions of Engagagement for the Appointment of an Architect.
- Small Works – Conditions of Appointment.
- Form of Appointment as Planning Supervisor – CDM Regulations.

16.5.2 Contracts between Client Team and Construction Team

The most important standard form of contract for the building industry exists between a client requiring construction activities and the builder or contractor who is prepared to execute them. The first modern type of formal contract was drafted during the late nineteenth century. It was prepared by the Royal Institute of British Architects (RIBA) and it became known as the RIBA Standard Form of Building Contract. In 1931 the RIBA was joined by the National Federation of Building Trades Employers (later known as Building Employers Confederation and now Construction Confederation) to review forms of contract, and this combination became known as the Joint Contracts Tribunal (JCT). In 1939 the RIBA Form of Contract was completely revised and replaced by the JCT Standard Form of Building Contract. However, the term 'RIBA Form' is still common parlance, probably because RIBA Publications produce the JCT forms.

Further revisions and editions to the JCT Form followed in 1963 and 1977, by which time the constituent bodies of the JCT had increased to include:

Royal Institute of British Architects
Building Employers Confederation
Royal Institution of Chartered Surveyors
Association of County Councils
Association of Metropolitan Authorities
Association of District Councils
Specialist Engineering Contractors Group
National Specialist Contractors Council
Association of Consulting Engineers
British Property Federation
Scottish Building Contract Committee

In 1980 a substantially new edition was produced. This became known as JCT 80 and it was published in various contract versions, including standard administration forms available for private sector and local authority clients as follows:

- Private edition with quantities
- Private edition without quantities
- Private edition with approximate quantities
- Local authority edition with quantities
- Local authority edition without quantities
- Local authority edition with approximate quantities

Between 1983 and 1998 the JCT 80 received a considerable number of amendments, culminating in the need to accommodate the requirements of the Housing Grants Construction and Regeneration Act – *The Construction Act 1998*. The volume of changes to JCT 80 now made it extremely difficult to use, so it was rationalised into an amalgamated edition published by the JCT Ltd under the title Standard Form of Building Contract 1998, i.e. JCT 98. The constituent bodies at this time were:

Royal Institute of British Architects
Construction Confederation
Royal Institution of Chartered Surveyors
National Specialist Contractors Council
Association of Consulting Engineers
British Property Federation
Scottish Building Contract Committee
Local Government Association

There are other styles of the JCT Standard Form of Building Contract:

- With contractor's design – with or without quantities for use where no architect or supervising officer is engaged by the employer.
- Intermediate form of building contract – for use where works are of a simple nature.
- Fluctuation clauses – for use with all versions of the local authorities and private editions.
- Management contract – prescribes all work to be conducted on a standard contractual base.
- Subcontract documents:
  - invitation to tender;
  - tender and agreement;
  - particular conditions;
  - standard form of employer/nominated subcontractor agreement;
  - standard form of nomination for subcontractor;
  - employer/specialist agreement, a warranty for named subcontractors who provide a design element;

– standard conditions of nominated subcontract;
– articles of agreement between contractor and nominated subcontractor.

- Sectional completion supplement – with quantities, approximate quantities and without quantities.
- Prime cost contract – replaces the former fixed fee contract. Used where the employer agrees to work commencing where only a price indication is given and the employer agrees to pay labour and material costs, plus a 'fixed fee' for the contractor's overheads and profit.
- Standard form of building works of a jobbing character – used by employers that place a number of small contracts (generally < £20,000) with various contractors.
- Measured term contract – used by employers who need regular maintenance and/or minor works to be carried out in a defined area over a specific time by one contractor.
- Minor building works – used where small-scale work is carried out for an agreed lump sum and an architect/supervisor has been appointed to represent the employer. Not suitable where a bill of quantities is used or the employer nominates subcontractors or suppliers.
- Formula rules for use with all standard forms of contract and subcontract. These are adjustments to accommodate for index changes in national wages and prices.
- Building contract for the homeowner or occupier – comprehensive document specifically for home improvements and domestic building work. Designed to enhance the professional relationship between house owners and reputable builders.

A building contract known as *A Form of Building Agreement 1982 (3rd edition 1998)* has also been prepared by the Association of Consultant Architects (ACA) who represent over 400 private sector architectural practices. This contract is shorter than the JCT contract and does not attempt to duplicate those areas of building legislation covered by common law. It should be used in conjunction with its own version of architect's instructions, interim certificates, final certificates and taking-over certificates.

The Chartered Society of Designers have similarly prepared their own form of contract. This is known as the *works agreement* and is intended to be used with its own standardised documentation covering instructions, directions, drawing issues, orders for nominated subcontractors, interim certificates, final certificates and practical completion details. The agreement is a

specialist short form of contract which provides a suitable legalised basis for carrying out interior design projects and makes provision for the nomination of suitable subcontractors.

Other forms of contract include several prepared by the Construction Confederation, e.g. for use where a contractor is to be responsible for both the design and construction of a project and there is no reference to independent consultants, or where a subcontractor has a design input. The Construction Confederation have also prepared contracts for use between a contractor and either nominated subcontractors or non-nominated subcontractors; and for a contractor who prefers to use subcontractors who supply labour only for the work involved.

Civil engineering work is contracted under the New Engineering Contracts (NEC) produced by the Thomas Telford organisation or the Civil Engineering Contractors Association (CECA) and the Institute of Civil Engineers (ICE) Conditions of Contract. The Chartered Institute of Building (CIOB) in partnership with Cameron McKenna, have produced the first Standard Form of Facilities Management Contract for application to all categories of private and public sector works. Government contracts (GC/Works) are widely used for government building and engineering works. Various forms exist within the following categories:

- GC/WKS/1 – Contract for building and civil engineering major works.
- GC/WKS/2 – Contract for building and civil engineering minor works.
- GC/WKS/3 – Contract for mechanical and electrical engineering works.

16.6 Fee accounts and certificates for payment

One important area of concern in most standard forms of contract involves payments for the services connected with the design and construction processes.

16.6.1 Fee accounts

Fees paid by the client to various members of the Design Team are usually governed by the conditions of engagement issued by the professional body which represents a particular design discipline. For example, the RIBA's practice documentation described earlier lists those services normally offered by an architect in accordance with the plan of work, and sets out indicative fees and stages of payment (Table 16.1, Fig. 16.18 and Fig. 16.19). For the large projects, which occupy long periods of time, it is becoming usual for designers' fees to be paid by instalments in accordance with a scheduled programme agreed with the client at the outset of negotiations. This programme can incorporate more frequent intervals between payments than provided by the work patterns given in the plan of work, and thereby enable the designer to receive more or less regular sums during a project's design and construction processes (see section 15.3).

Table 16.1   Classification of building types

Type	Class 1	Class 2	Class 3	Class 4	Class 5
Industrial	Storage sheds	Speculative factories and warehouses Assembly and machine workshops Transport garages	Purpose-build factories and warehouses		
Agricultural	Barns and sheds	Stables	Animal breeding units		
Commercial	Speculative shops Surface car parks	Multi-storey and underground car parks	Supermarkets Banks Purpose-built shops Retail warehouses Garage/showrooms	Department stores Shopping centres Food processing units Breweries Telecommunications and computer buildings Restaurants Public houses	High risk research and production buildings Research and development laboratories Radio, TV and recording studios
Community		Community halls	Community centres Branch libraries Ambulance and fire stations Bus stations Railway stations Airports Police stations Prisons Postal buildings Broadcasting	Civic centres Churches and crematoria Specialist libraries Museums and art galleries Magistrates/County courts	Theatres Opera houses Concert halls Cinemas Crown courts
Residential		Dormitory hostels	Estate housing and flats Barracks Sheltered housing Housing for single people Student housing	Parsonages/manses Apartment blocks Hotels Housing for the handicapped Housing for the frail elderly	Houses and flats for individual clients
Educational			Primary/nursery/first schools	Other schools including middle and secondary	University laboratories
Recreational			Sports halls Squash courts	Swimming pools Leisure complexes	Leisure pools Specialised complexes
Medical social services			Clinics	Health centres General hospital Nursing homes Surgeries	Teaching hospitals Hospital laboratories Dental surgeries

(Reproduced courtesy of RIBA Publications)

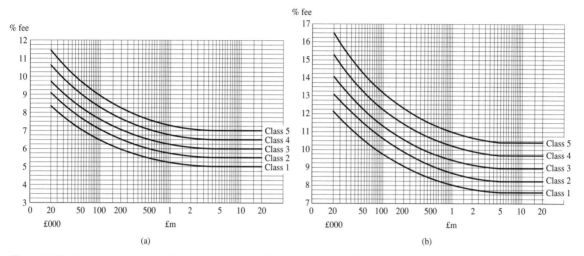

Figure 16.18 Indicative percentage fee scales for large works. (a) new works; (b) works to existing buildings. (Reproduced courtesy of RIBA Publications)

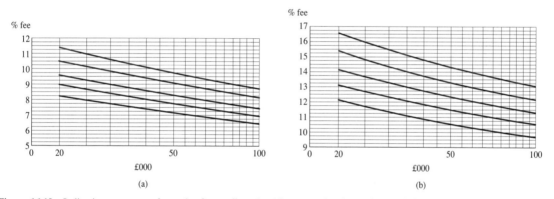

Figure 16.19 Indicative percentage fee scales for small works. (a) new works; (b) works to existing buildings. (Reproduced courtesy of RIBA Publications)

16.6.2 Certificates for payment

Depending on which standard form of building contract governs the construction of a particular project, it is also usual for payments by the client to the builder or contractor to be made at regular intervals. Normally the client's quantity surveyor prepares valuations (see page 133) at the periods stated in the contract (28 days or 14 days for large projects) in accordance with the given terms and conditions. Reports are made to the principal designer for guidance in preparing *interim certificates for payment*. These certificates give valuations of the work executed on site; materials delivered to the site but not used; nominated subcontractors' and suppliers' invoices; and other substantiating evidence of expenditure such as may result from daywork rates (see page 174) to cover designers' instructions not

already included in the contract documents. Thereafter, each interim certificate must contain an itemised statement of the principal sum due to the builder or contractor as well as the sums to be paid to subcontractors for their work. With most forms of contract there is provision for the retention of a small percentage of the total sum given in each certificate. This is normally 5 per cent (or 3 per cent where the estimated contract sum is £500 000 or more), and acts as a safety retainer in the event of subsequent default by the builder or contractor, subcontractor or supplier.

Once the construction work has been completed, a *final certificate of payment* can be issued to cover the agreed costs of the project. At this stage, usually only half the retention sum previously withheld is paid; the rest is kept for a period agreed in the contract during which it is assumed any defects will become apparent which are the sole liability of the builder or contractor, subcontractor or supplier. After this agreed *defects liability period* has elapsed (generally 3 months for small works and 6–12 months for larger works), the residue retention sum can be released in total, or in part, according to the expenditure necessary to correct any faults. The retention sum does not cover faults in design, or any work executed by the builder or contractor in total accordance with the instructions from the Design Team (i.e. as given in the contract documents). Failures arising from the content of the contract documents are subject to separate negotiations between client and designer, although the experience of the builder or contractor may be used in the event of a dispute. Where no quantity surveyor is employed to prepare financial statements, the builder or contractor submits a claim directly to the principal designer, who will then prepare the certificates for payment as already described.

After preparation by the principal designer, certificates of payment are issued initially to the builder or contractor, who will then submit them to the client for payment. This procedure arises because the contract is between client and builder; members of the Design Team are only acting as agents who administer the rules of agreement. Nevertheless, when the principal designer issues the certificate to the builder or contractor, the client and quantity surveyor (if involved) are automatically informed with a copy of the certificate. Under the conditions of the building contract, the client must pay the contractor within a stipulated period, normally 14 days. Appendix II illustrates some of the various standard forms which are available from the RIBA and the CSD to ease the administrative procedures briefly described above.

16.7 Programmes of work

Programmes of work for the design and construction of a building project involve the coordination of many complex human and material resources to ensure economy and efficiency. The more detailed the programme, the less likely will unforeseen circumstances upset either the intentions or the timetabling of a project, thereby reducing the chances of frustration and delays.

The three main areas requiring detailed programmes are as follows:

Design programmes (Fig. 16.20)   These can take the form of a *bar chart* and are often called pre-contract programmes. They are prepared by the Design Team in order to establish how long it will take from the beginning of the design process to start of construction work on site. Apart from providing valuable information to the client – particularly when

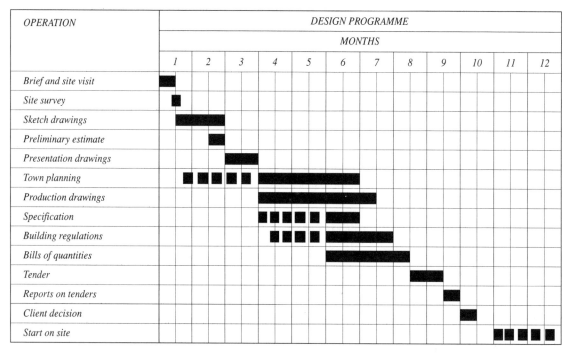

Figure 16.20   Design programme.

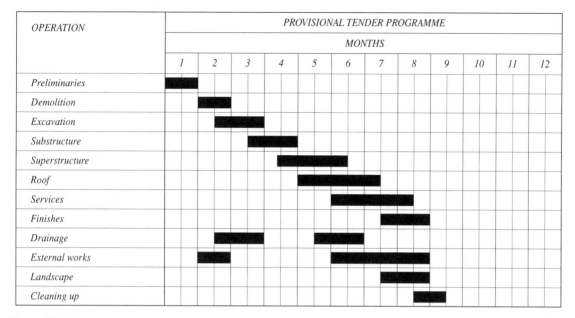

Figure 16.21   Tender programme.

complicated negotiations are required for the release of capital for building – they allow the Design Team to accurately assess the demands on staff and the effects which the new project may have on other work running concurrently.

Tendering programmes (Fig. 16.21)   Sometimes, on behalf of a client, the principal designer will request builders or contractors who are tendering for a project to submit an outline, or tender programme. This can also consist of a bar chart which indicates the approximate periods of time required for each major construction activity and, together with the total period indicated for the proposed project, provides a method of assessment concerning the organisational ability of the builder or contractor. The total period required for construction is very often the critical factor used when making a choice from a selection of tenders which have been priced competitively. Generally speaking, time delays are dearer for the client after work has commenced on site.

Contract programmes (Fig. 16.22)   Once the successful builder or contractor has signed the building contract, he or she must plan the project in great detail. Whereas the Design Team specifies the work to be done, the order in which it is executed is usually the responsibility of the contractor, and even the smallest project demands a close examination of available resources, assessments of time and the influences of subcontracted work.

A *method statement* should be prepared from information given by the contract documents. This takes the form of a schedule breaking down the intended construction work into operations, then labour and plant requirements (or availability) for each (Fig. 16.23). The information can then be converted into a draft programme which should be issued prior to a general meeting between the management members of all the parties concerned with construction activities. Having had time to consider the proposals, this meeting will be able to confirm working methods and agree amendments as necessary. The contractor can then prepare a 'final' contract programme.

If the contract stipulates a monetary penalty, *liquidated and ascertained damages*, for delays beyond the completion date required by the client, a prudent builder or contractor will ensure the work is finished earlier than the programmed finish. This potential time difference allows for the unforeseen eventualities which usually result in longer time periods for certain building operations. A contractor may find that the initial timetable determined by the programme soon becomes unrealistic because of delays caused by bad weather conditions, late delivery of materials, subcontractors not being nominated quickly, or instructions not being given on time by the designers. If such problems develop, the contract programme should be updated to indicate a revised plan of action, and allowance made to cover the builder or contractor against unjust claims for delay.

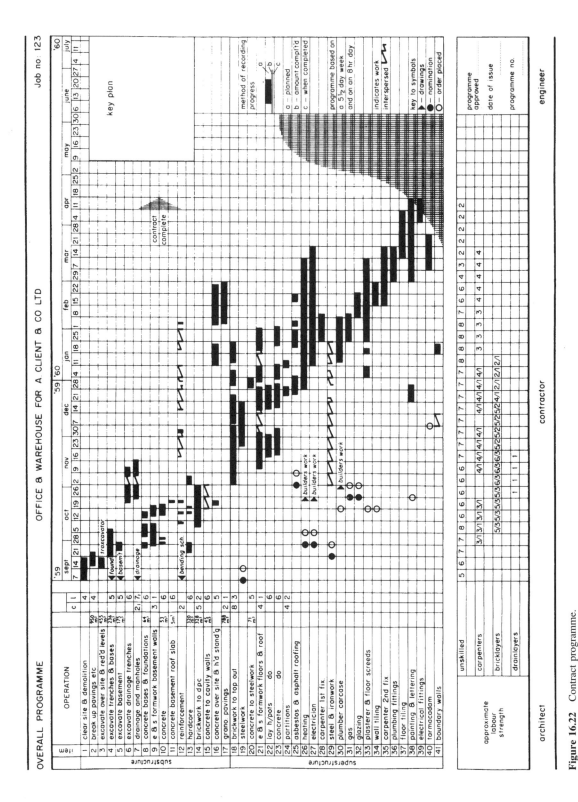

Figure 16.22 Contract programme.

Ref.	Item	Quantity	Method	Output	Labour	Plant	Time

Figure 16.23 Method statement.

Highly sophisticated programmes can be produced to provide a visual representation of the interrelationship between material and component deliveries, subcontractor and trade work. These can provide a more accurate programme and are based on the *critical path method* (CPM) (Fig. 16.24). This method is sometimes considered too complicated for practical use in small to medium-size projects because of the frequent reconstruction which may be necessary as a result of updating. Nevertheless, the critical path method reflects actual site techniques more accurately than bar charts; this is because the tasks performed on site do not divide naturally into 'design element' parts, or the sequential use of trades. Each task has its own breaks determined by the technology of production often peculiar to the particular site concerned, and a programme oriented towards reality should greatly improve the effective use of resources and progress of building work. No doubt the use of computer-produced programmes with the associated ease of updating by continuous 'feeding' of data will make the CPM more desirable for future building work.

16.8 Maintenance manuals

A completed modern building is often complicated to operate and maintain. Therefore, it is to the advantage of all those directly concerned with its eventual use that a *maintenance manual* is produced. This manual should be prepared by members of the Design Team, and perform a similar function to the operation and maintenance book for a car or any other machine.

For building, the maintenance manual can be conveniently divided into three parts:

Statement of the designers' intentions   This includes key drawings, schedules and photographs of the newly completed building; a description of its function; details of specialist designers; contract information; consents, licences and approvals obtained; details of Building Team; floor areas and permitted loadings; fire-fighting methods, precautions taken and means of escape; and a list of those maintenance

contracts which are absolutely necessary as well as those considered desirable.

Statement of the general maintenance required   This includes regular cleaning instructions and methods when applicable; general maintenance log sheets; and a schedule of fittings and components which require regular attention and/or replacement.

Statement of the mechanical services provided   This includes a maintenance guide giving timing and frequency of required operations; fittings requiring periodic replacement; and drawings which are required for an understanding of facilities provided.

16.9 Feedback data

It is important that those concerned with the design and construction processes of a building also prepare data which will prove invaluable in the event that they will be concerned with a similar building in the future. This information should include a complete set of contract documents; an appraisal of the contract procedures between designer, client, builder or contractor, and subcontractors; a detailed cost analysis; and an appraisal of design and technical criteria which incorporates details of operational procedures associated with the subsequent use of the building.

This information may also prove invaluable in the event of legal claims arising as a result of disputes concerning responsibility for subsequent faults in the performance of a building. Claims of this nature could arise many years after the completion of a building when, perhaps, most of the original Design and Construction Teams have dispersed.

Organisations requiring to retain this information for all or most of the projects for which they have been responsible will face a serious problem of storage, and even of retrieval when required for subsequent research. Large projects may have several thousands of large drawings and schedules, as well as other documentation. Storage problems can be overcome by one of the many microcopying techniques ranging from simple reduction to *35 mm slide transparencies*, to *35 mm or*

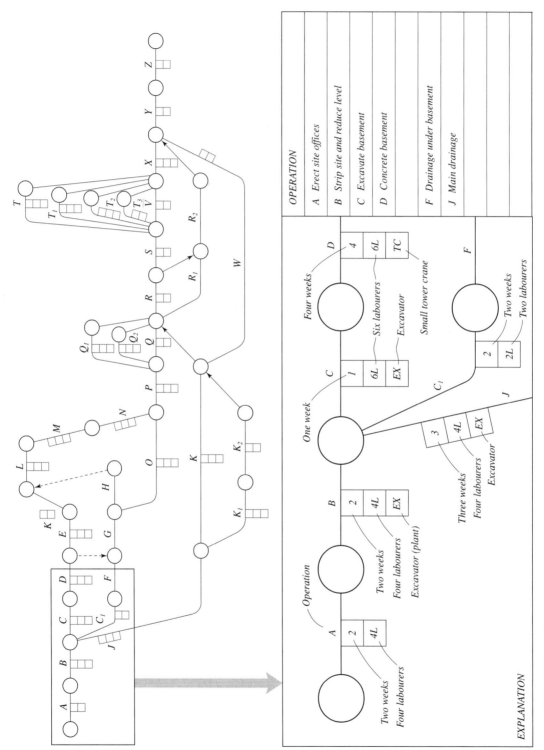

Figure 16.24 Critical path programme.

16 mm microfilm or *microfiche processes* and *storage on computer disks or CD-ROMs*. The microfiche process involves the photographic reduction of documents on transparent cards usually 150 mm × 100 mm in size. Each card carries up to 60 micro-images set in rows which can be displayed on an enlarging screen for easy reading.

The use of micro-reproduction techniques is inseparable from the systematic information storage method mentioned in section 16.1; it can provide a total package of information on a particular project in a very small storage area. The CI/SfB system could be used for classification and ease of retrieval (see Table 0 in Fig. 16.1).

Further specific reading

Mitchell's Building Series

Internal Components	Chapter 1 Component design
External Components	Section 2.8 Prefabricated component design: accuracy and tolerances
Materials Technology	Section 1.3 Specification of building materials
Structure and Fabric Part 1	Chapter 2 The production of buildings
Structure and Fabric Part 2	Chapter 1 Contract planning and site organization Chapter 2 Contractors' mechanical plant

Building Research Establishment

BRE Digest 53   *Project network analysis*
BRE Digest 166   *European product-approval procedures* Part 1
BRE Digest 167   *European product-approval procedures* Part 2
BRE Digest 172   *Production drawings*
BRE Digest 271   *Project information for statutory authorities*
BRE Digest 289   *Building management systems*
BRE Digest 335   *Electrical interference in buildings*
BRE Digest 397   *Standardisation in support of European legislation*
BRE Digest 424   *Installing BMS to meet electromagnetic compatibility requirements*
BRE Digest 450   *Better building/Integrating the supply chain*

Part C

An analysis of a building in terms
of typical construction methods

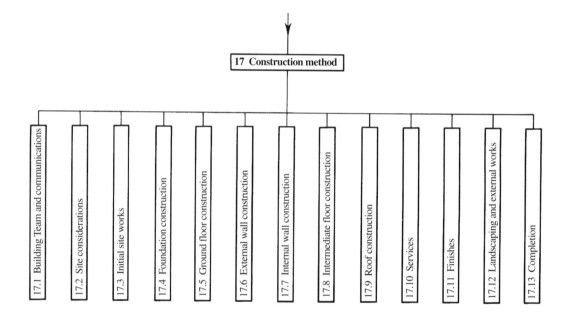

17 Construction method

17.1 Building Team and communications

17.2 Site considerations

17.3 Initial site works

17.4 Foundation construction

17.5 Ground floor construction

17.6 External wall construction

17.7 Internal wall construction

17.8 Intermediate floor construction

17.9 Roof construction

17.10 Services

17.11 Finishes

17.12 Landscaping and external works

17.13 Completion

17 Construction method

CI/SfB (A5)

The following description will give an outline of some of the various considerations, activities and processes which result in the erection of a simple building. A two-storey house has been chosen primarily because it is likely to be the most familiar form of building to those with limited experience of the construction industry. And a house is now the most common form of building in the United Kingdom which continues to be erected using conventional techniques derived from traditional craft skills. This creates a particular kind of 'domestic' aesthetic.

This is not to say that house-building techniques have not changed over the years. Indeed, the standards of performance requirements are continually being upgraded to provide increasingly sophisticated enclosures which reflect ideas of comfort, stability and aesthetic appeal, as indicated in Part A. It should also be borne in mind that the techniques to be described represent only one method of erection; there are many others. Furthermore, this chapter is not an exhaustive study on a particular construction method and should be regarded as a brief basic indication only. Further information can be obtained from the other volumes of the Mitchell's Building Series.

In order to help the designer to assess the effects of design decisions it is often useful for a sequence diagram to be prepared which shows the relationship of the activities involved. Figure 17.1 is a typical example where individual boxes define *activities* and circles contain the sequential *operation number*. Operations are usually sequential, but with good site management, organisation and coordination of trades, some activity overlap may be possible. This ensures continuity of work and an efficient construction process. A timescale can be applied to the diagram so that progress of construction is carefully monitored (see also section 16.7).

17.1 Building Team and communications
CI/SfB (A1g)

The Building Team and method of communication adopted for the creation of a two-storey house could in certain areas be less extensive than those described in Part B. Also, the number and type of personnel involved, and the resulting degree of communication, will obviously vary according to the problems to be solved. A large house with unusual, complicated or elaborate structural and/or environmental requirements may involve a correspondingly elaborate Building Team. The Client Team for the two-storey house to be considered here will most likely be a single individual or a couple who employ the services of a single principal designer (e.g. architect) from the Design Team. He or she may recommend the services of a quantity surveyor, although his or her skills should be sufficient to control expenditure accurately for such a small project. The designer's skills should also encompass sufficient expert knowledge of the drainage, plumbing, heating and electrical installation required, although he or she can recommend the services of subcontractors who will give recommendations and submit competitive estimates for these specialised areas.

The building work is best carried out by a general builder whose standard of work is familiar to the designer. Alternatively, contract documents in the form of working drawings, specification of works (and bills of quantities, if a quantity surveyor is employed), and the contract can be prepared by the designer in order that competitive tenders can be submitted by a number of general builders situated in the locality of the site for the proposed building.

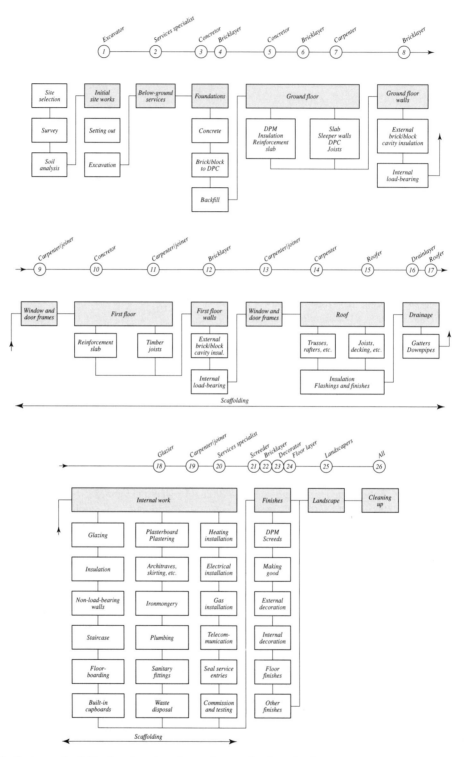

Figure 17.1 · Sequence of activities for a two-storey house of conventional construction.

It is important that the designer maintains overall control over the standard of work and materials used for the building work, especially when the builder employs a large number of subcontractors rather than his or her own workers. A form of contract which would be applicable for this project would be the JCT *Standard Form of Building Contract: Private edition without quantities* (or *private edition with quantities*).

The designer will be required to seek approval for the proposed scheme from the local authority's planning department, highways department, engineer's office and building control section. The house construction to be described will be governed by the Building Regulations 1991, 2000 and subsequent amendments.

17.2 Site considerations
CI/SfB (A3s)

17.2.1 Selection

The characteristics of a site (Fig. 17.2) are likely to have an important influence on the way a building is designed and it is therefore important that the designer is employed as soon as the desire for a house has been established by the client. As many sites as possible should then be inspected, and the relative merits compared before final selection. A badly chosen site can be one which offers few amenities, or few aesthetic advantages, and also one which involves the client in considerable additional building expenses arising from physical properties requiring special constructional solutions. It is also important that the local planning authority is consulted regarding any problems which may arise in connection with suitable zoning.

The early involvement of the designer is particularly important today, since those sites which provide highly suitable facilities are becoming scarce and land is increasingly having to be used which was previously avoided. For example, it may be necessary to select more remote sites than normal and, as outlined in Chapter 6, it will be essential to investigate the full range of climatic conditions as these will affect the final form

and appearance of the building fabric. Special consideration will influence the final form of building to be erected on a site which has poor subsoil or suffers from waterlogging, or a site to be reused following the demolition of existing structures where its physical properties may well have been altered during the life of the previous building.

17.2.2 Survey

An Ordinance Survey map (see Fig. 16.3) may give an initial guide to site features, but a detailed and accurate survey must be made to obtain more specific data. This should be done by a member of the Design Team; it will ascertain the precise shape and size of land involved, and the shape and position of any obstacles such as ponds, trees or buildings that exist within and around the boundaries of the site. It is also important to establish the exact path of the sun across the site, the prevailing direction of wind and rain, etc., and the way the site slopes.

Details will also be required about the privatised utilities (gas, electricity, telecommunications, water, drainage, etc.) and other services (cable television, refuse collection, maintenance contractors, etc.) as well as the transport routes. Greater details of these can probably be obtained from the local authority, which may also be able to supply information about desirable access points to the site, future road-widening schemes in the locality, preservation orders affecting buildings and/or views, rights of way and light, etc. Some planning authorities issue guidelines about the appearance of buildings to be erected in their locality. It may also be useful to make contact with the building control officer or approved inspector at this stage and discuss any general factors which may influence the construction methods applicable to a building to be erected on that particular site.

17.2.3 Soil investigation

Apart from the above-ground conditions, a survey must also include data about the *below-ground* conditions

Figure 17.2 Site selection.

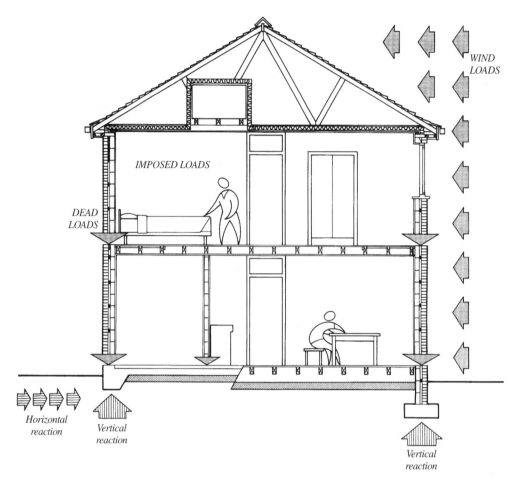

Figure 17.3   Combined building loads.

of the site. These include characteristics of the soil because this information influences the selection and design of an appropriate structural form for the proposed building.

The subsoil must safely support the combined building loads indicated in Figure 17.3 (explained in Chapter 5) and also ensure that unreasonable movements of the building do not occur. If the supporting soil is sufficiently resistant and its characteristics under load are likely to remain satisfactory, the problems of support and movement will be easily resolved. However, few soils other than rock can resist these concentrated loads and it is usually necessary to collect the resolved loads at their lowest point and transfer them to adequate bearing soil known to be available on a *particular* site (Fig. 17.4).

Building loads can be fairly easily calculated from known data, but the strength or resistance of the soil to them requires expert understanding and very careful analysis. Underneath the top layer of soil, which con-

sists of decaying vegetable matter and has little strength, lie bearing soils varying in strength according to type and consistency. Consideration must be given to the soil, including the particle sizes from which it is formed; moisture and chemical content; behaviour under loads; the presence of inherent or induced peculiarities; and the likely effects of variations in its normal undisturbed condition.

There are several methods employed to assess soil conditions; they can be used singly or in combination, depending upon the precise circumstances. Investigation can be done by reference to geological and topographical references or empirical knowledge from local residents, building inspectors, etc., and/or by physical exploration of the site. Physical exploration may be accomplished by trial holes dug into the ground or by taking test cores of the soil. Both techniques can provide a reliable source of information which costs as little as 1–2 per cent of the total building expenditure. It is important that *trial holes* are taken as near as possible to, but not on the actual line of, a proposed foundation in order to avoid the subsequent excessive consolidation of the relatively loosely backfilled soil by the constructed foundation (Fig. 17.5). Trial holes need to be formed to sufficient depth to expose the likely soil which will support the foundation. Most of this lies in the volume of soil beneath the foundation known as the *bulb of pressure* (Fig. 17.6). For lightly

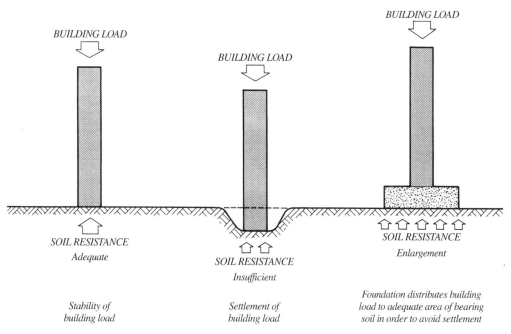

Figure 17.4   Method of transferring combined building loads to supporting soil.

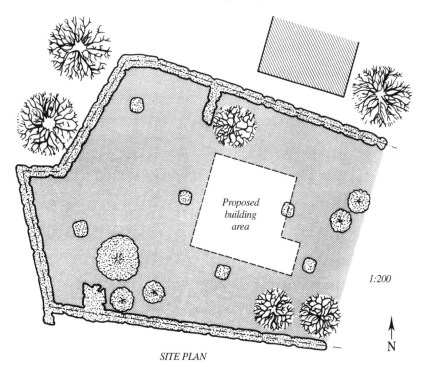

Figure 17.5   Suitable position for digging trial holes for soil investigation.

loaded buildings with simple foundations, this generally lies within 2 m below ground level. When necessary, investigations can be carried out beyond this depth by taking *sample cores* of soils, 150–200 mm in diameter, using special equipment, see BS 1377: *Methods of test for soils for civil engineering purposes*.

The general distribution of soil types in the United Kingdom is indicated in Figure 17.7; the soils include

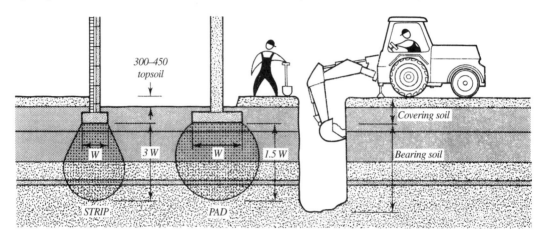

Figure 17.6   Influence of 'bulb pressure' on supporting soils.

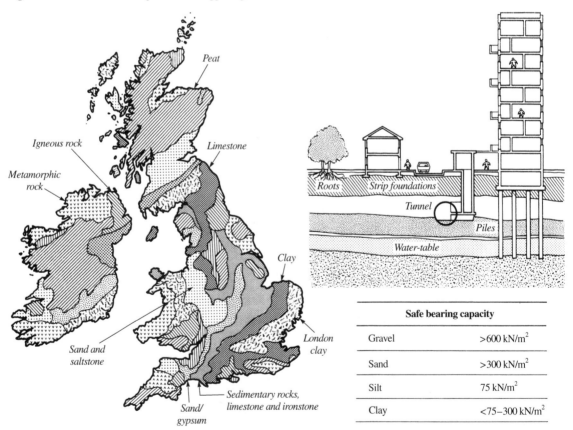

Safe bearing capacity	
Gravel	>600 kN/m²
Sand	>300 kN/m²
Silt	75 kN/m²
Clay	<75–300 kN/m²

Figure 17.7   Much simplified distribution of various types of supporting soils in the United Kingdom (see geological survey maps for greater detail). The quality of bearing soils will also vary with depth and various uses. Metamorphic rock – marble, limestone, sandstone and slate. Igneous rock – granite, pumice, basalt, and volcanic ash (tuff). Sedimentary rock – sandstone, mudstone, clay, chalk and gypsum.

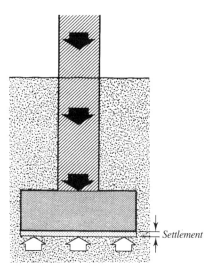

Figure 17.8 Initial consolidation of the supporting soil occurs during construction. Consolidation produces settlement and its effects are anticipated in the design.

peat, clay, silt, sand and gravel. Corresponding safe bearing pressures are also given.

The *steady* increase in loads resulting from the construction of a building causes initial consolidation of the soil beneath the foundation. The amount of this movement varies with the soil conditions and, in a correctly designed and formed foundation, causes little disturbance in the building fabric above. Unreasonable movements of foundations (Figs 17.8 and 17.9) can be caused by the use of poor materials (inadequate concrete mixes) and formation techniques, or commonly they are caused by earlier soil movements. These movements derive from either excessive consolidation of the soil particles under load or by other factors independent of building loads:

Swelling and shrinkage Clay soils are likely to swell or shrink with variations in moisture content caused by seasonal fluctuations in rainfall and the water-table, drought conditions, and the proximity of tree roots, boilers, kilns and furnaces.

Frost heave Ice lenses may form between the soil particles of chalks and silts, causing swelling.

Swallow- or sink-holes Chalk and limestones may contain pockets of weaker soil.

Slopes and landslips Clay soils may exhibit movement on sloping ground with a gradient steeper than 1:10.

Mining subsidence May occur when temporary supports in old or uncharted mines decay, causing the void to collapse.

Shock and vibration Soils near railways, motorways or active industrial plant may experience shock and vibration.

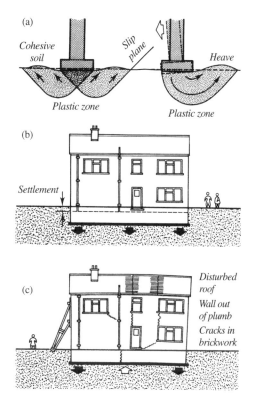

Figure 17.9 Foundations must be designed to prevent damage caused by unreasonable movements: (a) plastic failure of soil; (b) excessive settlement; (c) differential settlement.

17.3 Initial site works
CI/SfB (11)

17.3.1 Selection of plant

One of the first decisions the selected builder must make is whether it will be more economical to employ mechanical equipment (plant) or hand-excavation techniques for site work (Fig. 17.10). Plant should normally be used where it can be kept fully occupied, the ground is not too boggy, the excavation is not too 'fussy', there is room to manoeuvre, and the transport costs of plant (mechanical equipment) to and from the site can be paid from the savings over the cost of hand excavation. Regardless of whether the builder owns certain items of plant or hires it for certain periods, the choice whether it should be used or not is a matter for expert consideration. It is possible, however, for a designer to take the use of plant into consideration (perhaps in consultation with the builder) when designing a building so that maximum time saving can be obtained, which could cut costs.

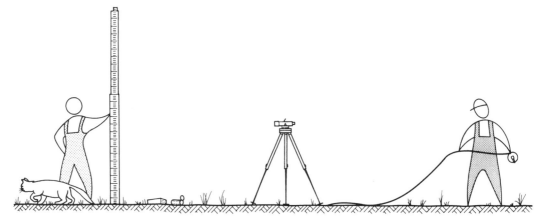

Figure 17.10   Initial site work: above-ground survey.

17.3.2 Setting out

Having cleared the site of debris, unwanted shrubs, etc., by using information from the designer's site plan, the builder can erect huts for workers (sanitary facilities, canteen, site office, etc.) and for storage of materials and components. These should be located in a position which will be unaffected by access points, circulation about the site and the subsequent building processes. The builder can then begin to set out the areas of the site which are required to be excavated. This is done by using a steel measuring-tape to establish the exact position and perimeter dimension of the building or other excavated area, then by providing *profiles* to give guidance during substructure operations. Profiles consist of timber pegs driven into the ground with timber crossbars fixed to them on which points are marked by nails to indicate lines of foundation and the walls above (Fig. 17.11).

The crossbar of the profile is fixed at a known height or datum which is established by a measuring-staff and an optical instrument known as a *level*. By periodically stretching a thin string or *line* between the tops of the crossbars during excavation and using a *boning rod* of known length, the amount of soil to be removed can be established (Fig. 17.12). It is always necessary to set up profiles away from the actual region

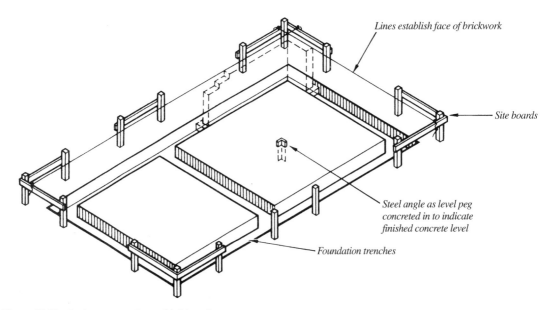

Lines establish face of brickwork

Site boards

Steel angle as level peg concreted in to indicate finished concrete level

Foundation trenches

Figure 17.11   Setting out trenches and brickwork.

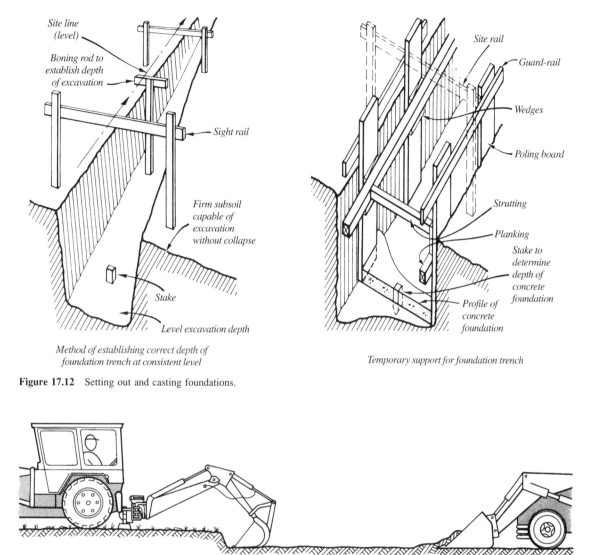

Site line (level)

Boning rod to establish depth of excavation

Sight rail

Firm subsoil capable of excavation without collapse

Stake

Level excavation depth

Method of establishing correct depth of foundation trench at consistent level

Site rail

Guard-rail

Wedges

Poling board

Strutting

Planking

Stake to determine depth of concrete foundation

Profile of concrete foundation

Temporary support for foundation trench

Figure 17.12   Setting out and casting foundations.

Figure 17.13   Clearance of topsoil over area of building before the foundations are excavated.

of excavation to allow for working room. The setting out of trenches for drainage can be accomplished in a similar manner, but as the bottom of these excavations should be laid to a fall, to ensure the pipes they support are self-cleansing, measurements down from the temporary line stretched between profiles need to be taken at frequent intervals.

17.3.3 Excavation

Excavation will consist of removing the topsoil from the area of the proposed building, then forming the trenches for the foundations (Fig. 17.13).

Topsoil exists on most sites to a depth of about 300 mm and is of considerable value for landscaping. But topsoil is highly water retentive, propagates plant life and is highly compressible, so it must be carefully removed from areas where it is not required and stored clear of building operations, ready to be reused. If the quantity involved is not all wanted for its site of origin, arrangements will have to be made for it to be removed and perhaps sold.

The type of excavation needed for the foundations depends on their category, chosen according to subsoil type and characteristics. Further details about the influence of these factors are given in section 17.4.3.

17.4 Foundation construction
CI/SfB (16)

17.4.1 Function

The resolved dead, imposed and wind loads may be transferred through the building by continuous walls, or through isolated piers, by columns or piers, or by combinations of these techniques. Because each of these structural systems results in foundation types which transfer the loads to the supporting soil in different manners, the selection of a foundation category (Fig. 17.14) should reflect the optimum relationship between building structure and the characteristics of the soil used as support. In other words, the aesthetic and technical factors resolving into a particular building form above ground are interdependent with the technical requirements of the foundation design below ground.

17.4.2 Design

The following steps are necessary to arrive at a correctly designed foundation which takes into account the relationship between soil conditions and building loads.

1. Assess soil conditions by suitable investigation (see section 17.2.3).
2. From investigations decide the permissible soil loading.
3. Determine the combined building loads (dead, imposed and wind) and allow for the self-weight of the foundation. Typical loads for a single- and two-storey house are indicated in Figure 17.15.

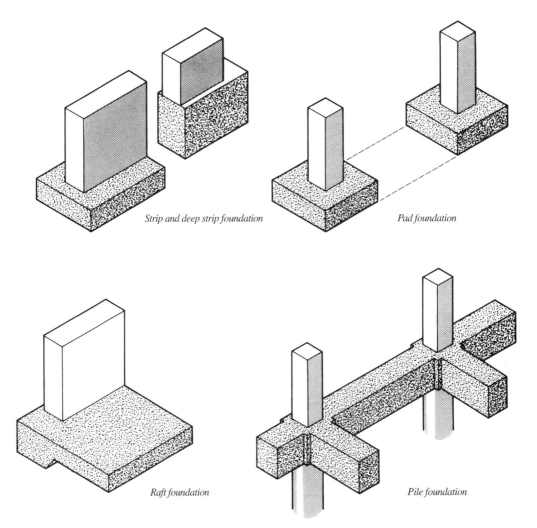

Strip and deep strip foundation

Pad foundation

Raft foundation

Pile foundation

Figure 17.14   Different types of foundation are used according to subsoil conditions and the form of building to be supported.

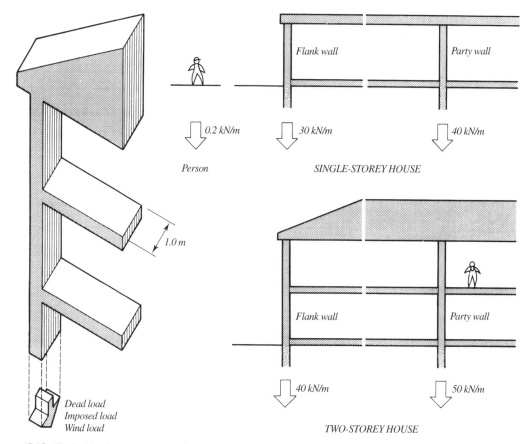

Figure 17.15 Typical loads on a two-storey house.

4. Establish the form and size of the foundation to suit loadings and correct for the self-weight allowances.
5. Check soil resistances against permitted stresses and estimate permissible movements.
6. Allow for necessary movement joints for differentially loaded parts of the building or other structural effects on the foundation.

17.4.3 Choice

For a two-storey house using continuous load-bearing masonry walls (brick, block or stone), a form of strip foundation will probably be most appropriate, and can consist of a horizontal strip *or* a vertical strip of mass concrete located beneath ground level. The choice between these two forms can be made during discussion between designer and builder as their structural significance is closely related to site conditions, plan shape of the building and the availability of appropriate resources. In both cases the depth of the foundation will depend upon the depth of adequate bearing soil, and the avoidance of certain phenomena which detrimentally affect the upper layers of the bearing soil: seasonal variations in moisture content, tree roots, etc. Generally, a depth of 1.0–1.2 m is adequate when a suitable subsoil is available. The width of excavation and hence the proposed strip foundation can be determined by calculation. Figure 17.16 shows a simple example, given analysis of the subsoil to indicate a safe bearing capacity of 80 kN/m², with a building load of 50 kN/m. Alternatively, simple field tests on the variety of soils represented in Table 17.1 can be used for guidance. For a wall of a building weighing 50 kN/m to be supported on firm clay the recommended width is given as 600 mm. Nevertheless, in order to avoid eccentric stresses in the foundation, consideration must be given to allowing enough tolerance for the bricklayer to build a perfectly straight wall exactly on the centreline of the concrete mass of the foundation formed within a trench which has been relatively crudely excavated by machine (or hand).

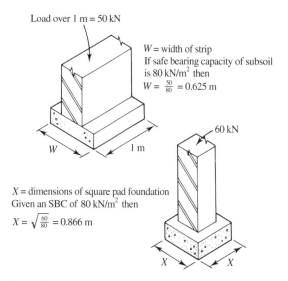

Load over 1 m = 50 kN

W = width of strip
If safe bearing capacity of subsoil
is 80 kN/m² then
$W = \frac{50}{80} = 0.625$ m

W 1 m

60 kN

X = dimensions of square pad foundation
Given an SBC of 80 kN/m² then

$X = \sqrt{\frac{60}{80}} = 0.866$ m

X X

Figure 17.16   Calculations for foundation design.

The *vertical strip foundation* is generally used in firm soils where costs can be reduced because they are self-supporting and do not require the trench support which the more loose soil types need. When safe soil bearing pressures and wall dimensions permit, the width of these foundations can be as little as 300 mm or the width of the smallest *mechanical* digging tool used for the excavation of the trench in which they are formed. (This is possible because the high vertical 'thickness' or sides of the deep strip foundation are provided with lateral restraint by the surrounding soil, which offsets the likelihood of the foundation overturning.) However, to avoid the possibility of the development of the eccentric stresses mentioned earlier, a tolerance dimension of 75 mm minimum should be allowed either side from the face of the wall above to the edge of the foundation. With the *horizontal strip foundation* form, allowances must be made for a bricklayer working *within* the trench (Fig. 17.17). A minimum of

Table 17.1   Strip foundation width relative to subsoil quality

Type of subsoil		Condition of subsoil	Field test applicable	Minimum width (mm) for total load of load-bearing walling of not more than (kN/m)					
				20	30	40	50	60	70
I	Rock	Not inferior to sandstone, limestone or firm chalk	Requires at least a pneumatic or other mechanically operated pick for excavation	In each case equal to the width of wall					
II	Gravel Sand	Compact Compact	Requires pick for excavation. Wooden peg 50 mm square in cross-section hard to drive beyond 150 mm	250	300	400	500	600	650
III	Clay Sandy clay	Stiff Stiff	Cannot be moulded with the fingers and requires a pick or pneumatic or other mechanically operated spade for its removal	250	300	400	500	600	650
IV	Clay Sandy clay	Firm Firm	Can be moulded by substantial pressure with the fingers and can be excavated with graft or spade	300	350	450	600	750	850
V	Sand Silty sand Clayey sand	Loose Loose Loose	Can be excavated with a spade. Wooden peg 50 mm square in cross-section can be easily driven	400	600	Foundations do not satisfy the regulations if the total load exceeds 30 kN/m			
VI	Silt Clay Sandy clay Silty clay	Soft Soft Soft Soft	Fairly easily moulded in the fingers and readily excavated	450	650	In relation to types VI and VII, foundations do not satisfy the regulations if the total load exceeds 30 kN/m			
VII	Silt Clay Sandy clay Silty clay	Very soft Very soft Very soft Very soft	Natural sample in winter conditions exudes between fingers when squeezed in fist	600	850				

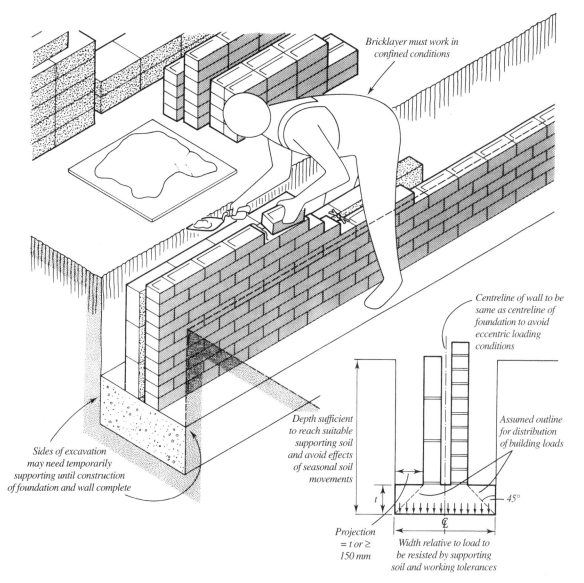

Bricklayer must work in
confined conditions

Centreline of wall to be
same as centreline of
foundation to avoid
eccentric loading
conditions

Depth sufficient
to reach suitable
supporting soil
and avoid effects
of seasonal soil
movements

Assumed outline
for distribution
of building loads

Sides of excavation
may need temporarily
supporting until construction
of foundation and wall complete

45°

t

Projection
= t or ≥
150 mm

℄

Width relative to load to
be resisted by supporting
soil and working tolerances

Figure 17.17 Horizontal strip foundation.

150 mm either side of the proposed wall, together with the structural consideration, may well determine the final width of foundation.

Generally, the chosen width of a vertical strip foundation may relate very closely to the width of the horizontal strip foundation, subject to similar loading conditions. However, the advantage of the deep strip foundation over the horizontal strip foundation is that it finishes just below ground level and does not limit the working space of the bricklayer (Fig. 17.18). Nevertheless, its use should be approached with caution as it is difficult to inspect the bottom of a narrow trench to

ensure consistency of soil before the concrete is placed, and also excessive soil movements can cause the narrow but high foundations to overturn unless adequate precautions are taken.

The thickness of the *horizontal strip* should never be less than 150 mm or the amount of projection from the face of the wall supported, whichever is the greater. This rule takes into account the general working tolerances applicable to the formation of an adequate amount of structural concrete for practical purposes. Unreinforced or mass concrete strip foundations can fail under the compressive loads from the building

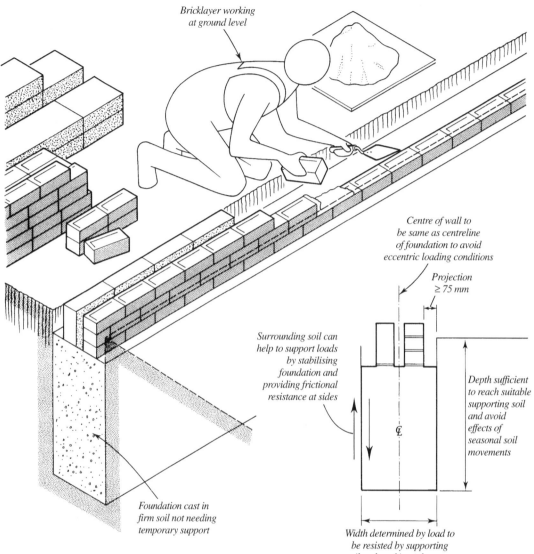

Bricklayer working at ground level

Centre of wall to be same as centreline of foundation to avoid eccentric loading conditions

Projection ≥ 75 mm

Surrounding soil can help to support loads by stabilising foundation and providing frictional resistance at sides

Depth sufficient to reach suitable supporting soil and avoid effects of seasonal soil movements

Foundation cast in firm soil not needing temporary support

Width determined by load to be resisted by supporting soil and working tolerances

Figure 17.18   Vertical strip foundation.

above as a result of excessive bending or from excessive shear stresses developing within the cross-section of the foundation. This latter form of failure generally occurs along the plane of maximum tensile shear which lies at an angle of about 45° to the horizontal. (In theory the concrete lying outside this plane does little in distributing loads to the supporting soil, but is nevertheless necessary for practical purposes in forming the foundation.) Therefore, if horizontal strip foundations are required to carry heavy loads or if the soil is weak, the corresponding increase in thickness of foundation (to maintain the 45° angle of spread; see Fig. 17.17)

could result in a large amount of concrete being used, which in turn increases the dead load on the soil. In such cases the foundations can be reduced in thickness and weight by using steel reinforcing bars to take up the tensile and bending stresses. These bars must be protected from corrosion by a covering of at least 40 mm concrete.

The bottoms of foundation trenches should be protected from the weather to avoid temporary shrinkage or swelling by leaving the last 100 mm or so unexcavated until immediately before pouring concrete. Alternatively, a blinding layer of concrete, about 50 mm thick, can be placed at the base of the completed excavation. The accurate thickness of concrete can be established prior to pouring by inserting timber stakes at suitable intervals in the base of the excavation. The stakes project from the excavation by the required amount and the concrete is levelled accordingly. Immediately before the pouring of concrete takes place, i.e. after excavation has finished, the bottom of the trench must be inspected and approved by a local authority building control officer, during his or her first formal site inspection.

17.4.4 Materials

The materials employed to make foundations must be strong and durable, particularly as it is not possible or practical to carry out maintenance on them once they have been installed in the ground. For this reason, and with few exceptions, they are formed from *good quality dense concrete*, consisting of cement, fine aggregate or well-graded coarse sand, and a coarse aggregate of natural gravel or crushed stone, all graded according to the designer's specification for the strength required.

Ordinary *Portland cement* is generally used in the concrete mixes, but where the ground conditions surrounding the foundation are likely to contain reactive chemicals, special forms of cement should be employed as an alternative. For example, sulphate solutions which occur naturally in groundwaters associated with clay soils, or the acidic solutions with peaty soils, attack and erode ordinary Portland cement. To avoid deterioration of the foundation, *sulphate-resisting cement* could be used. High alumina cement, or supersulphated cement (currently not available in the UK) are alternatives, but neither can be recommended since it is extremely difficult to achieve the essential quality of concrete under practical conditions. To avoid using any of these special cements (which are much more expensive than ordinary Portland cement) the foundation can be protected with an impermeable barrier wrapping such as rubber/bitumen-coated polythene sheet.

In certain cases the concrete may be reinforced with steel bars. Depending on loading conditions, these may be ordinary mild steel or high tensile steel bars, and when used they should be provided with a protective concrete cover of not less than 75 mm. In order to protect both steel and concrete from the soil during construction, a blinding layer or excavation coverage of concrete can be provided. This could be 50–100 mm thick, depending on site conditions, and if adopted, the protective concrete cover to reinforcing bars may be reduced to 40 mm.

17.4.5 Concrete strength and specification

Concrete is a composite of cement, fine aggregate (sand), course aggregate (stones) and water. It may also contain steel reinforcing rods to provide resistance to tensile forces. The quality of individual components can be varied considerably to suit a range of applications. BS 5328: *Methods for specifying concrete* provides details of acceptable cements, aggregates, mix applications and additives. Types of mix are defined as:

- Designed
- Prescribed
- Standard
- Designated

They have a numerical classification, i.e. 7.5, 10, 15, 20, 25, 30, 35, 40, 50, 55 and 60. Each represents the compressive strength of concrete in N/mm^2, 28 days after mixing and placing. *Designed* and *Prescribed* mixes can cover the full range, whilst *Standard* mixes are limited from 7.5 to 25 N/mm^2 and *Designated* mixes from 7.5 to 50 N/mm^2. The mix grade can be prefixed C, F or IT.

  C = characteristic compressive strength
  F = flexural strength
  IT = indirect tensile strength

A P suffix indicates a *Prescribed* mix, e.g. C30P is a *Prescribed* mix of 30 N/mm^2 characteristic compressive strength at 28 days.

Designed mix Concrete specified by its required performance. Variations are permitted for material content and water/cement ratios. Sample testing to ensure strength is essential.

Prescribed mix Concrete specified by constituent materials as mix proportions, i.e. a recipe of, say, 1:2:4 [20 mm] – 0.5 (1 part cement, 2 parts fine aggregate, 4 parts course aggregate [20 mm maximum size] and a water/cement ratio of 0.5). Only used where there is established evidence of its reliability.

Standard mix Suitable for general use in limited strength applications such as housing construction, where scale of operations does not justify full mix design procedures.

Designated mix Suitable for specific situations where the purchaser must provide the producer with full details of use and applications. The British Standard provides categories of mix to suit foundations, floors, paving, reinforced and prestressed concrete and general applications.

17.4.6 Construction

The sequence of the formation for the *vertical strip foundation* is indicated in Figure 17.19. As strip foundations are used in conjunction with continuous walls of small bonded units (bricks, blocks or stones), their top surfaces are laid to a true level surface. In this way the erection process of the walling unit will be made quicker and the wall will be more stable as the necessity of providing mortar packing of varying depth beneath each first course of wall unit is avoided. When a building is to be erected on a sloping site and the required bearing strata follows the ground slope, excavation costs and the amount of walling below ground can be reduced by forming the foundation in a series of steps, as shown in Figure 17.20.

Services entering a building below ground (water, electricity, gas, telephone, etc.) and those leaving a building (drainage, refuse systems, etc.) usually pass through the wall below ground in the case of horizontal strip foundation, or through the foundation itself when vertical strip foundations are employed (Fig. 17.21). This presents little problem in the former case as apertures can be left in predetermined positions by using precast reinforced concrete lintels or brick arches over openings, or by laying wall units in dry sand instead of mortar, so that they can be easily removed prior to the service being installed. If the service pipe can be positioned as the below-ground wall is being built, these processes can be simplified still further. However, when vertical or deep strip foundations are used, the positions for holes must be carefully predetermined and open-ended timber boxes or plastic pipes placed in position during the concrete-laying procedures. Whichever form of strip foundation is used, it is essential that the holes left are of adequate size to allow about 50 mm clearance all round the proposed services pipe, in order to avoid the possibility of its subsequent deflection or fracture should any settlement of the foundation occur. It is always best to avoid passing services beneath foundations unless precautions are taken to ensure localised settlement of the wall above does not occur at these points through the removal of foundation support (bearing soil).

After the concrete has been poured and has hardened, the construction of the wall above can commence and is subsequently taken to the height of the *damp-proof course* (see section 17.6). At this stage the local authority building control officer will make his or her second formal visit to the site and inspect the suitability and quality of the work and materials.

Once approval has been given, the rest of the *horizontal or shallow strip foundation trenches* can be

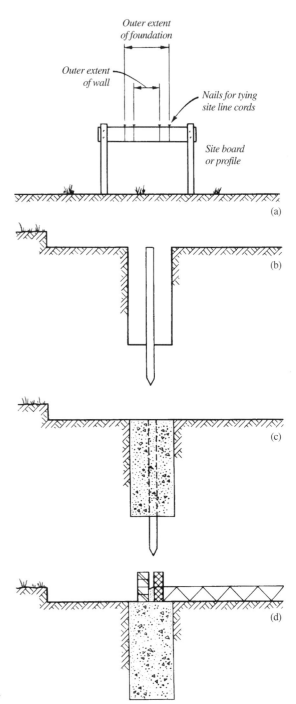

Figure 17.19 Foundation construction: (a) site board or profiles placed to establish position of future foundation (see Fig. 17.11); (b) the foundation is excavated and stakes are inserted to give a consistent level for the top surface; (c) the vertical or deep strip foundation is completed; (d) wall construction is begun just below natural ground level then hardcore is laid and consolidated.

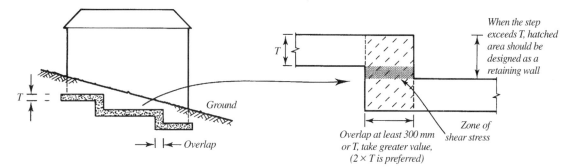

When the step exceeds T, hatched area should be designed as a retaining wall

Zone of shear stress

Overlap at least 300 mm or T, take greater value, (2 × T is preferred)

Figure 17.20 Stepped foundations.

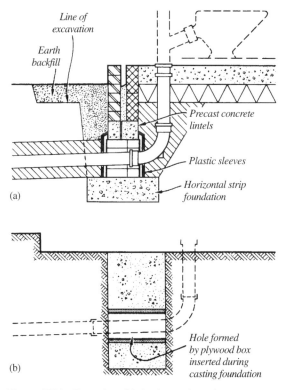

(a)

(b)

Figure 17.21 Formation of holes for services pipes: (a) horizontal or shallow strip foundation – formed by lintel construction; (b) vertical or deep strip foundation – formed by shuttering.

backfilled; ordinary excavated soil is generally suitable for this purpose on the *outside* face of the foundation; a denser material capable of greater consolidation is used for the backfill on the *inside* face of the foundation. This is because the backfill on the inside face of the foundation trench may be required to provide support for the ground floor construction.

Alternative forms of foundation to the horizontal or vertical strip may be necessary for a two-storey house when subsoil conditions are difficult. A *raft foundation* will be required when the soil is so weak that it is necessary to distribute the building loads within a wide volume of soil (Fig. 17.22); a *short-bore pile foundation* will be required when seasonal variations in moisture or vegetation are likely to affect the stability of the bearing soil lying relatively near the surface (Fig. 17.23). Both these forms of foundation, together with the pad foundation applicable to framed buildings, are described in detail in Mitchell's *Structure and Fabric Part 1*, Chapter 4. Sections 17.7.2 and 17.7.3 give details of internal wall foundations.

17.5 Ground floor construction
CI/SfB (16) + (43)

17.5.1 Function

Among other performance requirements, the ground floor construction must provide a level surface capable of supporting people, furniture, equipment, wheeled traffic, partitions, finishes, etc.; it must also provide environmental control against the passage of water (liquid and vapour) and heat. Various factors may affect the method of construction as follows:

- Existing ground levels
- Desired finished floor level(s)
- Required floor loading
- Bearing capacity of supporting soil
- Form of building foundation
- Availability and cost of suitable fill material

A ground floor construction which is continuously supported by the ground or through a fill material in contact with the supporting ground is called a *solid floor*; a construction supported by a foundation

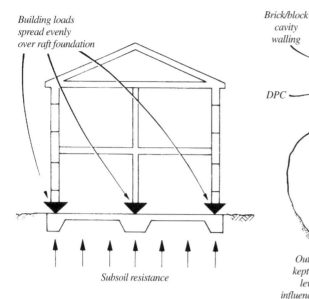

Figure 17.22   Typical details for raft foundation.

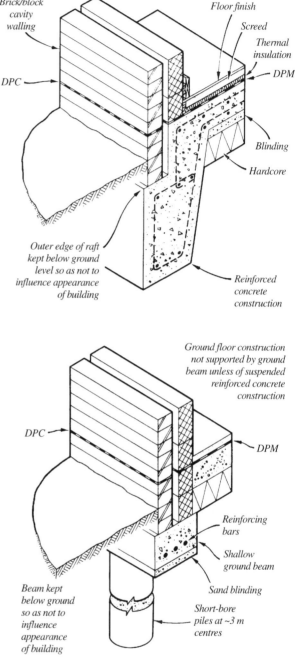

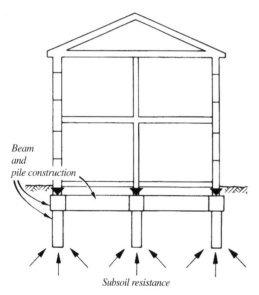

Figure 17.23   Typical short-bore pile foundation.

system at selected points is called a *suspended floor* (Fig. 17.24).

It is usual to commence some part of the floor construction (see page 208) as soon as the foundations and walls up to damp-proof course level have been com-pleted. This is because it is normally desirable to provide a firm and dry working surface for subsequent building operations, or for the storage of components and equipment requiring protection from moisture rising from the ground, e.g. windows, doors and frames. This

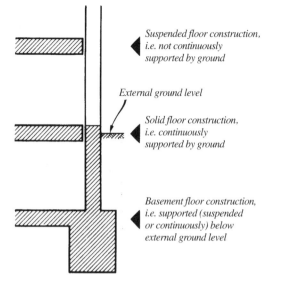

FACTORS TO BE CONSIDERED

Suspended floor construction,
i.e. not continuously
supported by ground

External ground level

Solid floor construction,
i.e. continuously
supported by ground

Basement floor construction,
i.e. supported (suspended
or continuously) below
external ground level

- *Method of support*
- *Type of main foundation*
- *Bearing capacity of ground*
- *Easy to waterproof*
- *Availability of alternative support*
  (hardcore or fill)
- *Sound insulation*
- *Thermal insulation*
- *Fire resistance*
- *Superimposed load*

- *Excavation costs*
- *Need for retaining walls*
- *Need for wall waterproofing*
- *Type of floor waterproofing*
- *Additional fire precaution requirements*
- *Availability of services, including drainage*

Figure 17.24   Forms of lowest floor construction.

provision is particularly important when the area of the building site is small and leaves little space for storage. However, it is also necessary that any prematurely completed work is suitably protected against damage during subsequent building work and that no materials are used in the construction which will be adversely affected by their temporary exposure to the environment, e.g. timber joists, boards, screeds and finishes.

Before the work to the ground floor commences, trenches must be dug and the internal below-ground service pipes for drainage, gas, water, telephone, television and electricity must be installed. These services may need special protection when they occur below the building (e.g. drainage pipes surrounded by concrete) and the trenches should be backfilled as solidly as practicable. It is important to supply the contractor with accurate information regarding the position of these services in order to avoid subsequent repositioning work.

17.5.2 Solid ground floor construction

The concrete ground floor slab must be laid to the correct level, and on a firm base to avoid the possibility of future movements which could cause cracking and produce an uneven top surface or moisture penetration from the subsoil below. The topsoil will have been removed from the formation area of the floor slab. The resulting reduced level should therefore be made good with a material or *fill* which is capable of full compaction so as to be strong enough to take the weight of the

floor slab and the loads it carries, as well as restricting or reducing the upward capillary movement of water from the soil. A material which most conveniently conforms to these requirements and is easy to use is gravel, but less expensive materials include brick, tile or concrete rubble (hardcore), or other waste products such as blast furnace slag, pulverised fuel ash and colliery shale. However, care must be taken when using these latter products to ensure that their unit size allows thorough compaction, and that they do not contain sulphates, which would attack the cement of the concrete slab under the foundation materials. When completed, the top surface of the fill material or hardcore should be dry and even, and at a level which best ensures the top surface of the concrete floor construction butts against the perimeter walling without the necessity for vertical damp-proofing techniques. This will avoid moisture penetration resulting from the lateral movement of water from outside (Fig. 17.25(a)).

The fill or hardcore must be fully compacted by mechanical hammers or rollers (depending on the area involved). In practice it is usually difficult to ensure enough consolidation when fill or hardcore depths greater than about 600 mm are required, even if compacted in the usual 150–200 mm layers. Therefore, in order to avoid the possibility of settlement occurring in the floor, a suspended floor construction will be more appropriate when the desired finished floor level requires more than 600 mm fill or hardcore (see section 17.5.3).

The concrete for the slab is mixed and poured so as to obtain a *minimum* thickness of 100 mm, although it

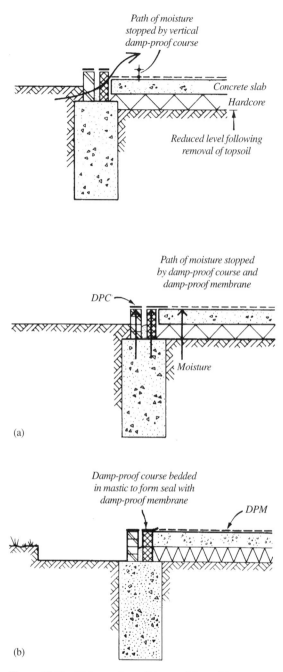

Figure 17.25   Waterproofing solid ground floor slab: (a) principles; (b) method.

separate foundations for this purpose are not necessary; the greater mass of concrete will offer greater resistance to rising moisture. If the subsoil conditions are relatively weak, the 150 mm slab should be reinforced with a mild steel mesh. This should be placed in the slab after an initial layer of concrete has been poured and then be covered by the subsequent and final pour. As concrete shrinks on drying, the reinforcement will also help in avoiding shrinkage hair-cracking. But if large areas of uninterrupted concrete slab are required, movement joints should be incorporated by casting the slab in approximately 3 m bays or using a modular dimension appropriate to the plan area.

The surface of the concrete slab can be finished in various ways, depending upon the position of the *damp-proof membrane*. This membrane consists of a sheet of polythene (1200 gauge), or heavy reinforced building paper (bitumen impregnated), or an *in situ* coating of hot-poured bitumen, or a minimum of three coats of cold-applied bitumen/rubber solution. The function of a damp-proof membrane (DPM) is to resist moisture in liquid and vapour form coming into contact with the internal surfaces of the floor construction and thereby causing damage. The concrete slab itself may not be able to resist this moisture – particularly when the water-table is close to the ground surface – because of the fine hair-cracks which develop during drying and, unless it contains special admixtures, concrete is not moisture vapour proof.

If either a sheet or an *in situ* coating damp-proof membrane is to be subsequently applied to the *top* of the concrete slab, as indicated in Figure 17.25(b), the top surface of the concrete must be trowelled smooth so that it is not punctured. The DPM must also be protected as soon as possible after it is laid by a sand and cement mixture (in the proportion 3:1) known as a *screed*. This screed should be a minimum thickness of 50 mm to avoid cracking and also trowelled smooth to a *perfectly level surface* in order to be satisfactory for the final floor finish, e.g. quarries, clay or PVC tile, carpet. The DPM, screed and subsequent floor finish cannot be placed until most of the other building works have been completed because they are all susceptible to damage which will be costly to rectify (see section 17.11).

Figure 17.26 indicates the alternative position for the DPM – *below* the concrete slab. In this case it is the fill or hardcore which must have the smooth surface to avoid sharp edges damaging the membrane; this is achieved by *blinding*, consisting of a 30–50 mm thick layer of sand. After the DPM has been positioned, rigid insulation is placed with edges returned and the concrete slab (reinforced if required) is poured very carefully to

is preferable to use a thicker slab of 150 mm when any doubt exists about either the full compaction of the fill or hardcore, or about the subsoil when it is less than very stable. The thicker slab will also assist in spreading the point loads from load-bearing partitions when

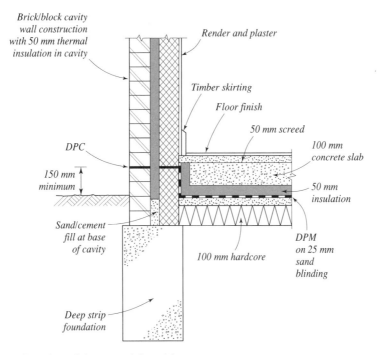

Figure 17.26   Damp-proof membrane below ground floor slab.

avoid damaging the insulation and DPM. The top surface of the slab should then be 'finished' with a rough surface by using a long plank of wood placed on edge and *tamping* across the wet surface. When dry, the resulting slightly irregularly raised surface texture will provide an excellent key for the screed. Because more positive bond is provided to the slab than when laid on the smooth DPM (Fig. 17.25(b)), the finishing screed can be of reduced thickness, say 30 mm. If the screed is laid while the concrete slab is still green, and therefore providing an even better key, the screed thickness can be reduced to 12 mm. Indeed, as an alternative to providing a screed, the green concrete slab can be finished to a very smooth surface by a light mechanical buffing process using a special machine known as a *power float*.

The designer must consult with the contractor concerning which of the above two positions for the DPM is most suitable. Both have advantages as well as disadvantages. Below-the-slab DPM leads to savings in both labour and materials, and, if correctly executed, the slab remains dry throughout the life of the building and helps to minimise heat losses as well as providing a thermal reservoir (heat storage). But it does mean great care must be taken during construction to ensure the DPM is not damaged. Unless a screed or some other finish (e.g. timber) is subsequently applied to the surface of the slab, it cannot be cast until quite late in

the building operations, otherwise the exposed surface will certainly be damaged. If a power-float finish to the slab is required, it must be temporarily protected until most building operations are complete.

The amount of heat which may be lost through a solid ground floor construction depends on several factors. The greatest concern for heat conservation derives from the extent of external wall bounding the floor because this juxtaposition will provide a direct path for heat loss to the external air. This problem may be overcome by using perimeter insulation material between the junction of the solid floor construction and the inside face of the external wall. This insulation should be waterproof and extend downwards either from the DPC/DPM position for 300–1000 mm, depending upon climatic conditions, or for the thickness of the slab before running horizontally beneath it.

Alternatively, insulating material can be placed directly on the DPM before the floor screed is applied, as indicated in Figure 17.27. This method of heat conservation is particularly useful when underfloor heating cables are to be inserted in the screed. However, care must be taken to ensure that the resulting *floating screed* (unattached) is of sufficient thickness to avoid cracking, and that it is separated from the insulation by a waterproof building paper to avoid saturation of the insulation and loss of the grout (cement/water) from

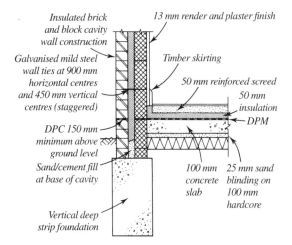

Insulated brick and block cavity wall construction

13 mm render and plaster finish

Galvanised mild steel wall ties at 900 mm horizontal centres and 450 mm vertical centres (staggered)

Timber skirting

50 mm reinforced screed

50 mm insulation

DPM

DPC 150 mm minimum above ground level

Sand/cement fill at base of cavity

100 mm concrete slab

25 mm sand blinding on 100 mm hardcore

Vertical deep strip foundation

Figure 17.27 Thermal insulation to ground floor slab.

the screed causing a 'cold bridge' link with the concrete slab below. The waterproof membrane will also act as a vapour control layer, and eliminate the possibility of interstitial condensation within the insulating material.

A minimum screed thickness of 75 mm is required and this should be reinforced with a light mesh reinforcement (chicken wire).

17.5.3 Suspended ground floor construction

This form of ground floor construction should be adopted when the desired finished floor level is above the supporting ground to the extent that the difference cannot conveniently be made up with fill or hardcore (see Fig. 17.24). The structural floor can be formed by timber joists (Fig. 17.28) or with reinforced concrete (cast *in situ* or by using precast concrete planks, see Fig. 17.32).

When *timber joists* are used, care must be taken to ensure they are not adversely affected by the potentially moist condition existing in the void beneath the floor. Precautions include the provision of a DPC immediately beneath the timber wall-plate which provides bearing for the timber joists through to the sleeper walls below; adequate *cross-ventilation* of the void by air from external perforated bricks circulating around the

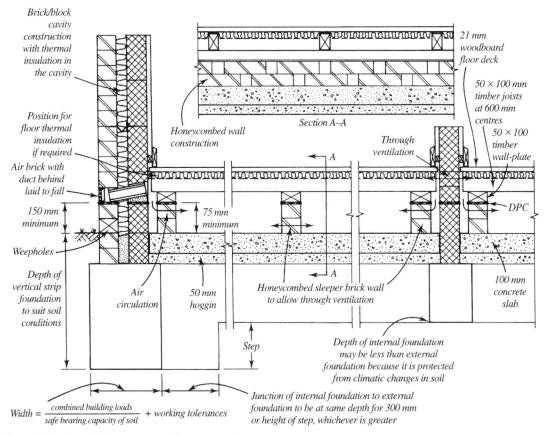

Brick/block cavity construction with thermal insulation in the cavity

21 mm woodboard floor deck

Honeycombed wall construction

Section A–A

Through ventilation

50 × 100 mm timber joists at 600 mm centres

50 × 100 timber wall-plate

Position for floor thermal insulation if required

Air brick with duct behind laid to fall

150 mm minimum

75 mm minimum

DPC

Weepholes

Depth of vertical strip foundation to suit soil conditions

Air circulation

50 mm hoggin

Honeycombed sleeper brick wall to allow through ventilation

100 mm concrete slab

Step

Depth of internal foundation may be less than external foundation because it is protected from climatic changes in soil

$$\text{Width} = \frac{\text{combined building loads}}{\text{safe bearing capacity of soil}} + \text{working tolerances}$$

Junction of internal foundation to external foundation to be at same depth for 300 mm or height of step, whichever is greater

Figure 17.28 Suspended timber ground floor construction.

timber joists and through gaps in the *sleeper walls*, which are of 'honeycomb' construction; and the use of structural timbers with not more than 18 per cent moisture content *at the time of fixing*. All timber should be impregnated with a preservative solution to counteract the possibility of fungal attack (dry rot) and act as a precautionary measure against the failure of the construction techniques to keep the moisture content of the timber below 18 per cent during its life in the building. Preservative-impregnated timber also proves unpalatable to the grubs of wood-eating insects (woodworm, house longhorn beetle, etc.).

For further comments regarding timber joists, and their sizes and spacing, see section 17.8.2 and Table 17.2.

The floor *decking* can be tongue-and-groove hardwood or softwood *boarding*, chipboard *sheets* or plywood *sheets* (Fig. 17.29). The narrow face width of boarding gives greater flexibility when providing a floor in an irregular or *non-modular* width of room, and also permits easier access to the void below, should it be required during the life of the building, e.g. for alterations or extensions of electric wiring contained within the void. However, sheet floor deckings are generally quicker to install, particularly when the spacing of the timber joists and sizes of rooms have been closely coordinated with dimensions of available plywood or chipboard sheets. Also, provided the sheets are allowed to acclimatise to the atmospheric conditions of the room prior to installation and provided they are correctly fixed, subsequent movements of the sheets are likely to have less effect on applied finishes than would the corresponding movements associated with softwood boarding. Boards as well as sheets should

Table 17.2 Span/depth for intermediate timber floor joists

Size of joist in millimetres graded GS or MGS	Maximum span of joist (m) for given dead load (kg/m²) supported by joist (excluding the mass of the joist)								
	Not more than 25			More than 25 but not more than 50			More than 50 but not more than 125		
	Spacing of joists (mm)			Spacing of joists (mm)			Spacing of joists (mm)		
	400	450	600	400	450	600	400	450	600
38 × 75	1.05	0.95	0.72	0.99	0.90	0.69	0.87	0.79	0.62
38 × 100	1.77	1.60	1.23	1.63	1.48	1.16	1.36	1.24	1.00
38 × 125	2.53	2.35	1.84	2.33	2.12	1.69	1.88	1.73	1.40
38 × 150	3.02	2.85	2.48	2.83	2.67	2.26	2.41	2.23	1.83
38 × 175	3.51	3.32	2.89	3.29	3.11	2.71	2.82	2.66	2.27
38 × 200	4.00	3.78	3.30	3.75	3.55	3.09	3.21	3.03	2.64
38 × 225	4.49	4.24	3.70	4.21	3.98	3.47	3.61	3.41	2.96
44 × 75	1.20	1.08	0.83	1.13	1.02	0.79	0.98	0.89	0.70
44 × 100	2.01	1.82	1.41	1.83	1.67	1.31	1.51	1.39	1.12
44 × 125	2.71	2.56	2.09	2.54	2.38	1.90	2.08	1.92	1.56
44 × 150	3.24	3.06	2.67	3.04	2.87	2.50	2.60	2.45	2.03
44 × 175	3.77	3.56	3.10	3.53	3.34	2.91	3.02	2.86	2.48
44 × 200	4.29	4.06	3.54	4.02	3.80	3.31	3.45	3.26	2.83
44 × 225	4.81	4.55	3.97	4.51	4.27	3.72	3.87	3.66	3.18
50 × 75	1.35	1.22	0.93	1.26	1.14	0.89	1.08	0.99	0.78
50 × 100	2.22	2.03	1.58	2.03	1.85	1.46	1.66	1.53	1.23
50 × 125	2.84	2.72	2.33	2.70	2.55	2.10	2.27	2.09	1.71
50 × 150	3.40	3.26	2.84	3.23	3.05	2.66	2.76	2.61	2.21
50 × 175	3.95	3.78	3.30	3.75	3.55	3.09	3.22	3.04	2.64
50 × 200	4.51	4.31	3.76	4.27	4.04	3.52	3.67	3.46	3.01
50 × 225	5.06	4.83	4.22	4.79	4.53	3.95	4.11	3.89	3.39
63 × 150	3.66	3.52	3.17	3.50	3.38	2.97	3.09	2.92	2.54
63 × 175	4.25	4.10	3.68	4.07	3.93	3.45	3.59	3.40	2.96
63 × 200	4.84	4.67	4.20	4.64	4.48	3.93	4.09	3.87	3.37
63 × 225	5.43	5.24	4.70	5.21	5.02	4.41	4.59	4.34	3.78
75 × 200	5.10	4.93	4.51	4.90	4.72	4.27	4.43	4.20	3.67
75 × 225	5.72	5.52	5.06	5.49	5.30	4.79	4.97	4.71	4.11

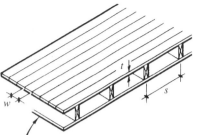

Boards	Finished thickness, t	Nominal width, w (mm)	Span, s	Approximate weight (kg/m^2)
Softwood t & g	16	100 and 150	505	7
	19		600	9
	21		635	10
	28		790	13
	47	150	2560	23
	60	150	3230	28
	72	150	3830	32

12.5 mm plasterboard
and plaster finish
= 12 kg/m²

Sheets	Thickness, t	Size, $(w \times l)$	Span, s	Approximate weight (kg/m^2)
Plywood t & g (flooring grade)	12	600 and 1200 × 2440 and 2750	400	8
	15		600	9
	18		700	11
	25		1200	15
Chipboard t & g (flooring grade)	18	600 and 1200 × 2400 and 2750	400	12
	19		450–500	13
	22		600	15

12 mm plasterboard
and plaster finish
= 12 kg/m²

Figure 17.29 Types of floor decking: dead weight or self-weight of flooring and ceiling can be used in conjunction with Table 17.2 to obtain suitable timber joist sizes.

be impregnated with a preservative solution in a similar manner as the structural timbers, although the sheets require special hard-wearing pre-treated flooring grades. When it is thought desirable to leave the top surfaces of the boards or sheets exposed, consideration must be given to the selection of suitable wood grain patterns and the effects of subsequent seals, stains or varnishes.

The *in situ* concrete slab used with the form of suspended floor construction, indicated in Figure 17.28, can be cast as soon as the walls have been completed up to DPC level in a similar manner to that described for solid floor construction. In this way the slab serves as a relatively dry, clean and level surface for storage and work by other trades, i.e. preparation of purpose-made door and window frames. The remainder of the floor construction can be executed as more of the building is completed.

If it is considered desirable for the floor decking to be installed before 'heavy' trade work is complete, a temporary layer of hardboard sheets must be installed to prevent damage to the surface of the boards or sheets. In any event, it is convenient if the services required to run beneath the floor decking can be installed before

all the boards or sheets are finally fixed to the supporting joists. Alternatively, predetermined areas of decking are left open to allow services to be fed through the void below. When the services are installed early, they can be left projecting just above the decking, pending subsequent connections and completion of the above-decking services.

Figure 17.30 illustrates a form of suspended ground floor construction where a concrete slab is not used. A slab cannot be considered as a water barrier since small cracks will develop through its thickness as a result of drying shrinkage. These gaps may be relatively large (2–3 mm) between two or more areas of slab cast next to each other on different days (daywork joints) and at the junction between a slab and the external wall. Furthermore, concrete is not moisture vapour proof. By using a sheet membrane (polythene) lapped and folded at edges and sealed with the wall DPC, a more positive barrier against free moisture and moisture vapour can be provided. This form of construction can best be executed without the use of intervening sleeper walls, which means that the structural floor joists must be larger to span the resulting longer distances. Ventilation

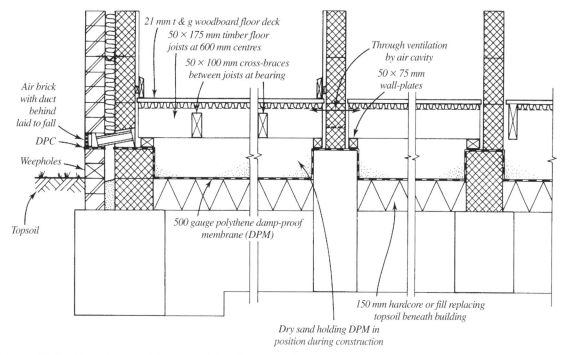

21 mm t & g woodboard floor deck
50 × 175 mm timber floor
joists at 600 mm centres

50 × 100 mm cross-braces
between joists at bearing

Through ventilation
by air cavity

50 × 75 mm
wall-plates

Air brick
with duct
behind
laid to fall

DPC

Weepholes

Topsoil

500 gauge polythene damp-proof
membrane (DPM)

150 mm hardcore or fill replacing
topsoil beneath building

Dry sand holding DPM in
position during construction

Figure 17.30   Alternative ground floor suspended timber construction.

of the floor void must be provided, as described earlier, and timbers should also be impregnated with a preservative treatment against fungal and insect attack.

This method of construction means that the dry working surface previously described is not available to the builder and he or she must take special measures to ensure the membrane barrier is not damaged during the installation of floor joists and decking. It causes some inconvenience to the builder and his or her opinion should therefore be sought before it is specified by the designer, even if considered essential to avoid excessive amounts of moisture occurring in the floor space beneath the decking. If the site is subject to waterlogging, it is probably best not to risk water standing freely on either concrete slab or membrane barrier. Figure 17.31 indicates a method of construction which does not employ either, but allows water to drain away during dry periods.

Generally speaking, a timber suspended floor provides better resistance to loss of heat to the surrounding ground by virtue of the better insulation properties of timber when compared with solid concrete. Additional thermal insulation can be provided, however, and this is usually draped over the timber joists prior to fixing the floor decking. Under these circumstances it is probably best not to provide a vapour barrier on the room side of the insulation (warmer) because it

would only serve as a trap for water which may be spilled on the upper surface of the floor decking and seep through joints to the spaces below. Seeping moisture from this source will soon dry out (as will any interstitial condensation) if the void is adequately ventilated.

Suspended ground floor constructions using *reinforced in situ or precast reinforced concrete* units allow for 25–40 per cent greater floor loadings than timber joists, and are capable of spans up to 6 m (Fig. 17.32). This may be particularly attractive when poor soil conditions require deep or complicated foundations and which should be formed as widely spaced as possible for reasons of economy. Also, concrete floor construction is not subject to dry rot or insect attack. It does, however, require formwork (permanent or temporary) for *in situ* construction, or lifting equipment (gantries or cranes) for precast concrete constructions. Precast constructions also require careful dimensional coordination to get the most from a minimum range of standard length and width components. As with solid floor construction, a suspended concrete floor construction can be provided with a timber floor finish (boards or sheets on preserved timber battens fixed to the concrete base), but does not provide the desirable springiness of a structural timber construction unless special sprung fixing devices are installed with the timber battens.

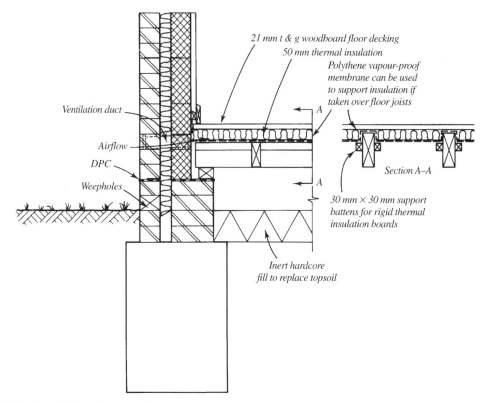

Figure 17.31 Second alternative ground floor suspended timber construction.

21 mm t & g woodboard floor decking

50 mm thermal insulation

Polythene vapour-proof membrane can be used to support insulation if taken over floor joists

Ventilation duct

Airflow

DPC

Weepholes

A

A

Section A–A

30 mm × 30 mm support battens for rigid thermal insulation boards

Inert hardcore fill to replace topsoil

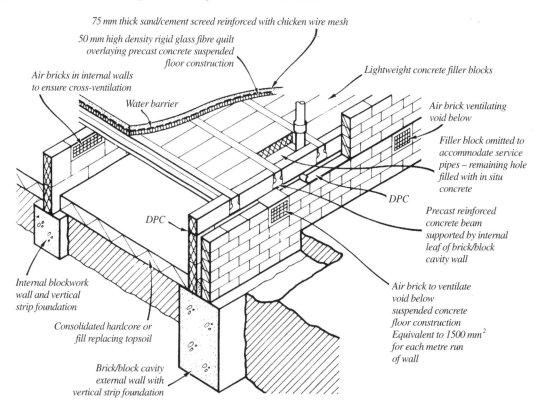

Figure 17.32 Suspended ground floor construction using precast concrete flooring units.

75 mm thick sand/cement screed reinforced with chicken wire mesh

50 mm high density rigid glass fibre quilt overlaying precast concrete suspended floor construction

Air bricks in internal walls to ensure cross-ventilation

Water barrier

Lightweight concrete filler blocks

Air brick ventilating void below

Filler block omitted to accommodate service pipes – remaining hole filled with in situ concrete

DPC

DPC

Precast reinforced concrete beam supported by internal leaf of brick/block cavity wall

Internal blockwork wall and vertical strip foundation

Consolidated hardcore or fill replacing topsoil

Brick/block cavity external wall with vertical strip foundation

Air brick to ventilate void below suspended concrete floor construction Equivalent to 1500 mm^2 for each metre run of wall

17.6 External wall construction
CI/SfB (21) + (31) + (41)

17.6.1 Function

The main function of an external wall involves providing the environmental control between the external and internal climates of a building. Also, depending upon the precise nature of the overall structural system, an external wall is usually required to support the combined dead, imposed and wind loads of the roof and floor construction, as well as its own combined loads, and transfer them safely to a foundation. However, apart from the technical requirements, considerations of appearance must be a critical part of the design since to a very large extent an external wall determines the architectural character and quality of a building.

17.6.2 Materials

Both technical and aesthetic criteria are influenced by the type and size of materials employed, standard of work and the effects of weathering and general durability resulting from design and construction methods. Therefore, external walls can take many forms since they can be made from a very wide range of materials used either singly or in combination with others:

- Small blocks of brick and/or concrete laid in mortar
- Dry-jointed units of timber, metal or concrete
- Homogeneous matrix of reinforced concrete
- Composite systems of timber, metal or concrete preformed units
- Rigid single sheets of glass, metal or plastics
- Flexible single sheets of rubber or plastics

When considering the construction of a dwelling in the United Kingdom, the first choice of material for an external wall is small blocks and bricks and/or concrete laid in mortar (Fig. 17.33). This choice derives partly from traditional attitudes, but also because of availability of materials and skilled work as well as the inherent performance requirements known to be satisfactorily fulfilled by this form of construction. Bricks and blocks are unlikely to cause the spread of fire from either outside or inside because they are non-combustible and highly resistant to collapse as a result of being subjected to fire. When used correctly, bricks and blocks are also highly resistant to weather penetration and can provide satisfactory standards of sound control and thermal insulation. However, the selection of the most suitable form would vary according to the location of the external wall relative to the disposition and siting of the dwelling; this involves a thorough analysis of the performance requirements outlined in Part A.

The majority of bricks used in the United Kingdom are manufactured from clay which, having been moulded to suitable dimensions when in a plastic condition, is burnt to a hardened and vitrified state. There are three technical varieties according to use: *common bricks* for general and ultimately unseen work; *facing bricks* specially made or selected to give an attractive appearance; and *engineering bricks* made to provide high levels of compressive strength *and* low levels of moisture absorption. The physical characteristics of the bricks are derived through the type of clay used and the manufacturing processes involved. Engineering bricks are of *special quality* and rely on a special form of clay for their production, whereas common and facing bricks may be of *ordinary quality* (used in normal

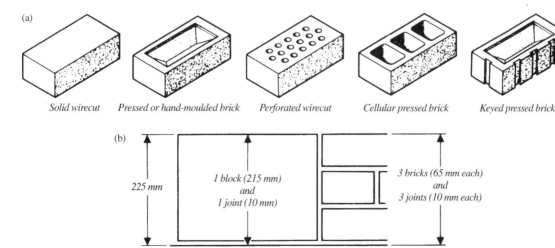

Figure 17.33 Bricks and brickwork: (a) types of brick; (b) relative course heights of brickwork and blockwork.

exposure condition) or of *internal quality* (used in protected position only) according to the combined strength and absorption characteristics of the clay used. Bricks can also be manufactured from shale (calcium silicate) or concrete.

Building blocks are larger than bricks and are manufactured from clay, plaster or concrete (see Fig. 17.47). However, concrete is the most suitable material for incorporation in an external wall construction since its use avoids the special measures which have to be taken with the other materials to ensure adequate weathering and thermal insulating properties. Concrete blocks are manufactured in several crushing strengths to BS 6073: 2.8, 3.5, 5, 7, 10, 15, 20 and 35 N/mm^2. The lower figures represent highly insulative blocks produced from lightweight aggregates but having load limitations. Blocks in the upper strength categories are produced from dense concrete for substructural and other high load-bearing situations. They are also manufactured in solid, perforated or hollow horizontal section.

Walls constructed from bricks and blocks rely for their strength on being laid in horizontal courses so that their vertical joints are staggered or *bonded* across the face of the wall (see Fig. 5.5). In this way the compression loads which may initially affect bricks or blocks can be successfully distributed through a greater volume of the wall. The bricks and concrete blocks are held together by an adhesive mixture known as *mortar*, thereby completing the structural and environmental enclosure. This mortar also serves the function of taking up any dimensional variations in the bricks or blocks, so that they can be laid in more or less horizontal and vertical alignment. Most mortars today usually consist of a water-activated binding medium of cement, and a fine aggregate filler such as sand, in the proportion of one part binder to three parts aggregate. The amount of water used to activate the binder in order for it to adhere to the fine particles of the filler and the face of the brick or concrete block must be kept to a minimum to avoid a sloppy, unworkable mix subject to excessive drying shrinkage. The use of an additive will aid workability and avoid an otherwise very dry, stiff mix which could be difficult to apply. Traditionally this was lime, but liquid plasticisers are more common today.

Mortar mixes on site should not be used when the air temperature is below 4°C, otherwise the water content may freeze and adversely affect the strength of the mortar. When such conditions arise, the contractor will have to stop bricklaying, or provide protective shelters around the building or over that part to be immediately constructed, and probably also provide artificial space heating sources to ensure continuity of work for the bricklayers. These precautions must be applied to all on-site 'wet' trade activities, including concrete laying and the application of plaster.

17.6.3 Brick/block cavity walls

Bricks and concrete blocks can be used as a walling material in a number of ways, as indicated in Figure 17.34. Traditionally, solid brick walls were considered sufficiently weather resistant in many areas of the United Kingdom, but under severe exposure conditions quite thick, solid brick walls may well permit water to penetrate to the inner face and would certainly not meet today's requirements for thermal insulation unless special precautions were taken. The most common form today derives from principles originated over 50 years ago, and uses the two materials together to create a *brick/block cavity wall construction*. This consists of an outer or external leaf of 102.5 mm thick brickwork (although this may be thicker to suit loading conditions), a separating air cavity 50–100 mm wide, and an inner or internal leaf of 100 mm thick concrete block (although this may be increased to suit loading conditions or thermal insulating requirements) to which a finish may or may not be applied.

This method of construction accepts that water may permeate through the relatively thin outer leaf of brickwork, but is then prevented by the air cavity from reaching the inner leaf of concrete blockwork which remains dry and therefore provides good thermal insulating properties, particularly when lightweight or aerated blocks are employed. These properties can be further increased by inserting a rigid thermal insulating material on the cavity face of the blockwork, but this must be done with great care to ensure that, whenever possible, a minimum cavity of 50 mm is maintained to provide the water barrier. Alternatively, a 50 mm cavity can be used and filled completely with either an insulating board (batt) or insulating foam. Complete cavity-filling techniques are easier for the bricklayer to implement, but can create problems of water penetration under conditions of severe exposure, owing to the possibility of mortar which has squeezed out of the joints in the bricks or blocks forming a link across the cavity. Figure 17.35 indicates the former construction method, which maintains a cavity barrier.

The relatively thin cross-section walling materials must be selected to withstand the effects of movements, mostly caused by dramatic temperature changes arising from winter to summer (seasonal) and, to a lesser extent, from night to day (diurnal). It is not uncommon for the annual temperature range of an external wall surface to be in the order of 90°C, although this will occur

Rain

215	solid brick wall
13	plaster
U value = 2.1 W/m^2 K	

102.5	brick outer leaf
50	cavity
102.5	brick inner leaf
13	plaster
U value = 1.7 W/m^2 K	

102.5	brick outer leaf
50	cavity
100	lw block inner leaf
13	plaster
U value = 1.0 W/m^2 K	

25	external skin
50	insulation
215	solid brick
13	plaster
U value = 0.55 W/m^2 K	

25	external skin
50	insulation
200	solid lw block
13	plaster
U value = 0.4 W/m^2 K	

102.5	brick outer leaf
50	cavity
150	lw block inner leaf
13	plaster
U value = 0.6 W/m^2 K	

102.5	brick outer leaf
45	cavity
25	insulation
100	lw block inner leaf
13	plaster
U value = 0.6 W/m^2 K	

25	external skin
100	lw block outer leaf
50	cavity
100	lw block inner leaf
13	plaster
U value = 0.57 W/m^2 K	

102.5	brick outer leaf
50	insulation
100	lw block inner leaf
13	plaster
U value = 0.45 W/m^2 K	

25	external skin
100	lw block outer leaf
50	insulation
100	lw block inner leaf
13	plaster
U value = 0.35 W/m^2 K	

Figure 17.34 Use of bricks and blocks to form external walls and a comparison of their approximate thermal insulation values.

quite slowly over the four seasons, thereby giving time for corresponding movements to happen gradually. The continually alternating wetting and drying created by weather processes can also cause problems of movements as well as frost and sulphate attack of the outer brick leaf. Open-pored bricks have a high rate of water absorption and may allow water to pass through their body. Although dense bricks have a low absorption rate and transfer little water through their body, large quantities of water running down the 'glass-like' wall face may be drawn in through capillary paths formed by fine hair-cracks between the mortar used for the joints and the brick (see Fig. 6.9). However, the design of the cavity wall is such that, once water has reached the inner face of the outer leaf, it can run down to a point of collection before being redirected to the outside through strategically placed *weepholes* (see Fig. 6.5). These can be provided by vertical joints (perpends) left open without mortar filling. The cavity must be kept clear of obstructions and damp-proof courses must be inserted to stop and redirect water at points where the cavity is unavoidably bridged, e.g. around door and window openings, air bricks, etc.

Water which may come up from the ground or down from the exposed top of a wall is also barred from those parts of the wall where it will cause damage or loss of environmental control by a damp-proof course appropriately linked to damp-proof membranes as described for floor and roof construction.

17.6.4 Construction

The brickwork and blockwork are normally laid in stretcher bond, although half bats or 'snap headers' may be used to achieve the appearance of Flemish or other bonds. The brick and block leaves must be held together by *wall ties* placed at intervals which will ensure that the two leaves develop a mutual stiffness and act in unison in resisting the combined building loads. This is achieved with the ties spaced at 900 mm maximum horizontally, 450 mm vertically and staggered. A maximum of 300 mm (normally 225 mm to suit block dimensions) vertical spacing is required at door and window openings (see Fig. 17.40). The design of wall ties also ensures that water entering the outer leaf is stopped before reaching the inner leaf by a *drip*, which occurs centrally over the cavity. The ties should be embedded within the horizontal mortar joints to a minimum penetration of 50 mm in order to provide adequate restraint to each leaf of the wall. They must be completely surrounded by the mortar of the joint, and ideally laid to a slight fall from the inner leaf down to the outer leaf of wall construction. A range of overall lengths is available to suit different cavity widths, and the ties are produced in non-corrosive metals and plastics to varying shapes. It is essential that ties are kept clear of mortar droppings, which could accumulate during construction, as they may negate the drip mechanism of the ties and thereby provide a bridge for water to reach the inner leaf. Any mortar droppings must therefore be removed daily as the work proceeds; temporary openings called 'coring holes' or 'clean outs' are provided at 1 m centres to allow their clearance from the bottom of the cavity.

Where the type of cavity wall is that indicated by Figure 17.35, the construction method is briefly as follows:

- The DPC material is *bedded* on a layer of mortar on each leaf of the brick/block cavity foundation wall. It is important that the DPC is the same width as the walling below and the lengths are adequately lapped at the ends, at least 100 mm or the leaf thickness if greater, in order to prevent rising moisture travelling past to the walling above.
- A layer of mortar is placed on the DPC to provide a level bedding for the cavity wall construction. The load from each leaf is then evenly distributed to the foundation wall, the DPC is protected from damage which may be caused by the rough brick or

block underface (bed face), and the bond between DPC, mortar and walling material ensures minimal horizontal water penetration by capillary attraction.

- The first row of wall ties is also positioned at 600 mm centres within the layer of mortar placed on the DPC. Polythene retaining discs on the ties should be spaced at a distance from the blockwork inner leaf to suit the thickness of thermal insulation material.
- The blockwork inner leaf is built up to one course above the next row of wall ties which should not have retaining discs fitted at this stage.
- Excess mortar oozing from joints between newly placed blocks should be removed from surfaces, and all mortar droppings must be cleaned from cavity and wall ties.
- Rigid glass fibre insulating boards or batts are placed between polythene discs on the lower wall ties and against the blockwork inner leaf. Care should be taken to ensure edges of the insulating boards are butted closely together before the polythene retaining discs of the upper row of wall ties are fitted to hold the insulation firmly against the inner leaf. If full cavity width insulation is specified, the retaining discs are omitted.
- The brickwork outer leaf is built up to the top level of the insulation batt, ensuring that mortar does not

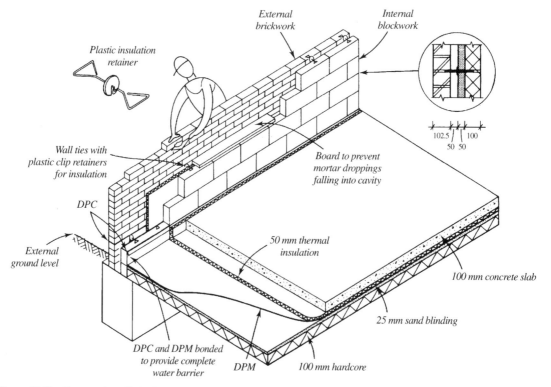

Figure 17.35   Construction of brick/block cavity wall.

drop into the remaining cavity. In practice this is difficult to achieve unless a special cavity batten is used, which is suspended on cords and collects the droppings as they fall. The batten is subsequently drawn up at each stage of wall construction to be cleaned.

- The blockwork inner leaf is built up and followed by the brickwork outer leaf and, as work progresses, the wall ties and insulating batts are positioned, ensuring at all times that the cavity, the wall ties and joints between batts are kept clean of mortar.

The usual procedure is to build the corners or the extremities of the wall to a height of about eight or twelve brick courses (600–900 mm), care being taken to ensure all edges are vertically plumb. The base of the corner is then extended along the wall and raked back as the work is carried up. The intermediate portion of the wall is built between the two corners, the bricks and blocks being kept level and straight by building their upper edges to a string line stretched between the corners of the building. Notwithstanding this building sequence, the whole of the walling should be carried up simultaneously; no part should be built higher than 900 mm to avoid the risk of unequal settlement on the foundations before the mortar has sufficiently set. The outside face of the mortar joints can be finished in a number of ways, as illustrated in Figure 17.36.

At the end of a day's work, or during rain, exposed areas of insulation, including edges, must be protected with polythene sheets. Without using special techniques, building work should not be carried out using wet mixes when the temperature is at 4°C or may soon approach it. Whenever necessary, newly completed work should be adequately protected overnight or during non-working days to avoid the detrimental effects of freezing.

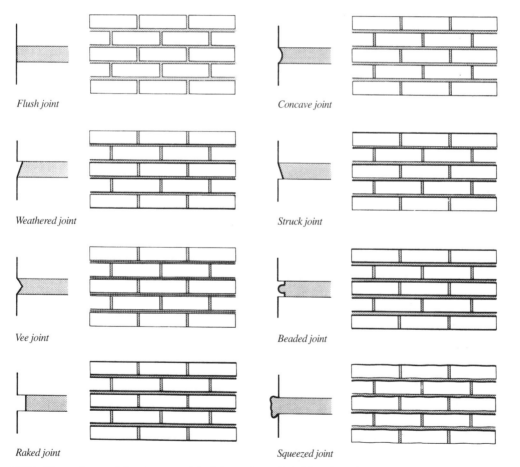

Flush joint

Concave joint

Weathered joint

Struck joint

Vee joint

Beaded joint

Raked joint

Squeezed joint

Figure 17.36 Types of mortar joint. A concave joint is better than a flush joint when bricks are uneven; struck, beaded or raked joints should not be used in exposed conditions because water will lie on the exposed edge of the bricks. The squeezed joint will also allow water to penetrate brickwork. Ordinary mortar joints can be raked back and repointed in coloured mortar to any of these profiles; coloured joints may have a dramatic effect on the overall appearance of the brickwork.

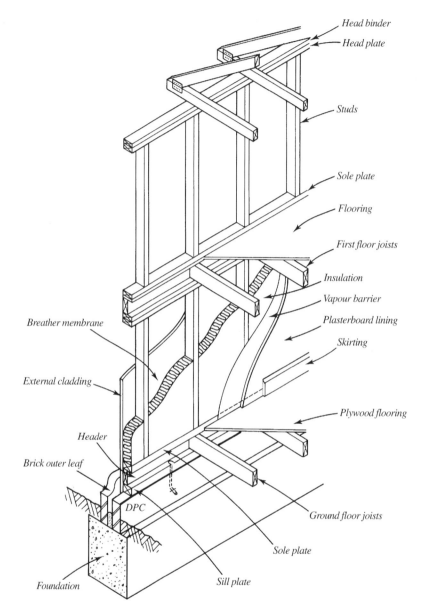

Head binder
Head plate
Studs
Sole plate
Flooring
First floor joists
Insulation
Vapour barrier
Plasterboard lining
Skirting
Breather membrane
External cladding
Plywood flooring
Header
Brick outer leaf
DPC
Ground floor joists
Sole plate
Foundation
Sill plate

Figure 17.37 Typical timber wall construction using platform frame.

Figure 17.37 illustrates an alternative construction method for an external wall using a timber frame technique incorporating thermal insulation, clad on both sides to provide a complete environmental enclosure. This construction technique is described fully in *Mitchell's Structure and Fabric Part 1*, Chapter 5.

17.6.5 Doors and windows

Doors and windows can provide the weak link in those performance requirements for an external wall concerned with weather exclusion, sound control, thermal comfort, fire protection and security (see Part A). In addition, their location, size, shape, proportion and material profiles profoundly influence the overall and detail appearance as well as the aesthetic aim of a building. The designer must carefully select the sizes of doors and windows relative to their location to ensure all these factors have equal value.

The frames for doors and windows can be 'built-in' openings left in the brick/block cavity wall construction, or can be positioned earlier so that the wall is constructed around them. Earlier positioning has many advantages when it comes to forming a weatherproof joint between wall construction and frame. However, the contractor must ensure that the frames are available on site at an early stage of the contract, and that they are adequately protected from physical damage as well as the weather during storage and after their positioning. It is important that the designer has ensured the horizontal lengths and vertical heights of wall construction conform with the overall dimensions and the distances between individual door and window frames (Fig. 17.38). Materials used for door and window frames include timber (softwood and hardwood), metals and plastics. Timber window and door frames should

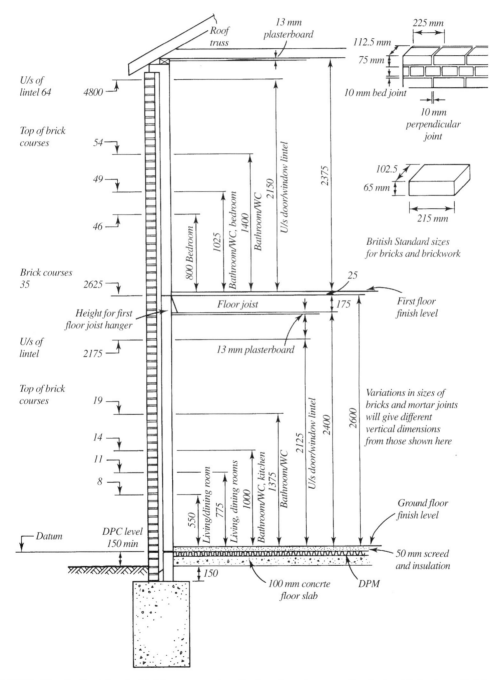

Figure 17.38 Coordination of vertical brick courses and positioning of openings as well as horizontal structure; all of them can eliminate wastage of materials and speed construction sequences (see also Figs 4.8 and 17.33).

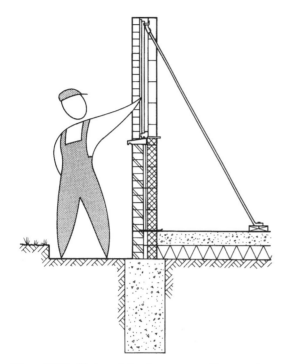

Figure 17.39 Timber window frame temporarily propped until lintel construction has been completed.

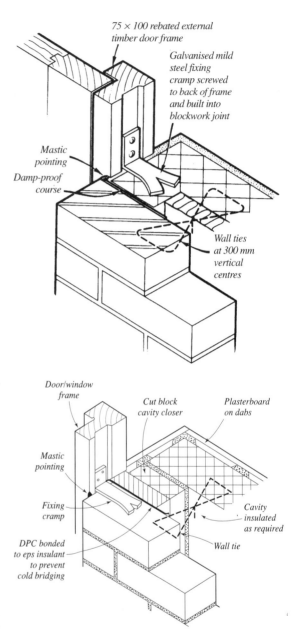

Figure 17.40 Alternative fixing methods for external door (or window) frame to brick/block cavity wall.

be treated in the factory with a preservative solution and, ideally, a primer so that all hidden surfaces are protected after building in. Both preservative and primer must be chemically compatible with subsequent paint systems. Alternatively, timber frames can be delivered to site having been treated with a decorative preservative stain which only requires recoating after installation.

Before any wall construction is built around the frames, they should be checked with a level to ensure perfect verticality, then they can be securely propped (Fig. 17.39). As the wall is constructed around the frames, they will become permanently fixed by building in right-angled galvanised mild steel lugs screwed to their side edges as the work proceeds (Figs 17.40 and 17.41). The cavity of the wall construction is closed by cutting the blockwork and making a 90° return to the inside face of the brick outer leaf. As this work is carried out, a vertical DPC is *bedded* between the two wall materials to form an effective water barrier. To ensure that the sides or jambs of the cavity wall construction are sufficiently robust to withstand the forces exerted by the door or window frames (wind pressures, slamming, etc.), the cavity wall ties around the opening are positioned at increased vertical centres of 300 mm or every three horizontal brick courses.

Once the cavity wall construction has reached the top of the door or window frame, a lintel construction must be formed to carry the walling above the opening.

Together with the number and proportion of openings, the method of forming a lintel has a critical influence on the appearance of the building. Furthermore, if not formed correctly, the lintel construction will also give rise to damp penetration problems as well as condensation due to occurrence of cold bridges. Lintels can be formed with reinforced concrete; with preformed steel section; or with a combination of these two which may or may not include a structural contri-

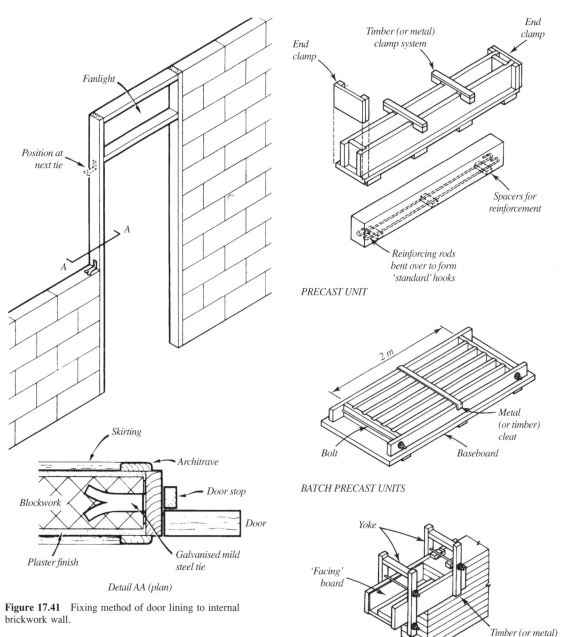

Detail AA (plan)

Figure 17.41   Fixing method of door lining to internal brickwork wall.

PRECAST UNIT

BATCH PRECAST UNITS

CAST IN SITU

Figure 17.42   Methods used for casting reinforced concrete lintels.

bution of the walling material itself. When considering the thermal insulation value of a door or a window opening, its effect on the total value provided by the whole external wall enclosure must be with regard to the thermal insulation provisions of the Building Regulations; see also section 8.3.

Reinforced concrete lintels can be formed *in situ* when the openings to be spanned are sufficiently large to make the bulk of the lintel too heavy to lift into position. Otherwise, they are generally preformed or precast by setting up a precast unit mould into which the

steel reinforcement is placed at the bottom on spacers to ensure adequate concrete cover. Concrete can then be poured into the mould. When many lintels of similar shape and size are required, a *batch precast unit* mould can be employed. This and the other methods of forming reinforced concrete lintels are indicated in Figure 17.42.

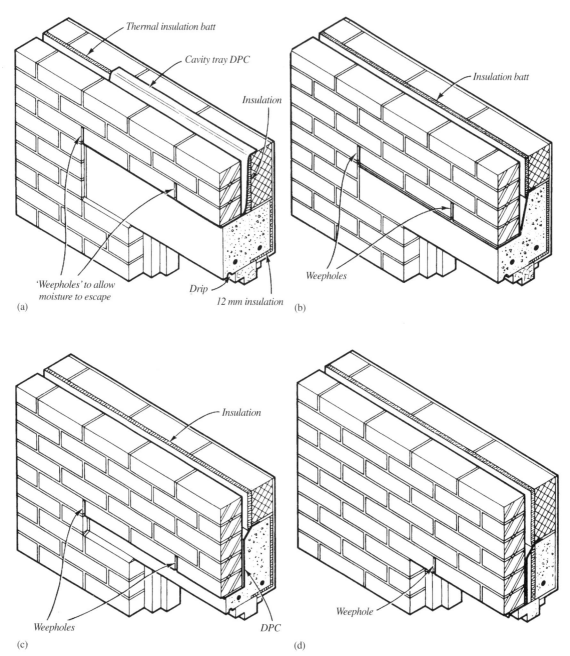

Figure 17.43 Lintel construction: (a) rectangular section: precast or *in situ* reinforced concrete lintel; (b) & (c) boot section: precast or *in situ* reinforced concrete lintel; (d) composite metal and precast reinforced concrete lintel; (*continued*)

The various forms of lintel construction are shown in Figure 17.43. Where thermal insulation standards have to satisfy current UK Building Regulations, the concrete examples are no longer applicable as excessive heat loss and cold bridging will occur through these units. Notice also that each variation considerably affects the building appearance.

(a) *Rectangular section precast or in situ reinforced concrete lintel*
The full depth of the concrete lintel shows on the face of the building. Because of its fully rect-angular section, a separate cavity tray must be provided in order that water percolating in the cavity wall is diverted through the weepholes to

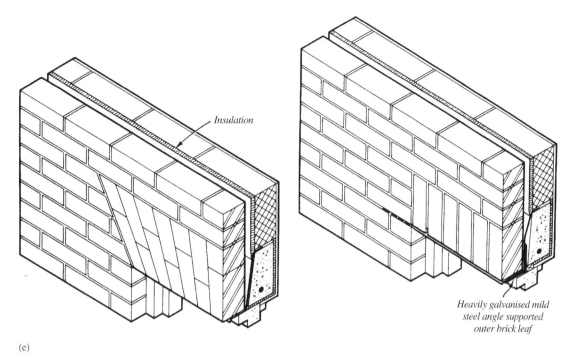

Insulation

Heavily galvanised mild
steel angle supported
outer brick leaf

(e)

Figure 17.43 Lintel construction (*continued*): (e) composite metal and precast reinforced concrete lintel with exposed brickwork arch construction. (*continued*)

the outside. The effects of heat losses (cold bridge and condensation) should be lessened by providing insulation material on the internal faces of the lintel, although this goes only a small way towards matching the total thermal insulation value of the wall supported, and 'pattern staining' is likely to occur (see section 8.3). The provision of a pelmet board for curtain hanging helps to hide the resulting dark staining. Depth of lintel depends upon the span.

(b) *'Boot' section precast or in situ reinforced concrete lintel*
Similar to (a) except that the lintel is shaped to reduce the amount of exposed concrete on the face of the wall. The sloping shape in the cavity provides support for the cavity tray, which would otherwise be liable to damage during construction, especially when the cavity is being raked clean of mortar droppings.

(c) *'Boot' section precast or in situ reinforced concrete lintel*
Similar to (b) except that the face depth of the externally exposed concrete is further reduced. To add to the visual effect of this lintel, the front edge or toe of the lintel can be slightly recessed back from the brick face to create a shadow.

(d) *Composite metal and precast reinforced concrete lintel*
This uses a precast concrete lintel to support the blockwork inner leaf and a preformed metal section to support the brickwork outer leaf (see Fig. 6.8). The metal lintel also acts as a cavity tray, but it is recommended that the lintel is heavily galvanised or painted with bituminous paint prior to being placed in position, in order to prevent corrosion. This composite lintel system maintains the cavity wall construction right down to the head of the door/window frame, and with the internally insulated concrete lintel goes some way towards maintaining a degree of thermal insulation at this point in the wall construction. Except for the narrow edge of the metal lintel, the supporting mechanism over openings is not visible and gives a different aesthetic from the exposed concrete lintels previously described.

(e) *Composite metal and precast reinforced concrete lintel with exposed brick arch construction*
Both diagrams show a brick lintel construction but the right-hand version would not actually provide structural support to the outer leaf unless reinforced internally. Both brick arch forms are in fact supported by the preformed non-ferrous or galvanised metal section.

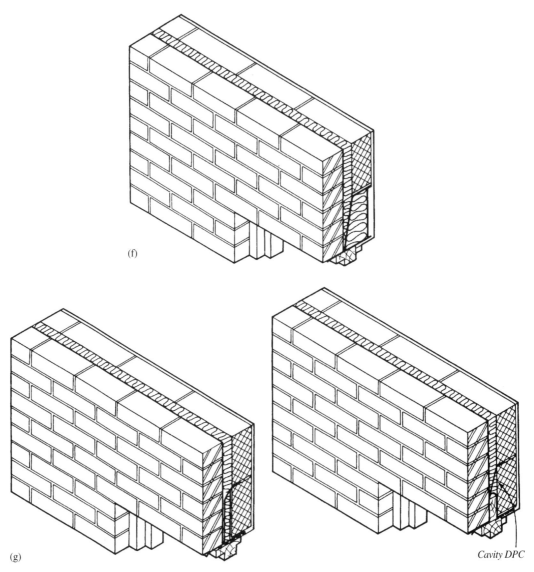

(f)

(g)

Cavity DPC

Figure 17.43   Lintel construction (*continued*): (f) preformed metal lintel; (g) preformed metal lintel and blockwork infill.

(f) *Preformed metal lintel*

The full advantage of preformed lightweight metal sections is taken in this example and there is no heavy precast concrete employed. The lintels can be supplied with the cavity filled with insulating foam, which gives thermal insulation properties corresponding to those of the supported wall. Again, the lintel is virtually unseen after construction is complete.

(g) *Preformed metal lintel with block infill*

Both these maintain the inner blockwork leaf as part of the lintel construction and, although giving a slightly reduced insulation value than the wall

construction, they provide a consistency of internal surfaces for fixings and finishes.

These forms of lintel construction would be suitable for doors and windows; alternative forms of sill construction for each of the two components are indicated in Figure 17.44. The threshold variation shown at (e) is to satisfy the Building Regulations, Part M: *Access and facilities for disabled people*. The principal entrance to all dwellings must not have a vertical projection in excess of 15 mm. This, along with a minimum clear opening width of 775 mm, is to ease wheelchair manoeuvrability.

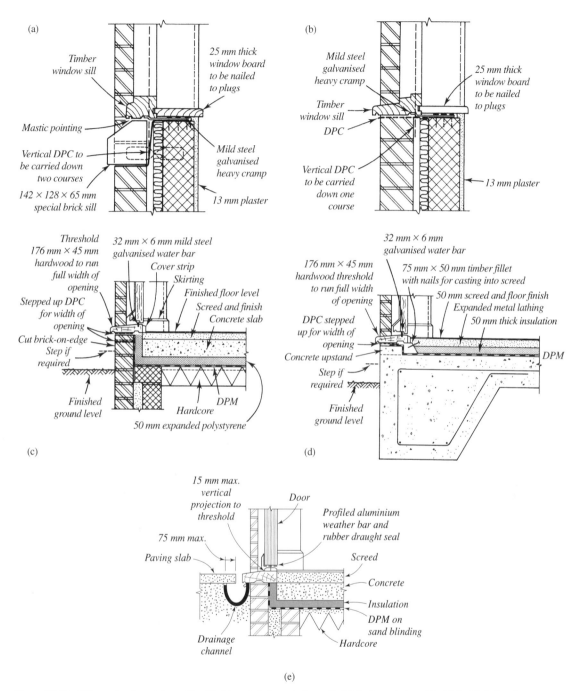

Figure 17.44 Sill construction: (a) combined brick and timber window sill; (b) timber window sill; (c) & (d) timber door threshold, with variation (e) for disabled access.

17.6.6 Scaffolding

As the work progresses beyond a height where it is reasonable for the bricklayer to lift materials from ground level, it will be necessary to erect scaffolding to support raised working platforms. The scaffolding generally consists of tubular steel or aluminium alloy connected by special fittings or couplings; the platform consists of scaffold boards of softwood. Figure 17.45 and some subsequent construction sequence figures

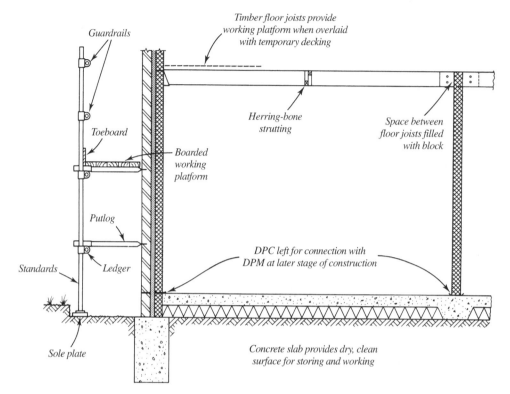

Figure 17.45 Construction sequence: first lift.

indicate *putlog scaffolding*, which consists of a single row of uprights set away from the wall for a sufficient distance to accommodate a working platform. The uprights (standards) are connected or coupled together with horizontal members (ledgers) which are tied back to the wall under construction by means of putlogs. Scaffolding platforms are provided at the required height by coupling putlogs to the ledgers, and their flattened ends are supported on the outer leaf of the cavity wall construction. After the scaffolding is no longer required, it will be dismantled and putlog holes filled or made good.

17.7 Internal wall construction
CI/SfB (22) + (32) + (42)

17.7.1 Function

Internal walls, or partitions, provide the physical space separation within a building which is necessary to isolate certain activities in order to provide privacy and security. They may be required to provide fire protection between the spaces they enclose as well as between the spaces and general circulation routes. A sufficient amount of thermal insulation and sound control may be necessary to permit some check on overall

performances. Internal walls must be robust enough to take various fixings for furniture and equipment, and be sufficiently durable to withstand the wear and tear associated with the activities of the building in which they are located. They should be of pleasing appearance, colour and texture which together should be compatible with the overall aesthetic character of the building of which they form a part.

Internal walls can be non-load-bearing or load-bearing. Load-bearing internal walls will assist in distributing the combined loads (dead, live and wind) of the building down to the foundation system then to the supporting soil. Also, they may sometimes be required to act as buttresses and provide lateral restraint to the external walls. By acting as sides of cellular forms (in conjunction with external wall and rigidly fixed floor/ceiling joists), the whole constructional system acts in structural unison in resisting the combined building loads (see section 5.3 and Fig. 5.4).

When planning the layout of internal walls, consideration must be given to the various components applied to the walls, floors and ceilings of the spaces they define. It is particularly important that, as with external walls, door openings are located so as to avoid excessive cutting and wastage of the bricks and blocks (see section 4.7).

17.7.2 Materials

Like external walls, internal walls may also be built from bricks or blocks, but rarely both together. Internal walls are seldom cavity walls unless they are used to separate two dwellings (i.e. party walls) or are designed as specific sound control barriers, e.g. walls to music rooms.

The internal walls of a house separate rooms in which the activities are relatively quiet. Also, at most, they are required to support only domestic loads from floors or roofs. Therefore, a single 102.5 mm thick brick (half brick), or 75–100 mm thick block is generally adequate and satisfies slenderness ratio requirements (see section 5.7). These thicknesses and materials are adequate in giving support for internal doors and frames, as well as fixings for cupboards, sanitary appliances, kitchen units, etc. They will also provide an adequate domestic standard of sound control, thermal insulation and fire resistance. For sound insulation specifically between bathrooms, bedrooms and other rooms, masonry walls will require a minimum mass of 120 kg/m^2, excluding finishing treatment.

As an alternative to bricks or blocks, internal partitions can be constructed from preserved timber using 50 mm × 100 mm vertical studs at 400 mm centres (to suit 1200 mm wide plasterboard cladding) braced at the midpoint by staggered timber noggins of the same cross-section (Fig. 17.46). The timber sections are planed prior to erection to ensure that the wall is of constant thickness with parallel faces to facilitate easy fixing of the plasterboard.

Another alternative is galvanised cold rolled steel channel profiles to BS 7364: *Specification for galvanised steel studs and channels for stud and sheet partitions and linings using screw fixed gypsum wallboards*, which can be used to provide a lightweight construction. Studs and channels are available in a variety of sizes and lengths, the smallest profile commencing with 50 × 32 mm channels for head and sole plates secured to the ceiling and floor. These are compatible with 48 × 32 mm stud sections which can be cut to length to slot into head and sole plates. Flat headed self-tapping screws are used as fixings. Channels 50 mm deep with a timber core are used around door openings. Plasterboard of 12.5 mm minimum thickness is secured to the framing with self-drilling/self-tapping countersunk head screws through the board and into the framing. As with all dry lining, taper edge boarding is preferred, with joints plaster filled, paper taped and plaster skimmed.

Stud partitions are generally non-load-bearing and are easy to construct, lightweight, adaptable and can be clad as well as infilled with various materials to give different finishes and properties. Stud partitions are of 'dry' construction and should not be commenced until the building has been made waterproof. Where sound insulation is critical, i.e. separating bedrooms and bathrooms from other rooms, the linings to each side of the studwork should be of at least two layers of minimum mass 9 kg/m^2 each. Two 12.5 mm plasterboard sheets each side will be adequate.

17.7.3 Construction

Brick and block internal walls can be erected directly from the ground floor solid concrete slab; from a sleeper wall forming part of a suspended ground floor construction; or, provided loads are not excessive, from some point between the main supports of a suspended floor (Fig. 17.47).

When subsoil conditions are less than stable or the loads on the partition are high, it will be necessary to provide separate foundations for internal walls (see Figs 17.26 and 17.27). The wall should incorporate a DPC, and when a separate foundation is not used, the DPC can be laid directly on the concrete slab and the first course of bricks or blocks bonded to it by a bed of mortar. This DPC should have been lapped with the DPC of the external cavity wall inner leaf, and must also extend a short distance either side of the internal wall so that eventually it can be stuck to the DPM laid on the concrete slab, at a later stage of construction. At junctions with the external wall, the partition will be *block bonded* into the inner leaf – it will be built into this leaf vertically every other block course, or be attached to the inner leaf with a proprietory system of structural profiles and ties.

First floor internal walls can also be constructed using bricks or blocks. Continuing upwards from the ground floor internal wall presents no particular problems, although when floor joists penetrate the partition care must be taken to ensure the brick or blockwork between is continued *solidly* to the underside of the wall above. If there is no wall below, a first floor non-load-bearing wall can be carried on double floor joists which have been bolted together at about 600 mm centres to ensure they act in unison. However, as the partition load may cause supporting joists to deflect slightly and therefore damage later finishes, it is probably best to insert a steel beam at this point (Fig. 17.48).

A non-load-bearing internal wall running *across* the timber joists can be carried by providing a timber sole plate at its base. Load-bearing partitions not continued from the floor below should always incorporate a rolled steel joist (RSJ) or a universal beam (UB) support in the intermediate timber floor construction.

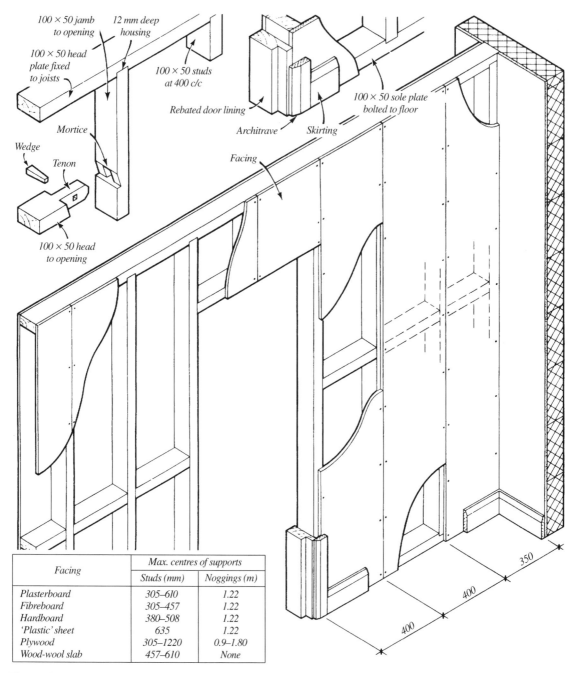

Facing	Max. centres of supports	
	Studs (mm)	Noggings (m)
Plasterboard	305–610	1.22
Fibreboard	305–457	1.22
Hardboard	380–508	1.22
'Plastic' sheet	635	1.22
Plywood	305–1220	0.9–1.80
Wood-wool slab	457–610	None

Figure 17.46   Stud partition details.

17.7.4 Door openings

Doors may be incorporated in the internal wall by building in standard *door linings* using galvanised metal ties similar to those used for the external door and window frames (see Fig. 17.41). The space over the door is often filled with a window or *fanlight* rather than solid brick or blockwork. The use of a fanlight avoids the need for a lintel in this position and also for cutting the bricks or blocks to bond between the walls either side of the opening. Although the use of fanlights may be economical, care must be taken to ensure that the

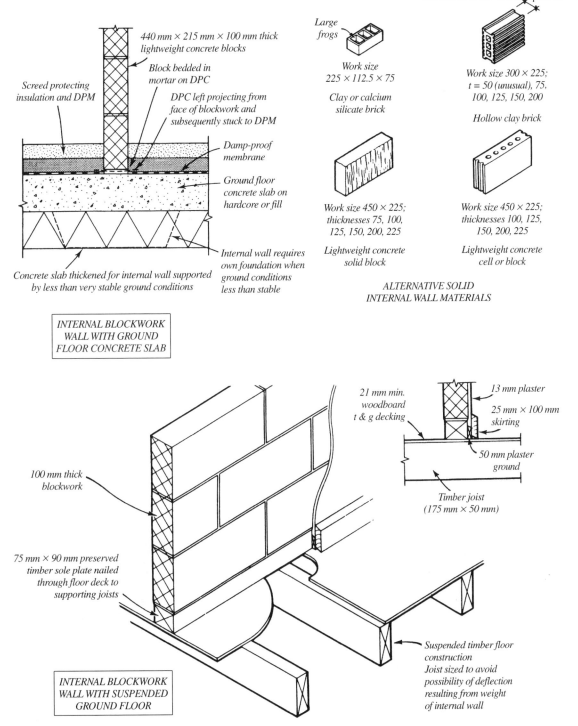

440 mm × 215 mm × 100 mm thick
lightweight concrete blocks

Block bedded in
mortar on DPC

Screed protecting
insulation and DPM

DPC left projecting from
face of blockwork and
subsequently stuck to DPM

Damp-proof
membrane

Ground floor
concrete slab on
hardcore or fill

Concrete slab thickened for internal wall supported
by less than very stable ground conditions

Internal wall requires
own foundation when
ground conditions
less than stable

**INTERNAL BLOCKWORK
WALL WITH GROUND
FLOOR CONCRETE SLAB**

Large
frogs

Work size
225 × 112.5 × 75

Clay or calcium
silicate brick

Work size 300 × 225;
t = 50 (unusual), 75,
100, 125, 150, 200

Hollow clay brick

Work size 450 × 225;
thicknesses 75, 100,
125, 150, 200, 225

Lightweight concrete
solid block

Work size 450 × 225;
thicknesses 100, 125,
150, 200, 225

Lightweight concrete
cell or block

**ALTERNATIVE SOLID
INTERNAL WALL MATERIALS**

21 mm min.
woodboard
t & g decking

13 mm plaster

25 mm × 100 mm
skirting

50 mm plaster
ground

Timber joist
(175 mm × 50 mm)

100 mm thick
blockwork

75 mm × 90 mm preserved
timber sole plate nailed
through floor deck to
supporting joists

Suspended timber floor
construction
Joist sized to avoid
possibility of deflection
resulting from weight
of internal wall

**INTERNAL BLOCKWORK
WALL WITH SUSPENDED
GROUND FLOOR**

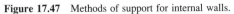

Figure 17.47 Methods of support for internal walls.

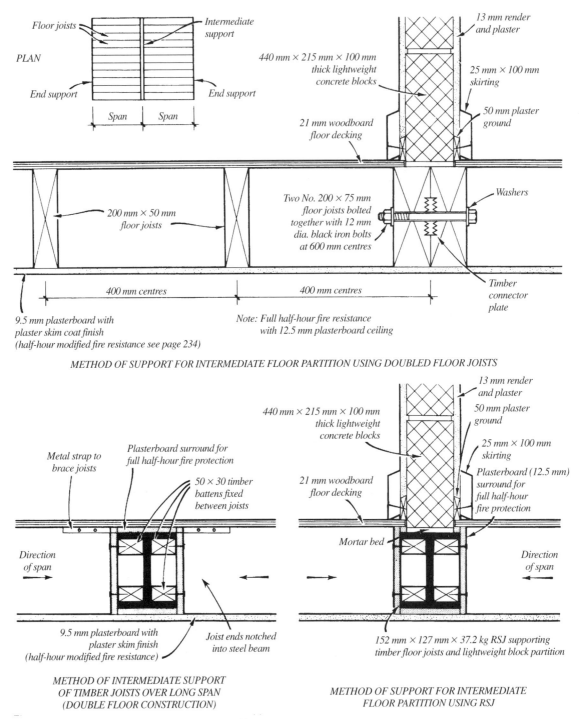

PLAN

Floor joists

Intermediate support

End support

End support

Span

Span

13 mm render and plaster

440 mm × 215 mm × 100 mm thick lightweight concrete blocks

25 mm × 100 mm skirting

50 mm plaster ground

21 mm woodboard floor decking

200 mm × 50 mm floor joists

Two No. 200 × 75 mm floor joists bolted together with 12 mm dia. black iron bolts at 600 mm centres

Washers

Timber connector plate

400 mm centres

400 mm centres

9.5 mm plasterboard with plaster skim coat finish (half-hour modified fire resistance see page 234)

Note: Full half-hour fire resistance with 12.5 mm plasterboard ceiling

METHOD OF SUPPORT FOR INTERMEDIATE FLOOR PARTITION USING DOUBLED FLOOR JOISTS

Metal strap to brace joists

Plasterboard surround for full half-hour fire protection

50 × 30 timber battens fixed between joists

Direction of span

9.5 mm plasterboard with plaster skim finish (half-hour modified fire resistance)

Joist ends notched into steel beam

METHOD OF INTERMEDIATE SUPPORT OF TIMBER JOISTS OVER LONG SPAN (DOUBLE FLOOR CONSTRUCTION)

440 mm × 215 mm × 100 mm thick lightweight concrete blocks

21 mm woodboard floor decking

13 mm render and plaster

50 mm plaster ground

25 mm × 100 mm skirting

Plasterboard (12.5 mm) surround for full half-hour fire protection

Mortar bed

Direction of span

152 mm × 127 mm × 37.2 kg RSJ supporting timber floor joists and lightweight block partition

METHOD OF SUPPORT FOR INTERMEDIATE FLOOR PARTITION USING RSJ

Figure 17.48 Methods of support for first floor partitions.

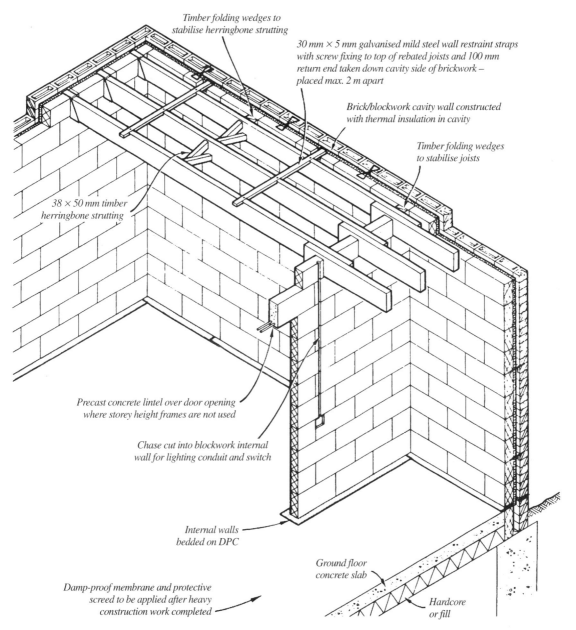

Timber folding wedges to
stabilise herringbone strutting

30 mm × 5 mm galvanised mild steel wall restraint straps
with screw fixing to top of rebated joists and 100 mm
return end taken down cavity side of brickwork –
placed max. 2 m apart

Brick/blockwork cavity wall constructed
with thermal insulation in cavity

Timber folding wedges
to stabilise joists

38 × 50 mm timber
herringbone strutting

Precast concrete lintel over door opening
where storey height frames are not used

Chase cut into blockwork internal
wall for lighting conduit and switch

Internal walls
bedded on DPC

Damp-proof membrane and protective
screed to be applied after heavy
construction work completed

Ground floor
concrete slab

Hardcore
or fill

Figure 17.49 Lateral bracing of walls by intermediate floor construction.

interrupted lengths of internal walls are adequately
stabilised at door openings. When an internal wall is
load-bearing, stability is enhanced by embedding intermediate floor or ceiling joists onto the top of the wall,
as shown in Figure 17.49. If the partition is non-
load-bearing and does not support floor or ceiling joists,
it is necessary to connect the top of the wall firmly
to the underside of joists with timber wedges. Often

the top of the brick or block partition is capped by a
50 mm deep timber wall-plate, which facilitates this
process by providing a level top surface. When a load-
bearing partition is not required to continue up to the
next floor, this timber plate also provides a level fixing
for the supported joists (see Fig. 17.45).

Brief comments regarding the fixing of door linings
are given in section 17.11.2.

17.8 Intermediate floor construction
CI/SfB (23) + (33) + (43)

17.8.1 Function

The construction of the external cavity wall and the internal walls of the house continues until the correct height is reached for the first floor timber joists to be put in position. The function of the *intermediate floor*, of which these joists form a part, is to provide a level surface with sufficient strength to support the dead loads of flooring and ceiling plus imposed loads of people and furniture. In addition, it must provide a degree of sound control and fire protection between the two levels. Often an intermediate floor must also assist in providing lateral restraint to heights of external and internal walls. This form of wall stability is shown in Figure 17.49 and may be achieved by the following means:

- At least 90 mm end bearing of joists
- Approved joist hangers
- Lateral restraint straps at no more than 2 m spacing

17.8.2 Materials

The materials used for the structure of domestic intermediate floors are reinforced concrete, or pressed metal joists, or timber joists. Although reinforced concrete construction (*in situ* or precast units) has many advantages associated with sound control and fire resistance, the materials are heavy and can involve complicated methods of assembly. Pressed steel joists do not have these advantages but are very lightweight; however, timber joist construction remains the most adaptable.

In working out the layout of timber joists, consideration must be given to their spacing, the clear spans and openings to be formed or 'trimmed' for the staircase, etc. Table 17.2 indicates typical spacings and spans. Variations in spans can be achieved for a given cross-sectional size of timber according to the actual stresses it can withstand. This capacity depends upon characteristics of the timber (structure, grain, knots, etc.) and is stress-graded to BS 4978: *Specification for softwood grades for structural use*, to provide known limits of loading which make the most economical use of timber.

The sizing of joists can be arrived at by calculation or by reference to tables in the Building Regulations or supplied by timber-interested organisations such as TRADA Technology Ltd, the Nordic Timber Council UK, etc. An approximate, although uneconomical, guide can be given by the formula $D = (S/24) + 50$, where D = depth of joist in mm and S = span in mm. This formula assumes that joists have a breadth of 50 mm and are used at 400 mm centre spacing.

Timber joists should be treated with a preservative against fungal and insect attack prior to final placing in the building. Joist lengths can be treated before delivery to site, but touching up on site will always be necessary as a result of cutting. Particular care needs to be taken with the ends of joists because they are most susceptible and are placed in vulnerable positions, such as adjoining the external wall construction.

The decking for timber intermediate floors can be tongue-and-groove hardwood or softwood boarding, or chipboard or plywood sheets. These are indicated in Figure 17.29. Tongue-and-groove profiles are preferred to butt-jointed boards because they reduce the effects of curling caused by thermal and moisture movement. In addition, the reduction of the amount of clear gaps is essential in obtaining a satisfactory resistance to the travel of flames and smoke during a fire. A timber tongue-and-groove board decking to floor joists acts in conjunction with a 9.5 mm plasterboard finish to form the ceiling to give the '*modified*' *half-hour fire resistance*. This means that for at least half an hour the floor must not collapse or allow the passage of flame or smoke, but the integrity and insulation requirement is reduced to 15 minutes (see section 9.2). Brief comments regarding the fixing of floor decking are given in section 17.11.2.

17.8.3 Construction

For the cavity wall construction already described, the level of the top of 175 mm × 50 mm joists should coincide with the level of 35 courses of brickwork (each brick 65 mm + 10 mm for bed joint) from the top of the DPC, as shown in Figure 17.38. This will provide a joist top level of 2625 mm and an underside joist level of 2450 mm: an allowance of 50 mm for screed will permit a 2400 mm clear storey height (excluding plasterboard ceiling), which is ideal for the storey-height door frames, as mentioned earlier (see Fig. 17.37). The ends of the joists are traditionally built into the inner blockwork leaf of the cavity wall because this prevents the end of the joists twisting as well as providing lateral restraint to the external wall by preloading (see section 5.3). However, the current tendency to use an increased amount of sapwood for structural timber, and also to avoid the risk of cold bridge or condensation, which might occur if the inner leaf is interrupted, means that it is probably wiser to use a *metal hanger* to support the joists. These are made of galvanised mild steel and can be supplied with a 'hooked' tail, which provides the necessary lateral restraint to the blockwork leaf of the cavity wall. The joists will be prevented from twisting by the shape of the hanger (see Fig. 17.49). In order to avoid cutting the blocks of the inner leaf so that the tail of the joist hangers align perfectly with the top of the joists, it will be necessary to provide one course of 140 mm high blocks (150 mm including joint) instead of the normal 215 mm high blocks (255 mm including joint). At the other end to the external wall, timber joists can be lapped over partitions where the bedded *wall-plate* mentioned earlier assists in spreading the loads taken by the joists, and provides fixing for them by cross or 'skew' nailing. Another method for accurate levelling of joists over partitions is to use a bearer bar cogged to the joists.

Having fixed the structure of the floor, a temporary decking of plywood can be laid over the joists in order to provide a working platform for subsequent building processes. The final finished decking should not be fixed until the building is weatherproof and all necessary work in the void of the floor has been completed. Similarly, a ceiling of plasterboard should not be fixed until electric conduits, etc., have been installed and there is little risk of subsequent damage occurring. The plasterboard for the ceiling is usually fixed at the same time as any other plasterboard, e.g. for stud partitions, ductwork, etc. (see section 17.11).

17.8.4 Lateral bracing of joists

When the timber joists span more than 2.4 m, it is necessary to provide lateral bracing or strutting in order to restrict the movements due to twisting, and vibrations which would damage ceiling finishes (see Fig. 17.49). The strutting can consist of 38 mm × 50 mm softwood battens or galvanised steel sections fixed diagonally between joists in a herringbone pattern. They should occur at midspan or at centres not less than 50 times the width of one joist. Alternatively, blocks of timber can be fixed between the joists at alternative centres for each space; they are easier to fix but provide less restraint. It is important that the restraint provided to the joists by strutting is continued through to the flanking walls. This is done by ensuring that the first joist next to the wall is positioned so as to allow a 40 mm gap into which timber *folding wedges* are driven on the line of the strutting.

17.8.5 Lateral bracing of walls

Hooked-tail joist hangers provide restraint to walls once in position and the joists have been securely nailed in position. Lateral bracing may also be required for walls which run parallel with the direction of span of the timber joists, particularly when the flanking wall is an external cavity wall. Bracing in this direction can be provided by galvanised mild steel straps fixed at not

more than 2 m centres, each strap being screwed to at least three joists and having a 90° return end built into the cavity wall construction (see Fig. 17.49).

17.8.6 Notches and holes in joists

Structural timbers used today can be stress-graded to give maximum economical depth/span ratios. It is therefore important that certain rules are observed should it be necessary to notch or drill holes in timber joists so that services (electrical cables, central heating pipes, etc.) can be accommodated within the floor space. For example, it is recommended that *notches* should occur within the middle half of the span of any joist, should be in the upper cross-section, and should never be more than one-eighth the depth of the joist. Holes should be situated within the middle two-thirds of the joist span and their diameter should not exceed one-sixth of the depth of the joist. In practice it may be very difficult to prevent workers on site from placing notches or holes anywhere in the joists; they tend to choose the most suitable position for their particular services. To overcome this problem, a prudent designer will oversize the joist sizes when services are required to be concealed in an intermediate floor space.

17.8.7 Openings in floor

Openings must be left in the intermediate floor so that staircases can be fitted and, perhaps, to allow service pipes which have been grouped together or central heating air ducts to pass from one floor to another. This means that the timber joists, which would normally span from one wall to another, would be stopped short and *trimmed* to allow for an opening. It is good practice for the designer to provide floor trimming layout drawings so that the carpenters can set out the joists for maximum usage. The joists forming the perimeter of the required opening will have to be at least 25 mm wider than the other joists, as indicated in Figure 17.50. The joists interrupted by the formation of the opening are trimmed and secured to a *trimmer joist* which in turn is carried by *trimming joists*. Methods adopted for fixing trimmed joists to a trimmer, and for fixing trimmers to trimming joists, will vary according to the expertise of the contractor's workers. Timber joints (tusk tenon and housed joints) can be skilfully carried out, or, as an alternative, fixing plates or joist hangers can be employed.

Typical details of a timber staircase to be incorporated in an opening are shown in Figure 17.51. These can be made on site, but it is more usual for prefabricated staircases to be fitted into a prepared opening.

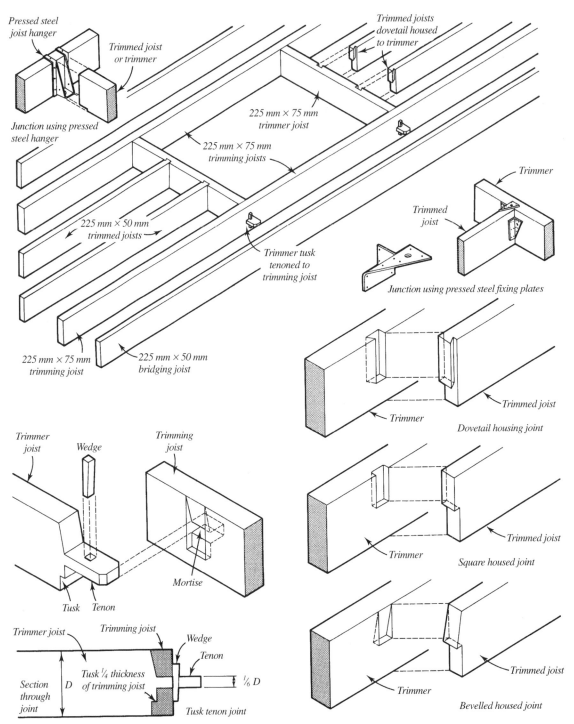

Figure 17.50   Trimming to timber floors.

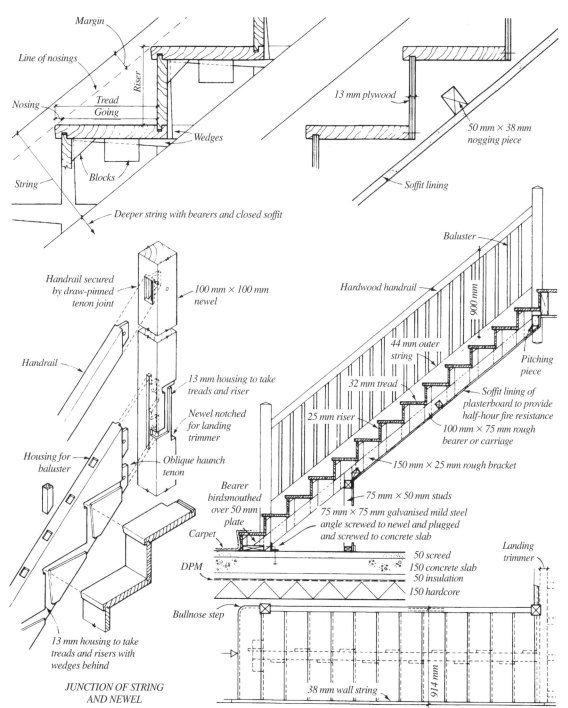

Figure 17.51 Typical timber staircase details.

17.9 Roof construction
CI/SfB (27) + (47)

17.9.1 Function

A roof is more vulnerable to the effects of wind, rain, snow, solar radiation and atmospheric pollution than any other part of a building. While avoiding the harmful results of these phenomena, it must also prevent excessive heat loss from the building in winter and be able to keep the interior cool during hot seasons. And in serving these functions, the roof construction is required to defy gravity by spanning horizontally between load-bearing elements (unless the roof and walls are combined in a tensile structure as described in section 5.6) while accommodating all stresses encountered, including movements due to changes in temperature and moisture content.

The roof of a building must also provide adequate defence for the occupants against the effects of fire and must limit its potential to spread. This is done by ensuring that the *external* cladding is able to withstand the *penetration* of fire from adjacent sources and for a defined period according to the proximity of the roof to a boundary. The roof cladding must also be able to limit the *spread of flame* over its surface.

Although hybrid variations occur, it is most convenient to consider roofs as being either *flat* if the external face is at not more than 10° to the horizontal, or *pitched* if the external face is at more than 10° to the horizontal. A *short span* roof is considered to be up to 7.5 m, *medium span* 7.5 m to 25 m, and *long span* over 25 m.

Apart from the purely technical factors, the choice between flat or pitched roof forms involves consideration of aesthetic appeal. When a flat roof is chosen, the buildings have a 'cut-off', or 'block' shape, but the appearance of the roof surface is relatively unimportant unless it is overlooked from higher levels or used as a terrace or for vehicular traffic. Conversely, the selection of a pitched roof will provide a dominant visual feature, the importance of which will depend on the viewing distance. Then the shape, colour and texture of the roof surface are very important factors.

17.9.2 Pitched roof construction

Of the many considerations involved in the selection of an appropriate form, the construction of a pitched

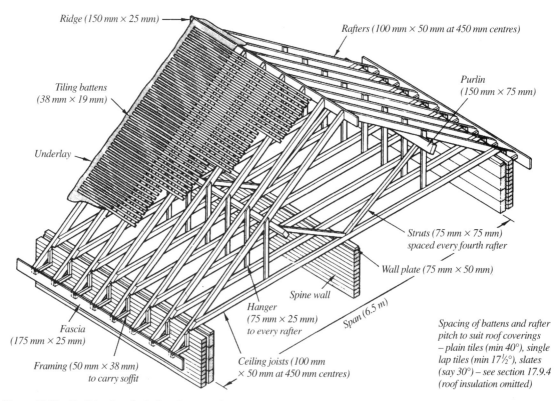

Figure 17.52 Traditional method of roof construction.

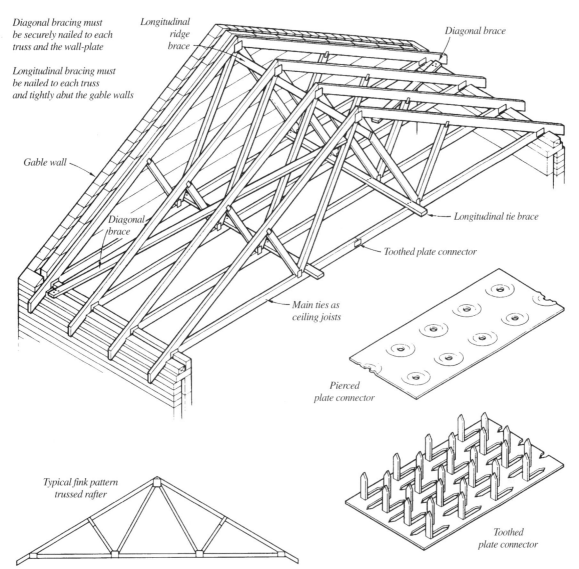

Diagonal bracing must be securely nailed to each truss and the wall-plate

Longitudinal bracing must be nailed to each truss and tightly abut the gable walls

Longitudinal ridge brace

Diagonal brace

Gable wall

Diagonal brace

Longitudinal tie brace

Toothed plate connector

Main ties as ceiling joists

Pierced plate connector

Typical fink pattern trussed rafter

Toothed plate connector

Figure 17.53   Trussed rafter roof construction.

roof is mainly dictated by the shape of the building and the span between supporting elements. The arrangement of the structural members forming the roof may follow similar principles, but their size, disposition and jointing methods will vary considerably. Figure 17.52 indicates a typical conventional domestic pitched roof construction. However, for simple spans without internal support walls, a method employing timber *trussed rafters* could be more economical (Fig. 17.53).

The modern trussed rafter roof indicates the development in empirical knowledge of triangulated roof forms. It involves the incorporation of more recent scientific

techniques resulting from the analysis of actual stresses induced in timber sections as well as at the critical points where two or more members are joined to act in unison. Instead of forming joints on site, using oversized sections for structural safety and convenience, the trussed rafter employs stress-graded timbers (section 17.8.2) joined when appropriate and with precision under factory-controlled conditions by galvanised mild steel *truss plates*. The resulting prefabricated, economical and lightweight truss rafters are delivered to the site at an appropriate time in the building programme. They are then hoisted into position, placed on

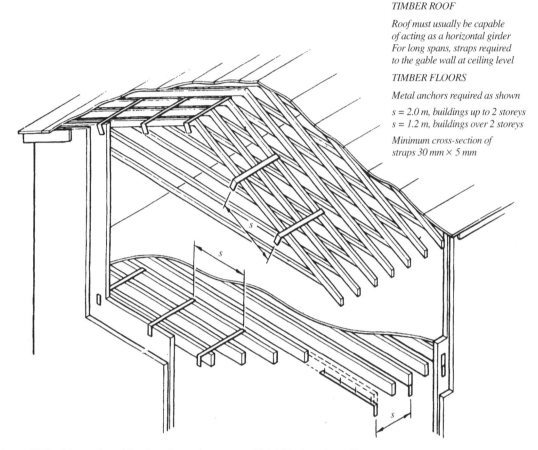

TIMBER ROOF

*Roof must usually be capable
of acting as a horizontal girder
For long spans, straps required
to the gable wall at ceiling level*

TIMBER FLOORS

Metal anchors required as shown

*s = 2.0 m, buildings up to 2 storeys
s = 1.2 m, buildings over 2 storeys*

*Minimum cross-section of
straps 30 mm × 5 mm*

Figure 17.54   Means of providing lateral restraint to external brick/block cavity walls.

the supporting wall and relatively rapidly fixed. However, it is particularly important that the light trusses are adequately protected from the elements during the period from when they leave the factory until they are finally covered by the external roof cladding. Should these trusses be allowed to become exposed to the weather for any length of time, the relatively thin sections of timber will become saturated, causing them to twist and loosen the truss plates. Twisting is still likely to occur after the roof construction is completed, particularly if the central heating of the building causes rapid drying.

17.9.3 Pitched roof bracing

Trussed rafters are spaced at close centres, usually 600 mm, to provide direct support for roof cladding and ceiling, which in turn helps in giving the necessary *lateral bracing* to the individual trussed rafter. Each rafter is securely nailed to the wall-plate already placed on a mortar bed on top of the cavity closure block, as indicated in Figure 17.56. The trussed rafters are lightweight and a wind tends to lift the whole roof construction, so it is important that this wall-plate is anchored to the supporting wall by galvanised mild steel straps at about 1.5 m centres (2 m maximum). When a flank wall of the house continues up above the lowest part of the rafter to form a gable wall, it is important that lateral bracing is provided for this otherwise free-standing wall by similar steel straps, as indicated in Figure 17.54. Finally, the whole roof system should be provided with wind braces running diagonally from a bottom corner to a top corner. This brace will prevent the roof timber from collapsing sideways as a result of laterally applied wind forces. Other timber braces are generally required to provide stability when trussed rafters are used, in order to prevent twisting and deflection from vertical alignment.

17.9.4 Pitched roof – batten spacing and sizing

Margin – the amount of tile or slate exposed.
Gauge – the spacing of timber battens for tile or slate fixing.

Plain tile batten gauge = [tile length (mm) – lap (mm)]
$$\div 2 = [265 - 65] \div 2 = 100 \text{ mm}$$

Single lap tile batten gauge = tile length (mm) –
lap (mm)

Example: concrete interlocking tile 381 mm long × 229 mm wide, laid to a 75 mm head lap: gauge = 381 – 75 = 306 mm

Slate batten gauge:

Head nailed = slate length (mm) – [lap (mm) +
25 mm] ÷ 2
Centre nailed = [slate length (mm) – lap (mm)] ÷ 2
Example: 610 × 305 mm standard *Duchess* slate laid to a 125 mm lap (note that slates are manufactured in a variety of sizes with established traditional names; see producers' catalogues for a full listing)
Head nailed batten gauge = 610 – [125 + 25] ÷ 2
= 230 mm
Centre nailed batten gauge = [610 – 125] ÷ 2
= 242 mm
Note: Head nailing is generally limited to smaller slates as wind lift and vibration may snap larger slates.

Timber batten sizes vary with type of tile or slate application and spacing of rafters:

Plain tiles: 38 mm × 19 mm (rafters up to 450 mm c/c)
38 mm × 25 mm (rafters up to 600 mm c/c)
Single lap tiles: 38 mm × 25 mm (rafters up to
450 mm c/c)
50 mm × 25 mm (rafters up to
600 mm c/c)
Slates: 50 mm × 25 mm (rafters up to 600 mm c/c)

Batten nails: about 3 mm diameter × 65 mm long to penetrate the rafter at least 40 mm.

17.9.5 Pitched roof cladding

On a pitched roof, gravitational forces move rainwater down the slope and, aggravated by wind currents, try to move it inwards (see Fig. 6.6). Because the roof has a relatively rapid degree of run-off, the external covering usually consists of relatively small *lapped* units such as tiles or slates, which are dry jointed and supported independently at intervals by timber battens. As the slope of the roof decreases, the resistance to the inward flow of water becomes less. Therefore, the amount of water penetration likely depends on the angle of slope and the amount of horizontal overlap provided by the slates or tiles. The movement of water around the tiles or slates depends on the *angle of creep*: the bigger the overlap (which corresponds to the size of cladding unit), the less the pitch can be (Fig. 17.55). Slates and tiles are complemented as a water barrier by the independent waterproof membrane of sarking felt laid under the battens, as shown in Fig. 6.6. This membrane prevents wind and rain, driven through gaps in the cladding, from penetrating the roof void. It also prevents rain penetration where tiles are broken. However, these traditional bituminous felts do not have the water vapour permeability required to reduce moisture levels and condensation which can occur in highly insulated roofs. In conjunction with ventilated roofs (see Figs 17.57 and 17.59) a bonded polypropylene-based breather membrane underlay is preferred, as this is capable of permitting air circulation through the roof without perforation.

Working from the scaffolding platform at roof level, the sarking felt or underlay is unrolled and laid across the rafters starting at the lowest point (eaves). The sarking is placed over a tilting fillet and lapped over the fascia board to form a small projection or drip. It is important that the sarking is allowed to sag slightly between rafters in order to form the space which permits the water permeating around the slates or tiles to be caught by the sarking and transferred by gravity to the eaves drip and rainwater gutter. The sarking is nailed into position before the next horizontal length is placed so as to permit an overlap to be formed of at least 150 mm. All sides of the roof are continued upwards until the sarking is taken over the top of the roof or *ridge*, except where gaps are left for continuity of roof-void ventilation through purpose-made ridge tiles (see Fig. 17.59). Vertically sloping edges of the roof or *hips* are overlayed with strips of sarking to ensure a complete enclosure. Slate or tile battens are then fixed at horizontal centres to suit the type and size of cladding unit.

When cladding to the pitched roof is to be in slates, each horizontal row is laid starting again from the eaves. Each slate is butt-jointed at the side and overlapped at the head so as to form *two thicknesses* of slate over each nail hole as protection, making in all three thicknesses of material at the overlaps. The side butt joint should be left slightly open so that water will drain

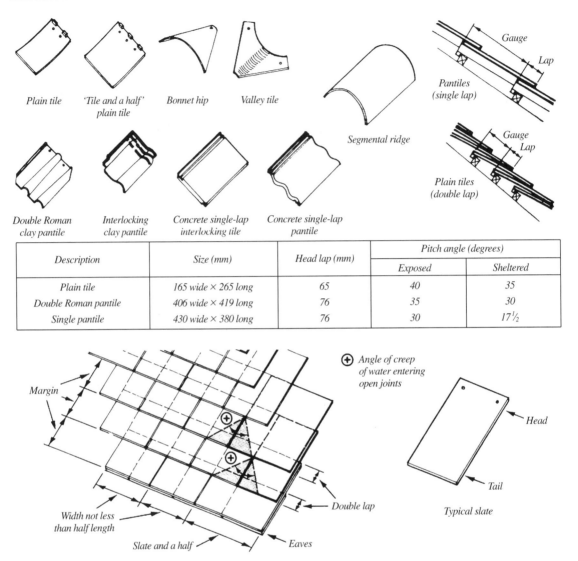

Description	Size (mm)	Head lap (mm)	Pitch angle (degrees)	
			Exposed	Sheltered
Plain tile	165 wide × 265 long	65	40	35
Double Roman pantile	406 wide × 419 long	76	35	30
Single pantile	430 wide × 380 long	76	30	17½

Figure 17.55 The angle of creep depends on the pitch of the roof. The steeper the pitch, the smaller the angle; this can be used to determine the minimum acceptable size for a tile or slate.

quickly. Special lengths of slate are required at eaves and sloping edges or *verges*. Each slate is fixed twice by yellow metal, aluminium alloy, copper or zinc nails. The slate should be holed so that breaking away of the edges of the hole (spalling) will form a countersinking for the nail-heads. This is best done by machine on site so that the holes can be correctly positioned by the fixers.

Clay plain tiles are fixed in a similar fashion except that each tile is retained on the battens by nibs, and, unless the roof is very exposed, they need only be nail-fixed every fourth or fifth course.

On completion of the external waterproofing system to the roof (Fig. 17.56), the rainwater gutters and drain-pipes should be installed to prevent the water collected by the roof surface from cascading down the face of the walls below and causing later problems. Work can also commence to the interior of the building, hitherto left until a dry interior 'climate' could be ensured.

17.9.6 Thermal insulation and ventilation

Pitched roofs using slate or tile external cladding can be either *warm roofs* or *cold roofs* according to the

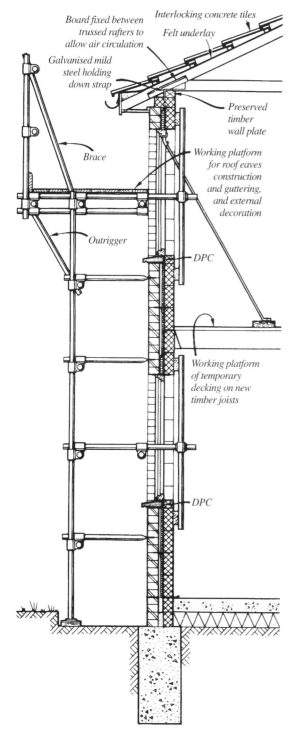

Board fixed between trussed rafters to allow air circulation

Interlocking concrete tiles

Felt underlay

Galvanised mild steel holding down strap

Brace

Preserved timber wall plate

Working platform for roof eaves construction and guttering, and external decoration

Outrigger

DPC

Working platform of temporary decking on new timber joists

DPC

Figure 17.56 Construction sequence: final lift.

positioning of the insulation (Fig. 17.57). An effective vapour control layer must always be incorporated on the warm side of the insulation.

Warm roof construction allows the roof space to catch the rising warm air from the rooms below, or even to be heated when it forms a room itself. Also, the roof timbers are protected from external solar heat gains which could cause excessive thermal movement and crack ceiling finishes, etc. The thermal insulating material is placed over the rafter prior to fixing the sarking. As it may be difficult to form the sag in the sarking necessary to allow permeating water from the roof cladding to run to the eaves, an alternative constructional system is required. This involves the incorporation of vertically fixed *counter battens* over the sarking and insulation on the lines of the rafters so that the space is created between the underside of the horizontal slate or tile battens (Fig. 17.58).

A similar system of counter battening is also necessary when boarding has to be fixed to the top face of the rafter to provide additional resistance to wind penetration. This boarding also gives the roof greater internal stability as well as greater defence against illegal entry. (There is easy access for a burglar into a dwelling through the roof; tiles can be lifted and the sarking felt cut away.)

The advantage of cold pitched roof construction lies in *energy conservation*; the roof space is unheated, and most of the heat in the rooms below is used solely for those spaces. However, thermal movements caused by solar gain within the roof space are likely to provide problems of movement and special cover strips will be required at junctions between the ceiling and the walls below. Having completed the outer cladding of the roof and made it waterproof, the builder is more able to accommodate insulating material over and/or between the ceiling joist later in the building programme, just prior to the finishes. However, care must be taken to ensure that the thermal insulation is placed *over* water cisterns so that the lost heat from the rooms below is at least used to prevent them freezing. Water pipes, including overflow pipes, must be separately lagged if they are located within the cold roof space.

As no vapour control layer system can be guaranteed to be entirely effective, it is very important that the space above the thermal insulation is adequately ventilated to remove the last vestige of moist air which could cause condensation. This is particularly important when 'cold roof' construction is employed and the Building Regulations (Approved Document F) require the provision of cross-ventilation by voids equivalent to a 10 mm gap at the eaves. However, insulation can

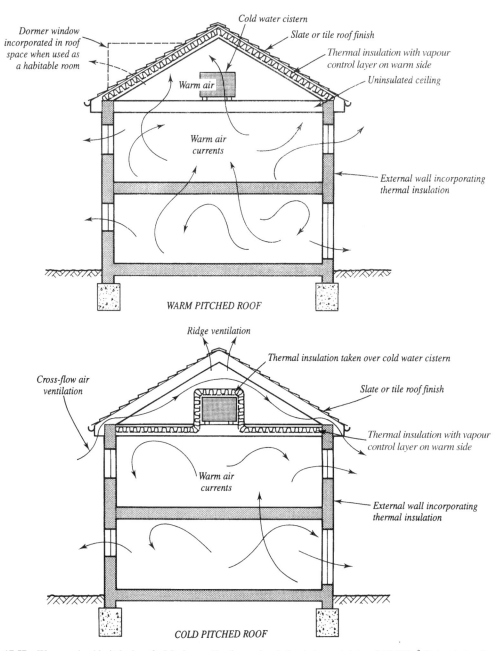

Figure 17.57 Warm and cold pitched roofs. Maximum *U* values – insulation between joists = 0.16 W/m² K, insulation between rafters = 0.20 W/m² K.

prevent this gap from being effective unless a board or patent profiled plastic spacer is inserted between rafters, as shown in Figure 17.59. Eaves ventilation gaps will also require a mesh screen to prevent insect penetration. Moreover, the effectiveness of ventilation can be more easily achieved by establishing the method

most suitable relative to the airflow required, determined by wind speed, the availability of a free flow of air and the plan shape of the roof space. Nevertheless, the Building Research Establishment (Digest 270) has concluded that 'air speeds required to avoid condensation (0.1 m/s) are much lower than those which cause

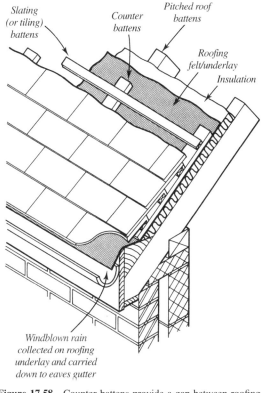

Figure 17.58   Counter battens provide a gap between roofing felt and slating battens (see also Fig. 6.6).

deterioration in performance of insulation (1.2 m/s)'. See also BS 5250: *Code of practice for control of condensation in buildings*, and Figure 17.59.

The use of tiles or slates provides a roof cladding which gives good protection against the effects of an external fire; they will not allow flames to penetrate into the roof space or allow flames to travel along their surface.

17.9.7 Flat roof construction

The structure of a flat roof is similar to intermediate floor construction, except that it must be waterproofed externally, thermally insulated, and, as lighter loads are usually required to be carried, the timber joist sections can be reduced in size. Because gravity cannot carry

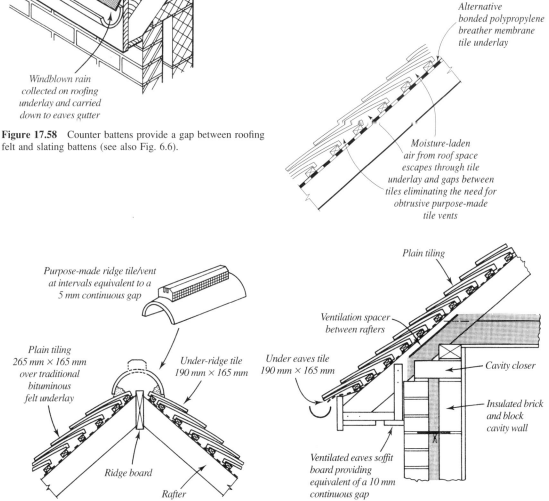

Figure 17.59   Ventilation of roof space.

the rainwater away, a *water barrier* must be provided, consisting of continuous impervious membrane such as jointed metal sheets, asphalt, multi-layer roofing felt or single-layer plastics. These have to be laid on a *firm decking to falls* to ensure that water drains away to the water collection points. On a timber-joist flat roof the falls are generally provided by *firring pieces*. Figure 17.60 illustrates typical details; particular care is required by the designer when detailing, and by the builder when constructing, the junctions between a 'flat' roof and any upstand, such as walls or projecting service pipes. When inadequately formed, these points usually provide sources of water penetration and, therefore, failure of the roof system. It should also be borne in mind that flat roof construction generally forms the major source of building failures in the United Kingdom; faults are attributed to poor design, inadequate materials and low standards of work.

Flat roof construction can also provide a 'warm' roof or a 'cold' roof, and various systems are shown in Figure 17.61. The jointed metal sheets and asphalt provide adequate protection against the effects of an external fire, but multi-layer roofing felts and single-layer plastic systems must be finished with a coating of stone chippings to provide the equivalent protection. Stone chippings also give protection against solar radiation and mechanical damage, and may therefore be incorporated on asphalt membranes.

17.10 Services
CI/SfB (5–) + (6–)

17.10.1 Below-ground drainage

Below-ground drains for a house are installed in trenches formed at appropriate falls by excavation techniques similar to those used for foundation trenches (Fig. 17.62). If the builder needs to hire an excavator or a specialist subcontractor for this purpose, it will be more economical to form all trenches, both foundation and drainage, during the same period of building work. Lengths of pipes can be installed and connected together up to convenient points where the above-ground drainage commences, such as gullies and WC outlets. The trenches can then be backfilled so as not to cause inconvenience to subsequent site and building work. Alternatively, the builder may prefer to carry out the majority of excavations for drainage (other than that affecting service pipes through foundation walls and ground floor slabs) once the bulk of the work on the building itself has been completed. In this way the installation of the drains, and subsequent backfilling processes, becomes part of the general work relating to external landscaping. Similarly, the connection to the public sewer can be delayed until most of the site drainage work is complete; or connection can be made at a very early stage in the building programme by forming

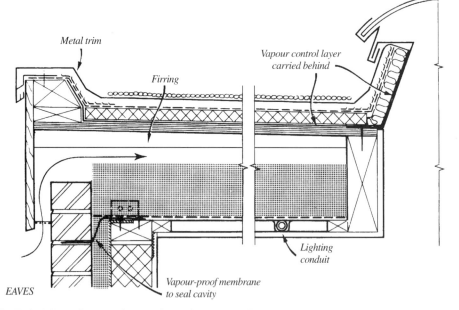

Figure 17.60 Typical flat roof construction. Details are for a cold roof construction where the thermal insulation and vapour control layer are held against the underside of structural timber joists by battens placed at right angles. In order to minimise the perforations in the vapour control layer, electrical lighting conduits are not installed in the structural zone of the roof.

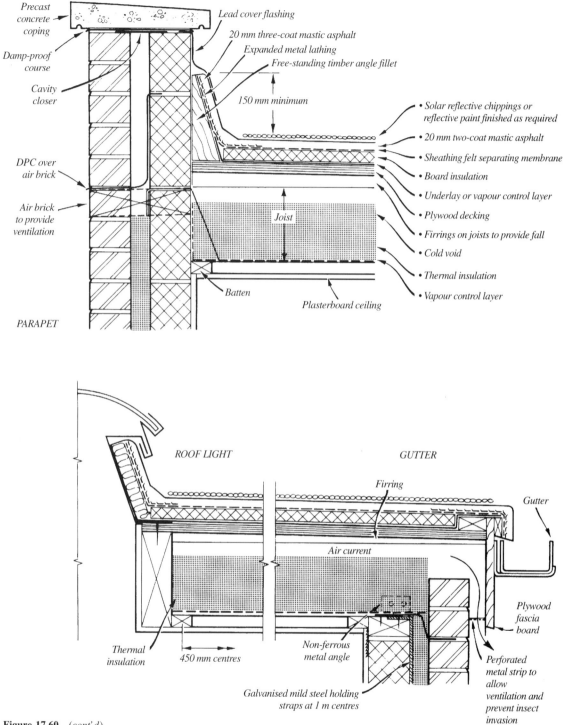

Precast concrete coping

Damp-proof course

Cavity closer

DPC over air brick

Air brick to provide ventilation

PARAPET

Lead cover flashing

20 mm three-coat mastic asphalt

Expanded metal lathing

Free-standing timber angle fillet

150 mm minimum

Joist

Batten

Plasterboard ceiling

• Solar reflective chippings or reflective paint finished as required

• 20 mm two-coat mastic asphalt

• Sheathing felt separating membrane

• Board insulation

• Underlay or vapour control layer

• Plywood decking

• Firrings on joists to provide fall

• Cold void

• Thermal insulation

• Vapour control layer

ROOF LIGHT

GUTTER

Firring

Air current

Gutter

Thermal insulation

450 mm centres

Non-ferrous metal angle

Galvanised mild steel holding straps at 1 m centres

Plywood fascia board

Perforated metal strip to allow ventilation and prevent insect invasion

Figure 17.60   (cont'd)

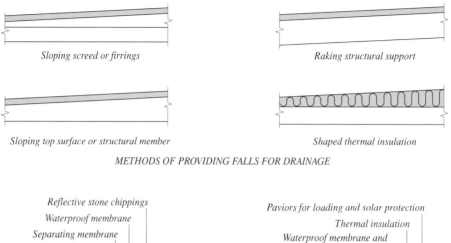

Sloping screed or firrings

Raking structural support

Sloping top surface or structural member

Shaped thermal insulation

METHODS OF PROVIDING FALLS FOR DRAINAGE

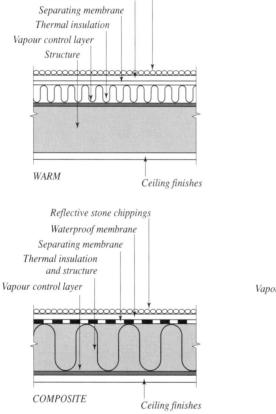

Reflective stone chippings
Waterproof membrane
Separating membrane
Thermal insulation
Vapour control layer
Structure

WARM   Ceiling finishes

Paviors for loading and solar protection
Thermal insulation
Waterproof membrane and
vapour control layer
Separating membrane
Structure

WARM   Ceiling finishes

Reflective stone chippings
Waterproof membrane
Separating membrane
Thermal insulation
and structure
Vapour control layer

COMPOSITE   Ceiling finishes

Reflective stone chippings
Waterproof membrane
Separating membrane
Structure
Thermal barrier
Vapour control layer

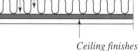

Ventilation
of void

COLD
                Ceiling finishes

Figure 17.61   Warm and cold flat roof construction.

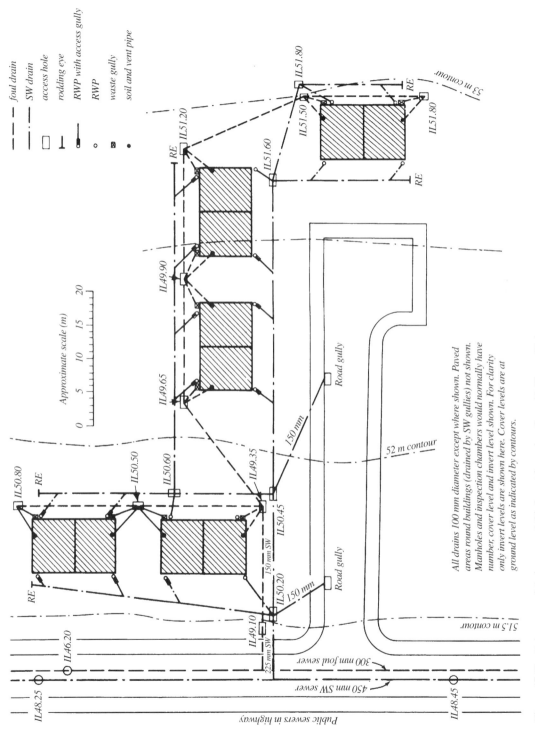

Figure 17.62 Below-ground drainage: read this layout in conjunction with Figure 11.4.

All drains 100 mm diameter except where shown. Paved areas round buildings (drained by SW gullies) not shown. Manholes and inspection chambers would normally have number; cover level and invert level shown. For clarity only invert levels are shown here. Cover levels are at ground level as indicated by contours.

Legend:
- foul drain
- SW drain
- access hole
- rodding eye
- RWP with access gully
- RWP
- waste gully
- soil and vent pipe

Approximate scale (m) 0 5 10 15 20

the link drain between the sewer and a manhole on site. This means of drain access must be within 22 m of the sewer, where a saddle connection is permitted, as shown in Figure 11.4. The manhole (inspection chamber if less than 1 m deep) will also accommodate any difference in levels which may exist between the two drains. Local authority permission must be sought when connections to a public sewer are required, and the work should be carefully planned to cause minimum inconvenience to the public. Normally the local authority form the connection using their own appointed contractor.

Drainage 'tails', left for future connections with above-ground drainage, must be temporarily sealed against the leakage of smells and gases from the main sewer, and/or the possibility of them becoming blocked by debris. Backfilling drainage trenches must be done with great care; the exact method employed for casing the pipes will depend to a large extent on local conditions.

It will be necessary for the local authority to inspect and approve all drainage work and the subsequently installed plumbing work. Although they are looked at during their installation for general approval, the final tests on drainage by the inspector generally take place after the trenches have been backfilled. Consequently, it is desirable for the builder to give the drains a preliminary water test *before* backfilling commences, in order to avoid the trouble and expense of opening up again at a later date if leaks become evident as a result of the local authority test.

17.10.2 Above-ground drainage and plumbing

The drainage above ground can be installed as soon as practicable; usually this means after the basic structural work has been completed (walls, floors and roof). Care needs to be taken to ensure that all necessary holes have been formed in the correct position before the drainage is installed. Once the interior of the building has been made 'dry' by completion of the external enclosure, plumbing work can also commence.

Sanitary fittings connecting to the internal drainage system are not usually installed until after the basic finishes of the building have been completed. This is because it is often easier for operatives, such as plasterers, to carry out their work on walls and ceilings unhampered by pipes and fittings, saving on the 'cleaning down' processes after they have finished. Plumbing and heating pipes are generally left until after plastering for similar reasons. However, when the fittings and pipes are to be surrounded with duct enclosures or other built-in features, or where finishes of a less 'messy' nature than wet plastering processes are

used, it is often more convenient to install them as soon as the internal drainage is complete. These matters should have been considered during the design stage of the project, and normally form a subject of discussion between designer and builder during the progress of the work on site.

17.10.3 Water

The water supply pipe is taken from the water company's stop valve on the distribution main to another valve and a meter at the site boundary. From here the underground service continues into the building, terminating with a combined stop and drain valve at a convenient point (usually under the kitchen sink) – see Figures 11.1 and 17.63. From this point the rising main serves drinking-water facilities and a cold water storage cistern installed at high level.

17.10.4 Gas and electricity

The gas pipe and electricity service cable are brought independently from their mains position outside the site to a point of termination by a meter (Fig. 17.64). (The entry of the gas connecting pipe should not be through the same duct as the electrical cabling duct.) This work is carried out by the gas and electricity companies to a programme agreed with the builder.

Unless separately ducted, the work of installing internal gas pipes and electricity cables from the meter positions should be carried out before basic finishes to the building are commenced. This work may also be carried out by the relevant authorities, but is more usually carried out by specialist subcontractors, a satisfactory price for the work having been given through competitive tendering procedures. Once the gas and electrical services have been installed to the layout requirements stipulated by the drawings and specification of the designer, the work must be inspected by the representatives of the two companies for compliance with safety regulations.

17.10.5 Telecommunication systems

Like the other services described, the telephone, computer and cable television terminals may have localised common supply points from which connections are made to a building (Fig. 17.65). Alternatively, they can be supplied from remotely located exchanges. These installations are carried out externally and internally by the provider's engineers, although the site electrician's work can include internal distribution to telephone sockets.

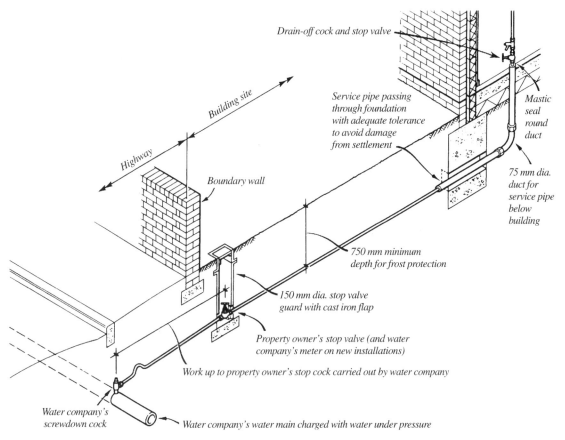

Drain-off cock and stop valve

Building site

Service pipe passing
through foundation
with adequate tolerance
to avoid damage
from settlement

Highway

Boundary wall

Mastic
seal
round
duct

75 mm dia.
duct for
service pipe
below
building

750 mm minimum
depth for frost protection

150 mm dia. stop valve
guard with cast iron flap

Property owner's stop valve (and water
company's meter on new installations)

Work up to property owner's stop cock carried out by water company

Water company's
screwdown cock

Water company's water main charged with water under pressure

Figure 17.63 Water supply.

17.11 Finishes
CI/SfB (4–)

As soon as the constructional processes reach the stage where the structural elements of the house (walls, floors and roof) have more or less been finished, and the weatherproofing completed by window/door glazing and roof claddings, etc., the work on the interior can commence in earnest. Initially, this will involve the erection of dry construction components such as stud partitions, and internal door frames and linings as described earlier. At about the same time, the service trades – gas, water, plumbing, heating and electrical – will be continuing their installations from the previously incorporated terminal points. In close association with all these activities, the building contractor will wish to make a start on the coverings for the internal surfaces of walls, floors and ceilings, as well as building the staircase, cupboard fitments and ducting work. Therefore, whereas previous trades were working relatively independently of each other and progress relied on consecutive working

programmes, the processes now required are much more interdependent and the degree of organisation required increases. Builders and subcontractors must work closely together to ensure that the correct labour skills and adequate material resources are available, so as to avoid the accumulation of unreasonable delays with their consequent frustrations and financial penalties.

17.11.1 Plasterboard and plastering

Stud partitions and joisted ceilings are usually faced with *plasterboard* as it is reasonably strong, flexible in use, and contributes towards fire protection. Plasterboard consists of a plaster core encased in and firmly bonded to specially prepared durable lining paper. A grey lining paper indicates that it is meant to receive a coating of wet plaster and ivory that can be decorated direct. Plasterboard used without wet plaster (except in joints) is called *dry lining*.

Plasterboards are available in a number of standard sizes to conform with the centres of structural studs

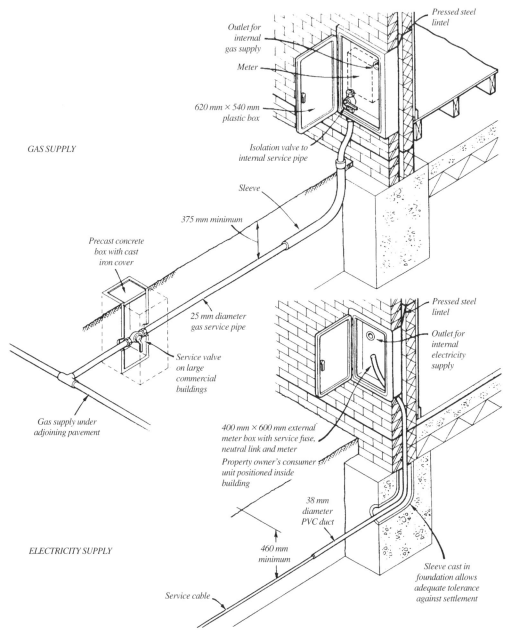

GAS SUPPLY

Outlet for
internal
gas supply

Meter

620 mm × 540 mm
plastic box

Pressed steel
lintel

Isolation valve to
internal service pipe

Sleeve

375 mm minimum

Precast concrete
box with cast
iron cover

25 mm diameter
gas service pipe

Service valve
on large
commercial
buildings

Gas supply under
adjoining pavement

Pressed steel
lintel

Outlet for
internal
electricity
supply

400 mm × 600 mm external
meter box with service fuse,
neutral link and meter

Property owner's consumer
unit positioned inside
building

38 mm
diameter
PVC duct

ELECTRICITY SUPPLY

460 mm
minimum

Sleeve cast in
foundation allows
adequate tolerance
against settlement

Service cable

Figure 17.64   Gas and electricity supply.

or joists, and are supplied in various thicknesses, i.e.
6, 9.5, 12.5, 15, 19 and 25 mm. A metallised vapour
check backing is an option on 9.5 and 12.5 mm boards.
Thermal-check boards with a bonded expanded polystyrene (eps) laminate are also available in various overall thicknesses up to 55 mm. For fire-resisting purposes,
a 12.5 mm thickness of plasterboard applied to each
side of 75 × 38 mm (minimum) timber stud partition

framing at 600 mm (maximum) spacing will achieve a
half-hour fire resistance. The same fire-resisting standard
for a floor is achieved with 38 mm (minimum) width
joists at 450 mm (maximum) spacing, a 12.5 mm
plasterboard ceiling and 21 mm (minimum) tongued
and grooved woodboard decking (see section 17.8.2).

Plasterboards are secured by galvanised or sheradised
nails at 150 mm maximum spacing. Inevitably, some

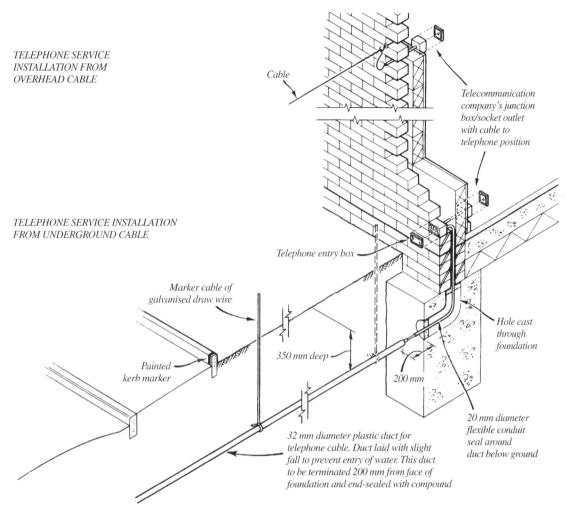

TELEPHONE SERVICE INSTALLATION FROM OVERHEAD CABLE

Cable

Telecommunication company's junction box/socket outlet with cable to telephone position

TELEPHONE SERVICE INSTALLATION FROM UNDERGROUND CABLE

Telephone entry box

Marker cable of galvanised draw wire

350 mm deep

Painted kerb marker

Hole cast through foundation

200 mm

32 mm diameter plastic duct for telephone cable. Duct laid with slight fall to prevent entry of water. This duct to be terminated 200 mm from face of foundation and end-sealed with compound

20 mm diameter flexible conduit seal around duct below ground

Figure 17.65   Telephone cable supply.

cutting will be necessary, particularly around door openings, electric light switches and power points, and when splays (to form bulkheads around a staircase) are incorporated in the design of the stud wall and/or ceiling. It is essential that all cross-joints in the plasterboards are staggered and that they are covered with a *jute scrim* or paper tape, which forms a seal and helps to prevent subsequent cracking of future finishes caused by thermal movements. It is also good practice to form the joints at right angles to the support of the sheets, and for each joint to be backed by a timber batten or *noggin*.

When all the plasterboard is in place, the plastering can commence to the plasterboard as well as the internal blockwork surfaces. A plaster finish fulfils the function of camouflaging irregularities in the backing wall as well as exposed conduits on the surface. It also provides a sufficiently hard surface to resist damage by impact and a surface smooth enough to be suitable for direct decoration. Gypsum plasters are suitable for this, and the choice of mix, type and number of coats will depend upon the characteristics of the background and its surface texture. Generally, the lightweight concrete blocks used for the internal leaf of the external cavity wall and for the internal partitions, and the surface of plasterboard, provide a suitable key for the direct application of plasters without further preparations. Although one-coat plasters are also available for blockwork, it is more usual to use a two-coat process. The undercoat of cement and sand render is applied using a wooden float or rule which is worked between 'dots' or 'runs' of plaster to give a true level surface. This undercoat is about 10 mm thick and is scratched before drying to provide a suitable key for the finishing coat. The thin finishing coat is applied to a depth of about 3 mm and finished with a steel float to provide a glass-like surface.

Junctions between backing walls of different materials (plasterboard adjoining block walls) require special treatment if cracking is to be avoided as a result of subsequent differential movements. Jute scrim reinforcement can be used at these points, although a much better detail can be adopted by using alloy edge-reinforcing strips, as indicated in Figure 17.66. These can also be used on the corners of walls to prevent damage by impact. Plaster reinforcing details between roof ceiling joists and the wall below are also indicated in Figure 17.66.

When a 'warm roof' construction is used, it will be necessary to install the vapour control layer prior to the fixing of the plasterboard lining for the ceiling. Although plasterboards are available with a thin metal-lised foil bonded to their back face for this purpose, with the high degree of insulation required it is wise to incorporate an additional independent vapour control layer. This is because the metallised foil will probably be damaged during erection and will be pierced for electric light fittings. The independent vapour control layer is best incorporated as part of the insulation, which can be laid once the ceiling and electrical work has been completed. It is important that the warm roof insulation is taken *over* any water cisterns in the roof space and that any water pipes are suitably lagged to prevent the possibility of freezing.

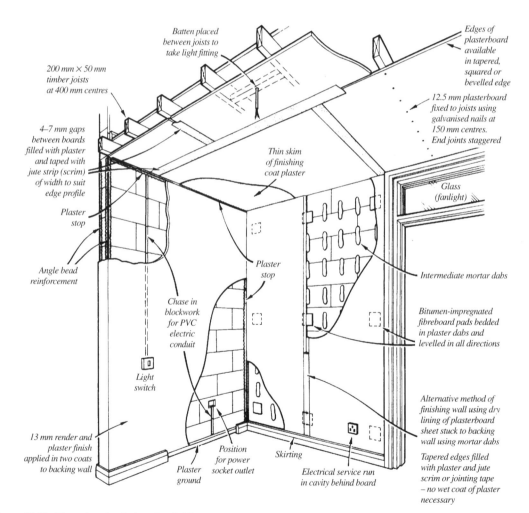

Figure 17.66   Plasterboard and plastering finishes.

17.11.2 Joinery

As soon as practicable, the upper surfaces of the horizontal structural elements should be made available by fixing the decking in an order which allows other work to be carried out from them. The staircase, which links these surfaces, should also be constructed in parallel, and door linings as well as window sills, etc., must be fixed so that plasterboard linings and plastering can continue without hindrance.

The intermediate floor decking can be *tongue-and-groove* (t & g) hardwood or softwood boarding, or chipboard or plywood sheets (see Fig. 17.29). When narrow face width boarding is used, it will be necessary for the boards to be cramped securely together before nailing to the joists in order to close the joints between each, and lessen the effect of shrinkage once the central heating has been switched on. Areas requiring future access for maintenance and alterations to electric wiring, junction boxes, etc., can be accommodated by removing the tongues from the perimeter board affected and using screws to form a removable panel. The use of t & g chipboard, or plywood sheets (say 1.2 m × 2.4 m), will speed the laying of floor decking provided their standardised dimensions are coordinated with the centres of the structural joists and the overall sizes of individual rooms.

Access areas can be provided by removable screw-down panels in a similar manner to hardwood or softwood boarding. Flooring grades of chipboard sheets need particular attention if subsequent bowing of the floor deck is to be avoided as a result of moisture absorption. It is recommended that the boards are allowed to acclimatise to the moisture levels of the room in which they are to be used before being fixed, and that movement gaps are left around the perimeter of the floor at junctions with walls.

Door linings and window boards are normally included in the 'first fixing' programme of the joinery. Linings are carefully set up to be plumb, then screwed to the plugs already built into blockwork or to the double stud and head timbers of stud partitioning. They should be of a width which allows for the thickness of plasterboard and/or plaster. Internal window boards are levelled and fixed direct to the blocks using cut nails or screws or, depending on the precise detail, they can be nailed to timber packing pieces already fixed to the top course of the blockwork opening. Sometimes tiled sills are required, so the tiles are bedded in mortar direct to the blockwork. 'Second fixing' joinery, which includes items such as doors, architraves, skirtings, shelves, cupboard units and ironmongery (locks, handles, etc.), should not be installed until just prior to final finishing.

When considering the design of joinery items, it is important to select appropriate sizes for timber sections to allow for the reduction in size of the original rough or sawn (unwrot) standard sections caused by their being smooth (wrot). The amount of reduction will vary according to species of timber and the original sawn size – the range for softwoods is from 1.5 mm of *each* face to be planed for up to 22 mm sections, to 6.5 mm of each face for sections over 150 mm. Joinery sections are therefore specified by using the prefix '*ex:*' (out of) against the unwrot size of timber to be used, e.g. an ex: 100 mm × 25 mm softwood skirting produces a wrot size of 94 mm × 19 mm. For further clarification, see section 5 of Mitchell's *Internal Components*.

After the main construction work has been completed, but before the 'second fixing' of the joinery, the ground floor damp-proof membrane (DPM) and insulation can be installed if they have not already been incorporated under the solid concrete slab. After the DPM has been laid on the top surface of a concrete floor slab, rigid insulation is placed before screeding, as detailed in Figure 17.27. If the final floor finish is to be applied by the builder (plastic tiles, wood strip, parquet, etc.), the DPM, insulation and screed should not be begun until the very last period of the construction programme.

17.11.3 Decorations

Once the construction work has been completed, those parts of the house which require decoration can be appropriately prepared (Fig. 17.67). The more care that is taken with preparatory work, the better the finished result will be.

Sometimes the externally decorated areas will have been prepared and finished during an earlier stage of the building programme. This involves bringing in the painters and decorators out of sequence relative to the work to be done inside the building. This procedure may still be economical since it allows the scaffolding to be removed much earlier in the contract, and may mean that the external landscaping can proceed earlier and unhampered by construction activities. However, it is important that allowance is made for protecting the newly finished work while any construction work is still being carried out in the interior of the building. Protection is particularly important for window and door openings, especially as they are often supplied to site in some prefinished form (primed or preservative decorative staining to timber window and door frames, or anodised aluminium sections). They will require protection against mechanical damage arising

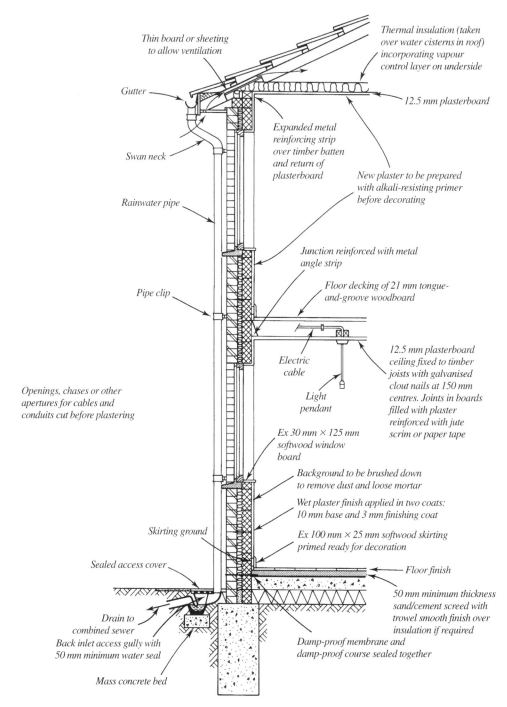

Thin board or sheeting
to allow ventilation

Thermal insulation (taken
over water cisterns in roof)
incorporating vapour
control layer on underside

12.5 mm plasterboard

Gutter

Expanded metal
reinforcing strip
over timber batten
and return of
plasterboard

Swan neck

New plaster to be prepared
with alkali-resisting primer
before decorating

Rainwater pipe

Junction reinforced with metal
angle strip

Floor decking of 21 mm tongue-
and-groove woodboard

Pipe clip

Electric
cable

12.5 mm plasterboard
ceiling fixed to timber
joists with galvanised
clout nails at 150 mm
centres. Joints in boards
filled with plaster
reinforced with jute
scrim or paper tape

Light
pendant

Openings, chases or other
apertures for cables and
conduits cut before plastering

Ex 30 mm × 125 mm
softwood window
board

Background to be brushed down
to remove dust and loose mortar

Wet plaster finish applied in two coats:
10 mm base and 3 mm finishing coat

Ex 100 mm × 25 mm softwood skirting
primed ready for decoration

Skirting ground

Floor finish

Sealed access cover

50 mm minimum thickness
sand/cement screed with
trowel smooth finish over
insulation if required

Drain to
combined sewer
Back inlet access gully with
50 mm minimum water seal

Damp-proof membrane and
damp-proof course sealed together

Mass concrete bed

Figure 17.67 Construction sequence: finishes.

from loading material to upper levels from scaffolding platforms.

Before any of the woodwork is to be painted on site, it should be carefully rubbed down, all nails punched

below the surface and any necessary filling done. When choosing a particular manufacturer's priming, undercoat and finishing coat, it is essential to check that they are chemically compatible with any preservative treatment used on the joinery. The outside junction between door and window frames with the brickwork should be sealed with a mastic.

Internally, the walls to be decorated should receive their final preparation. This process includes the removal of temporary fixing nails, the repair of damaged plasterwork, and the smoothing of rough surfaces generally. In a new building there may be a small amount of movement due to initial structural settlement, and to the drying out of structural timbers, blockwork, etc. Drying out may persist for some time, and is likely to cause some cracking in plaster finishes, particularly during the first heating season. Plastered blockwork walls need a 'drying out' period after their completion because of the amount of water used in their construction. Their initial decoration should therefore be regarded as temporary. Certain emulsion paints allow the walls to breathe and residual moisture to escape without causing excessive or speedy breakdown of the surface film. It is normal to choose a neutral colour for the temporary finishing emulsion. Blemishes will be less obvious, and the occupants of the dwelling will be able more easily to appreciate the quality of space and light around them before selecting the colours and textures of more permanent wall finishes which will compliment the chosen style of furnishings.

17.12 Landscaping and external works
CI/SfB (90)

The external landscaping around the building provides the vital aesthetic and physical link between outside spaces and inside spaces (Fig. 17.68). For this reason the design and construction of landscaped areas requires just as much detailed consideration as applies to the building itself. Apart from purely enhancing the *appearance and quality* of the building with which it most closely associates, as well as that of the surrounding environment, the landscaping will also serve in providing the following practical functions:

- Convenient methods of circulation for foot and vehicular travel
- Shelter from wind, snow, rain and unwanted sunshine
- Visual and/or aural privacy
- Security from unlawful access to the site and building

The required function and subsequent design of the building should maximise the natural landscaping of the site as far as is possible. Beyond this, consideration must be given to the need for the planting of trees and shrubs, the formation of paths, driveways and roads, the erection of fences and walls, and the construction of shelters for storage or amenity. Most of this work will be executed after the construction of the main building because the areas of site will become progressively free as storage areas, etc., become redundant. Also, newly completed landscaped areas are less likely to be damaged because the majority of building construction will be completed. (An exception might occur on large building contracts when the long period of construction can be used to allow planting to mature in advance of the finally completed buildings.)

It would be economically desirable if most of the earth landscape features can be formed from the topsoil saved as a result of excavation of the building area of the site (see section 17.3.3). If there is an insufficient amount, the contractor will be required to import additional quantities from an outside source. Existing trees and shrubs which have been retained and protected during construction work will also provide considerable saving against planting new ones.

It is important that detailed thought is given to the selection of the correct species of new trees or shrubs. These will be small during their infancy and subsequent growth may result in undesirable amounts of overshadowing and the curtailment of desirable vistas. It is particularly necessary to analyse the effects of future root growth relative to the type of soil (including the proximity of water sources and the influences of pruning) and their proximity to the building (Fig. 17.69). Building closer to a tree than its mature height is not recommended, and, where several trees occur, the safe distance is considered to be $1\frac{1}{2}$ times the mature height of the trees. Some guidance on precautionary measures for foundation dimensions in the proximity of trees is shown in Figure 17.70.

The hard surface landscaping areas created by paths, driveways and roads on the site must be designed to retain their interest through all the changing seasons of the year. Natural stone pavings and gravels can be incorporated in a way which permits mosses and lichen to spread over areas of their surfaces to unify them with the natural landscape. Alternatively, they can be designed and constructed to create a separation rigidly defining dry and durable circulation paths. Hard surfaces may also require a system of rainwater collection to ensure that they are relatively free from the effects of rainwater. Large expanses can be drained by continuous 'monsoon type' drains which are revealed at the surface only by a continuous gap. Other materials considered suitable for large hard surfaced areas include tarmacadam, hot rolled asphalt and *in situ* concrete. Care must be taken to ensure that they do not give a monotonous appearance or crack excessively owing to dimensional movements. Accordingly, these materials can be formed in bays using another material at the

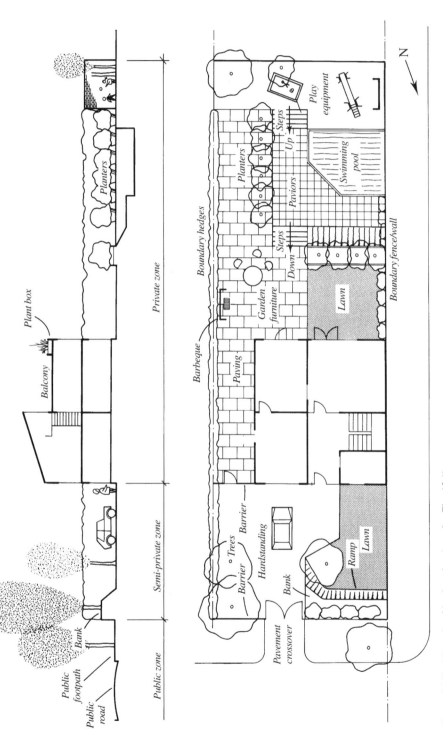

Figure 17.68   The role of landscaping (see also Fig. 12.1).

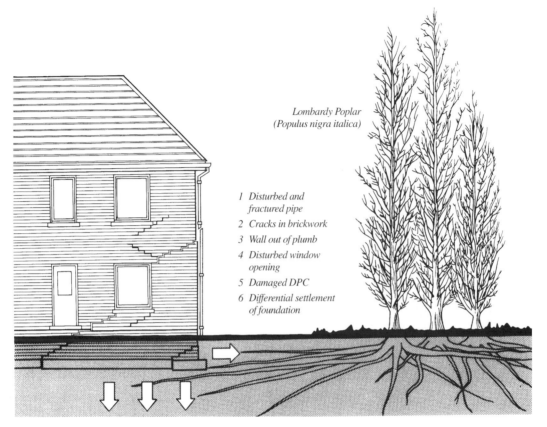

1 Disturbed and
   fractured pipe
2 Cracks in brickwork
3 Wall out of plumb
4 Disturbed window
   opening
5 Damaged DPC
6 Differential settlement
   of foundation

Lombardy Poplar
(Populus nigra italica)

Figure 17.69 Danger of planting trees too close to a building erected on conventional foundations.

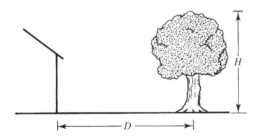

Tree	Depth of foundation trench (m)						
	$D/H = 1/10$	$D/H = 1/4$	$D/H = 1/3$	$D/H = 1/2$	$D/H = 2/3$	$D/H = 3/4$	$D/H = 1$
Poplar elm and willow	*Not acceptable*	2.8	2.6	2.3	2.1	1.9	1.5
All others	*Not acceptable*	2.4	2.1	1.5	1.5	1.2	1.0

Figure 17.70 Proximity of trees.

joints which helps relieve monotony and create the opportunity for an aesthetically acceptable movement joint. Precast concrete paving slabs, brick pavers, clay tiles or cobbles can also be used with effect.

The provision of screening and boundary fences/walls has an important influence on the overall quality of the site and building. Timber can be used in a variety of ways which result from construction techniques ranging from driving poles for fixing preformed panels, to sophisticated joinery incorporating hardwood sections to give maximum visual appeal. Brick-built walls may also be pleasing, and can form an 'introduction' at the perimeter of the site to the aesthetic of the brick-faced building within. However, in order for this ideal to be maintained for the life of the building, it will be necessary to ensure that the effects of weathering on a fully exposed boundary wall remain similar to the walls of the building which are subjected to less hazardous conditions. The choice of brick and mortar should be primarily influenced by the exposure condition of the boundary wall. Typical details are indicated in Figure 17.71.

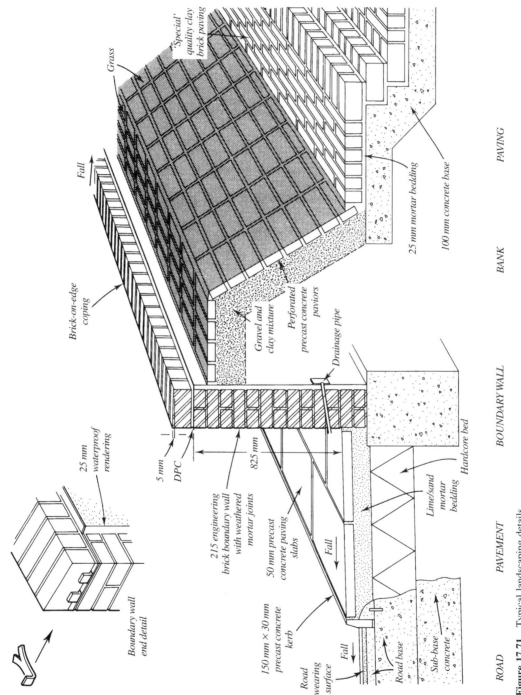

Figure 17.71 Typical landscaping details.

Special' quality clay brick paving

Grass

Fall

Brick-on-edge coping

Gravel and clay mixture

Perforated precast concrete paviors

25 mm mortar bedding

100 mm concrete base

PAVING

BANK

Drainage pipe

25 mm waterproof rendering

5 mm

DPC

825 mm

215 engineering brick boundary wall with weathered mortar joints

50 mm precast concrete paving slabs

Boundary wall end detail

BOUNDARY WALL

Hardcore bed

Lime/sand mortar bedding

Fall

150 mm × 30 mm precast concrete kerb

Road wearing surface

Fall

Road base

Sub-base concrete

PAVEMENT

ROAD

17.13 Completion
CI/SfB (A8)

Once all work has been completed the contractor will be required to clear away all surplus materials and debris, remove plant/huts, etc., and leave the building and site in a clean and workable condition. The designer will inspect the work, ensure the satisfaction of the local authority inspectors, and prepare the necessary maintenance manual before passing the completed project to the client (see section 16.8).

Further specific reading

Note: The Building Regulations Approved Document supporting Regulation 7 – *Materials and workmanship* – applies throughout.

17.1 Building Team and communication

Mitchell's Building Series

External Components	Chapter 2 Prefabricated building components
Structure and Fabric Part 1	Chapter 2 The production of buildings
Structure and Fabric Part 2	Chapter 1 Contract planning and site organization
	Chapter 2 Contractors' mechanical plant

17.2 Site considerations

Mitchell's Building Series

Structure and Fabric Part 1	Section 4.1 Foundations: soil and soil characteristics
	Section 4.2 Foundations: site exploration
Structure and Fabric Part 2	Section 3.1 Foundations: soil mechanics

Building Research Establishment

BRE Digest 63	*Soils and foundations* Part 1
BRE Digest 64	*Soils and foundations* Part 2
BRE Digest 67	*Soils and foundations* Part 3
BRE Digest 202	*Site use of theodolite and surveyor's level*
BRE Digest 234	*Accuracy in setting-out*
BRE Digest 247	*Waste of building materials*
BRE Digest 298	*The low-rise building foundations: the influence of trees in clay soils*
BRE Digest 318	*Site investigation for low-rise building: desk studies*
BRE Digest 322	*Site investigation for low-rise building: procurement*
BRE Digest 348	*Site investigation for low-rise building: the walk-over survey*
BRE Digest 363	*Sulphate and acid resistance of concrete in the ground*
BRE Digest 381	*Site investigation for low-rise building: trial pits*
BRE Digest 383	*Site investigation for low-rise building: soil description*
BRE Digest 411	*Site investigation for low-rise building: direct investigations*
BRE Digest 412	*Desiccation in clay soils*
BRE Digest 427	*Low-rise buildings on fill* (Parts 1–3)
BRE Digest 447	*Waste minimisation on a construction site*

Building Regulations

A1   Loading
A2   Ground movement
C1   Site preparation and resistance to contaminants
C2   Resistance to moisture

17.3 Initial site works

Mitchell's Building Series

Structure and Fabric Part 1	Chapter 2 The production of buildings
	Section 11.1 Timbering for excavations
Structure and Fabric Part 2	Chapter 2 Contractors' mechanical plant
	Section 11.2 Timbering for excavations

Building Research Establishment

BRE Digest 179	*Electricity distribution on sites*
BRE Digest 274	*Fill Part 1: Classification and load-carrying characteristics*
BRE Digest 275	*Fill Part 2: Site investigation, ground movement and foundation design*
BRE Digest 276	*Hardcore*
BRE Digest 298	*The low-rise building foundations: the influence of trees in clay soils*
BRE Digest 318	*Site investigation for low-rise building: desk studies*
BRE Digest 322	*Site investigation for low-rise building: procurement*
BRE Digest 343	*Simple measuring and monitoring of movement in low-rise buildings* Part 1: Cracks
BRE Digest 344	*Simple measuring and monitoring of movement in low-rise buildings* Part 2: Settlement, heave and out-of-plumb
BRE Digest 361	*Why do buildings crack?*
BRE Digest 381	*Site investigation for low-rise building: trial pits*
BRE Digest 383	*Site investigation for low-rise building: soil description*

Building Regulations

A1   Loading
A2   Ground movement
C1   Site preparation and resistance to contaminants
C2   Resistance to moisture

17.4 Foundation construction

Mitchell's Building Series

Materials	Chapter 8 Concretes
Materials Technology	Chapter 3 Concrete
Structure and Fabric Part 1	Chapter 4 Foundations
Structure and Fabric Part 2	Section 3.2 Foundation design
	Section 3.3 Foundation types

Building Research Establishment

BRE Digest 63   *Soils and foundations* Part 1
BRE Digest 64   *Soils and foundations* Part 2
BRE Digest 67   *Soils and foundations* Part 3
BRE Digest 240   *Low-rise buildings on shrinkable clay soils* Part 1
BRE Digest 241   *Low-rise buildings on shrinkable clay soils* Part 2
BRE Digest 242   *Low-rise buildings on shrinkable clay soils* Part 3
BRE Digest 251   *Assessment of damage in low-rise building*
BRE Digest 274   *Fill* Part 1: Classification and load-carrying characteristics
BRE Digest 275   *Fill* Part 2: Site investigation, ground improvement and foundation design
BRE Digest 276   *Hardcore*
BRE Digest 298   *The low-rise building foundations: the influence of trees in clay soils*
BRE Digest 313   *Mini-piling for low-rise building*
BRE Digest 315   *Choosing piles for new construction*
BRE Digest 325   *Concrete* Part 1: Materials
BRE Digest 326   *Concrete* Part 2: Specification, design and quality control
BRE Digest 352   *Underpinning*
BRE Digest 412   *Desiccation in clay soils*
BRE Digest 427   *Low-rise buildings on fill* Parts 1, 2 and 3

Building Regulations

A1   Loading
A2   Ground movement
C1   Site preparation and resistance to contaminants
C2   Resistance to moisture

17.5 Ground floor construction

Mitchell's Building Series

Environment and Services	Chapter 5 Heat
Internal Components	Chapter 4 Raised floors
Materials	Chapter 2 Timber
	Chapter 3 Boards, slabs and panels
	Chapter 8 Concretes
Materials Technology	Chapter 3 Concrete
	Chapter 12 Timber
	Chapter 13 Artificial composites: timber-based boards and slabs
Structure and Fabric Part 1	Section 8.1 Ground floor construction
Finishes	Chapter 4 Ceramic materials
	Chapter 7 Composites

Building Research Establishment

BRE Digest 33   *Sheet and tile flooring made from thermoplastic binders*
BRE Digest 54   *Damp-proofing solid floors*
BRE Digest 145   *Heat losses through ground floors*
BRE Digest 163   *Drying out buildings*

BRE Digest 276   *Hardcore*
BRE Digest 323   *Selecting wood-based panel products*
BRE Digest 325   *Concrete* Part 1: Materials
BRE Digest 326   *Concrete* Part 2: Specification, design and quality control
BRE Digest 364   *Design of timber floors to prevent decay*
BRE Digest 371   *Remedial wood preservatives: use them safely*
BRE Digest 373   *Wood chipboard*
BRE Digest 378   *Wood preservatives: application methods*
BRE Digest 380   *Damp-proof courses*
BRE Digest 393   *Specifying preservative treatments: the new European approach*
BRE Digest 394   *Plywood*
BRE Digest 416   *Specifying structural timbers*
BRE Digest 429   *Timbers: their natural durability and resistance to preservative treatment*

Building Regulations

A1   Loading
C2   Resistance to moisture
L1   Conservation of fuel and power in dwellings
L2   Conservation of fuel and power in buildings other than dwellings

17.6 External wall construction

Mitchell's Building Series

Environment and Services   Chapter 5 Heat

External Components   Section 2.8 Prefabricated component design: accuracy and tolerances
Chapter 3 External glazing
Chapter 4 Windows
Chapter 7 External doors

Materials   Chapter 6 Bricks and blocks
Chapter 15 Mortars for jointing
Chapter 16 Sealants

Materials Technology   Chapter 4 Masonry construction
Chapter 5 Glass

Structure and Fabric Part 1   Chapter 5 Walls and piers
Chapter 9 Fireplaces, flues and chimneys

Structure and Fabric Part 2   Section 4.1 Solid masonry walls
Section 4.3 Cavity load-bearing walls

Finishes   Chapter 2 Polymeric materials
Chapter 4 Ceramic materials

Building Research Establishment

BRE Digest 108   *Standard 'U' values*
BRE Digest 230   *Fire performance of walls and linings*
BRE Digest 236   *Cavity insulation*
BRE Digest 246   *Strength of brickwork and blockwork walls: design for vertical load*
BRE Digest 273   *Perforated clay bricks*
BRE Digest 281   *Safety of large masonry walls*
BRE Digest 294   *Fire risk from combustible cavity insulation*

BRE Digest 297   *Surface condensation and mould growth in traditionally built dwellings*
BRE Digest 304   *Preventing decay in external joinery*
BRE Digest 338   *Insulation against external noise*
BRE Digest 362   *Building mortar*
BRE Digest 369   *Interstitial condensation and fabric degradation*
BRE Digest 371   *Remedial wood preservatives: use them safely*
BRE Digest 377   *Selecting windows by performance*
BRE Digest 379   *Double glazing for heat and sound insulation*
BRE Digest 380   *Damp-proof courses*
BRE Digest 394   *Plywood*
BRE Digest 401   *Replacing wall ties*
BRE Digest 404   *PVC-U windows*
BRE Digest 410   *Cementitious renders for external walls*
BRE Digest 420   *Selecting natural building stones*
BRE Digest 440   *Weathering of white external PVC-U*
BRE Digest 441   *Clay bricks and clay masonry* (Parts 1 and 2)

Building Regulations

A1   Loading
B4   External fire spread
C2   Resistance to moisture
D1   Cavity insulation
F1   Means of ventilation
J1   Air supply
L1   Conservation of fuel and power in dwellings
N1   Glazing – Protection against impact
N3   Glazing – Safe opening and closing of windows, etc.
N4   Glazing – Safe access for cleaning windows, etc.

17.7 Internal wall construction

Mitchell's Building Series

Environment and Services	Chapter 6 Sound
Internal Components	Chapter 2 Demountable partitions
Materials	Chapter 2 Timber
	Chapter 6 Bricks and blocks
	Chapter 15 Mortars for jointing
Materials Technology	Chapter 4 Masonry construction
	Chapter 12 Timber
	Chapter 13 Artificial composites: timber-based boards and slabs
Structure and Fabric Part 1	Section 5.6 Partitions
Structure and Fabric Part 2	Section 4.6 Cross-wall construction
	Section 4.8 Compartment, separating and other walls

Building Research Establishment

BRE Digest 293   *Improving the sound insulation of separating walls and floors*
BRE Digest 320   *Fire doors*

BRE Digest 333   *Sound insulation of separating walls and floors* Part 1: Walls
BRE Digest 337   *Sound insulation: basic principles*

Building Regulations

A1   Loading
B2   Internal fire spread (linings)
B3   Internal fire spread (structure)
E1   Protection against sound from adjoining dwellings or buildings, etc.
E2   Protection against sound within a dwelling, etc.
E3   Reverberation in the common internal parts of buildings containing dwellings, etc.
L1   Conservation of fuel and power in dwellings
N1   Glazing – Protection against impact
N2   Glazing – Manifestation of glazing

17.8 Intermediate floor construction

Mitchell's Building Series

Environment and Services	Chapter 6 Sound
Internal Components	Chapter 8 Stairs and balustrades
Materials	Chapter 2 Timber
	Chapter 3 Boards, slabs and panels
Materials Technology	Chapter 12 Timber
	Chapter 13 Artificial composites: timber-based boards and slabs
Structure and Fabric Part 1	Section 8.2 Upper floor construction
	Chapter 10 Stairs
Structure and Fabric Part 2	Chapter 6 Floor structures
	Section 8.1 Stairs
Finishes	Chapter 4 Ceramic materials
	Chapter 7 Composites

Building Research Establishment

BRE Digest 33   *Sheet and tile flooring made from thermoplastic binders*
BRE Digest 208   *Increasing the fire resistance of existing timber floors*
BRE Digest 293   *Improving the sound insulation of separating walls and floors*
BRE Digest 323   *Selecting wood-based panel products*
BRE Digest 334   *Sound insulation of separating walls and floors* Part 2: Floors
BRE Digest 337   *Sound insulation: basic principles*
BRE Digest 373   *Wood chipboard*
BRE Digest 378   *Wood preservatives: application methods*
BRE Digest 393   *Specifying preservative treatments: the new European approach*
BRE Digest 394   *Plywood*
BRE Digest 416   *Specifying structural timber*
BRE Report 128   *Guidelines for the construction of fire-resisting structural elements*

Building Regulations

A1   Loading
B2   Internal fire spread (linings)

B3   Internal fire spread (structure)
E1   Protection against sound from adjoining dwellings or buildings, etc.
E2   Protection against sound within a dwelling, etc.
E3   Reverberation in the common internal parts of buildings containing dwellings, etc.
L1   Conservation of fuel and power

17.9 Roof construction

Mitchell's Building Series

Environment and Services	Chapter 5 Heat
External Components	Chapter 8 Roofings
Materials	Chapter 2 Timber
	Chapter 3 Boards, slabs and panels
	Chapter 9 Metals
	Chapter 11 Bituminous products
Materials Technology	Chapter 12 Timber
	Chapter 13 Artificial composites: timber-based boards and slabs
Structure and Fabric Part 1	Chapter 7 Roof structures

Building Research Establishment

BRE Digest 108   *Standard 'U' values*
BRE Digest 144   *Asphalt and built-up felt roofings: durability*
BRE Digest 163   *Drying out buildings*
BRE Digest 180   *Condensation in roofs*
BRE Digest 270   *Condensation in insulated domestic roofs*
BRE Digest 284   *Wind loads on canopy roofs*
BRE Digest 295   *Stability under wind loads of loose laid external roof insulation boards*
BRE Digest 312   *Flat roof design: the technical options*
BRE Digest 323   *Selecting wood-based panel products*
BRE Digest 324   *Flat roof design: thermal insulation*
BRE Digest 336   *Swimming pool roofs: minimising the risk of condensation using warm-deck roofing*
BRE Digest 338   *Insulation against external noise*
BRE Digest 371   *Remedial wood preservatives: use them safely*
BRE Digest 372   *Flat roof design: waterproof membranes*
BRE Digest 373   *Wood chipboard*
BRE Digest 378   *Wood preservatives: application methods*
BRE Digest 393   *Specifying preservative treatments: the new European approach*
BRE Digest 394   *Plywood*
BRE Digest 416   *Specifying structural timber*
BRE Digest 419   *Flat roof design: bituminous roofing membranes*
BRE Digest 429   *Timbers: their natural durability and resistance to preservative treatment*
BRE Digest 439   *Roof loads due to drifting of snow*

Building Regulations

A1   Loading
B2   Internal fire spread (linings)
B3   Internal fire spread (structure)

B4   External fire spread
C2   Resistance to moisture
F2   Condensation in roofs
H3   Rainwater drainage
L1   Conservation of fuel and power in dwellings

17.10 Services

Mitchell's Building Series

Environment and Services	Chapter 7 Thermal installations
	Chapter 8 Electric lighting
	Chapter 9 Water supply
	Chapter 10 Sanitary appliances
	Chapter 11 Pipes
	Chapter 12 Drainage installations
	Chapter 13 Sewage disposal
	Chapter 15 Electricity and telecommunications
	Chapter 16 Gas
Internal Components	Chapter 2 Demountable partitions
	Chapter 3 Suspended ceilings
	Chapter 4 Raised floors
Structure and Fabric Part 1	Chapter 9 Fireplaces, flues and chimneys
Structure and Fabric Part 2	Chapter 7 Chimney shafts, flues and ducts

Building Research Establishment

BRE Digest 248   *Sanitary pipework* Part 1: Design basis
BRE Digest 249   *Sanitary pipework* Part 2: Design of pipework
BRE Digest 254   *Reliability and performance of solar collector systems*
BRE Digest 289   *Building management systems*
BRE Digest 292   *Access to domestic underground drainage systems*
BRE Digest 308   *Domestic unvented hot-water systems*
BRE Digest 335   *Electric interferences in buildings*
BRE Digest 339   *Condensing boilers*

Building Regulations

B3   Internal fire spread (structure)
F1   Means of ventilation
G1   Sanitary conveniences and washing facilities
G2   Bathrooms
G3   Hot water storage
H1   Foul water drainage
H2   Cesspools, septic tanks and settlement tanks
H3   Rainwater drainage
J1   Air supply
J2   Discharge of products of combustion
J3   Protection of building
L1   Conservation of fuel and power in dwellings

17.11 Finishes

Mitchell's Building Series

Internal Components	Chapter 2 Demountable partitions
	Chapter 3 Suspended ceilings
	Chapter 4 Raised floors
	Chapter 5 Joinery
	Chapter 6 Doors
	Chapter 7 Ironmongery
	Chapter 8 Stairs and balustrades
External Components	Chapter 4 Windows
	Chapter 7 External doors
Materials	Chapter 5 Ceramics
	Section 13.6 Thermoplastics
Finishes	All chapters

Building Research Establishment

BRE Digest 33   *Sheet and tile flooring made from thermoplastic binders*
BRE Digest 163   *Drying out buildings*
BRE Digest 197   *Painting walls* Part 1: Choice of paint
BRE Digest 198   *Painting walls* Part 2: Failures and remedies
BRE Digest 280   *Cleaning external surfaces of buildings*
BRE Digest 301   *Corrosion of metals by wood*
BRE Digest 304   *Preventing decay in external joinery*
BRE Digest 323   *Selecting wood-based panel products*
BRE Digest 387   *Natural finishes for exterior wood*
BRE Digest 407   *Timber for joinery*
BRE Digest 422   *Painting exterior wood*
BRE Digest 429   *Timbers: their natural durability and resistance to preservative treatment*

Building Regulations

B2   Internal fire spread (linings)
G2   Bathrooms

17.12 Landscape and external works

Mitchell's Building Series

Materials	Chapter 4 Stones
	Chapter 6 Bricks and blocks
	Chapter 8 Concretes
Materials Technology	Chapter 3 Concrete
	Chapter 4 Masonry construction

Building Research Establishment

BRE Digest 298   *The low-rise building foundations: the influence of trees in clay soils*

Appendix I

Working drawings and schedules

This list should not be taken as a definitive checklist of items to be included on working drawings/schedules. It is only a guide as to the types of information which may be required. Drawings and schedules may be supplemented by information given in specifications, bills of quantities, sample boards and three-dimensional representations.

1.00 General information on each drawing/schedule
1.01 Name of project and location
1.02 Project reference and drawing number
1.03 Name of designer/consultant and address of contract, including telephone number
1.04 General description of content of drawing
1.05 Cross-reference to other relevant drawings, schedules or bills of quantities/specification
1.06 Scale(s) employed on drawing(s)
1.07 Date of drawing when completed
1.08 Space for date of issue stamp
1.09 North point when appropriate
1.10 Amendment panel, including date of revision
1.11 General instructions panel
1.12 Subtitles for each individual item shown
1.13 Key to non-standard conventions

2.00 Site plans
2.01 District and address
2.02 Boundaries and site lines
2.03 Adjoining buildings, road and paths (existing and proposed)
2.04 Rights of way to be maintained or temporarily obstructed
2.05 Existing features (trees, mounds, fences, hedges, ditches, etc.)
2.06 Existing datum, levels, and contours
2.07 Existing drainage: sewers, cesspools, septic tanks, soakaways and watercourses, including levels
2.08 Existing gas, water, electricity, telecommunication services, including levels
2.09 Means of access
2.10 Setting out of proposed building and site, including levels
2.11 Proposed routes for rubbish disposal and fire-fighting vehicles
2.12 Proposed landscape feature, including levels
2.13 Alteration to existing features (boundaries, trees, landscape)
2.14 Alterations to existing drainage, gas, water, electricity and telecommunication services, including points of connection to proposed buildings
2.15 Location of proposed drainage, gas, water, electricity and telecommunication services

3.00 Foundation plans
3.01 Grid/modular planning lines
3.02 Datum level for excavation
3.03 Indication of existing foundations, earthworks, etc., to be removed
3.04 Dimensions of new foundations to define shape, including vertical steps
3.05 Levels of top and/or underside of foundations
3.06 Positions of walls relative to foundations
3.07 Positions of services to be installed below ground level
3.08 Location and size of holes left through foundations for service pipes
3.09 Drain and manhole foundations and levels
3.10 Typical details of excavations

4.00 Floor plans
4.01 Grid/modular planning lines
4.02 Datum level
4.03 Finished floor levels and changes of level
4.04 External dimensions
(a) changes of direction, openings, etc.
(b) overall of building
4.05 Internal dimensions
(a) changes of direction, openings, etc.
(b) overall of areas
4.06 Room/space names and areas if required
4.07 Wall dimensions and description of materials, thickness and jointing
4.08 Location of movement joints
4.09 Air bricks and ventilation grilles
4.10 Layout of horizontal and vertical service ducting: dimensions and description of services carried (gas, water, electricity, telephone, TV and cable)
4.11 Fireplaces: dimensions and/or description of openings, hearth, projections, etc.
4.12 Flues: dimensions and/or description of type, position, size, lining, etc.
4.13 Mat wells and other changes in floor surface
4.14 Overhead floor/roof construction: materials, size, spacing and direction of span, etc.
4.15 Overhead rooflights, ventilation cowls, etc.
4.16 Staircases: dimension and/or description of direction (up/down), treads (and number), rise/going, flight width, handrails, landings, balconies, etc.
4.17 Doors and door frames/linings: dimensions and/or description cross-referenced to schedules (including ironmongery)
4.18 Windows and window frames: dimensions and/or description cross-referenced to schedules (including ironmongery)

4.19 Hatches/access panels and frames: dimensions and/or description cross-referenced to schedules (including ironmongery)

4.20 Door (and window) direction of opening, number, and manufacturer's reference if applicable

4.21 Vertical damp-proof courses and membranes

4.22 External and internal wall and floor finishes, including materials, thickness, jointing, position and techniques

4.23 Internal ceiling finishes where to be linked in with wall components

Depending on the size of the project, the following items may be included, either on the main drawings or on separate 'specialist' drawings *cross-referenced* with main drawings.

4.24 Built-in furniture, cupboards, units, etc.: dimensions and/or description cross-referenced to schedule/detail drawings

4.25 Heating units/radiators, etc.: dimensions and/or description cross-referenced to schedule

4.26 Sanitary fittings: dimensions and/or description cross-referenced to schedule

4.27 Specialist equipment, fire hoses, alarms, mechanical plant

4.28 Electrical installation: description and dimensioned position of intake, consumer unit (main switch/fuses), power socket outlets, lighting points and switches, emergency lighting, etc.

4.29 Gas installation: description and dimensioned position of intake, meter and supply points

4.30 Water installation: description and dimensioned position of intake, valve controls, rising main, supply points, and cold water tank, draindown points, etc.

4.31 Heating installation: description and dimensioned position of boiler (equipment, model, size and rating), hot water cylinder, expansion tank, heating units, draindown points, etc.

4.32 Drainage installation: description and dimension of outfall, manholes, drainpipes, soil stacks, wastes, vent pipes, gullies, branch drains from fittings to include details of material, size, gradient and direction of flow when appropriate

4.33 Rainwater drainage installation: description and dimensions of outfall, manholes, drainpipes, rainwater pipes and gullies, surface water gullies and direction of falls, etc.

4.34 External works and landscaping: steps, ramps, paving, grass, planting, trees and built features, such as terraces, playgrounds (including equipment), amenities, etc.

4.35 Details, dimensions and/or description of external fuel storage, oil-tanks and sheds, garages, car ports, etc.

5.00 Roof plans

5.01 Grid/modular planning lines

5.02 Datum level

5.03 Dimensions
(a) changes of direction, openings, etc.
(b) overall of building

5.04 Description of roof structure: decking, furrings, and integral thermal insulation – vapour control layer and ventilation

5.05 Screeds: composition, maximum and minimum thicknesses

5.06 Roof finishes: materials, sizes, thickness, gauge, pitch, etc.

5.07 Direction of falls

5.08 Rooflights and ventilation cowls: description and dimensions cross-referenced to schedule if applicable

5.09 Ventilation pipes, rainwater pipes and gutters: description and dimensions

5.10 Parapet, coping, eaves, ridge and upstands: description and dimensions

5.11 Lift-motor and mechanical plant rooms: description and dimensions

5.12 Location and description of cantilever beams, cradle fixing position, etc., for window cleaning

5.13 Location of areas unsafe for maintenance workers without additional precautions; areas safe for maintenance workers only; areas safe for foot traffic and/or heavier loads

5.14 Access positions: staircases, traps, ladders, etc.

5.15 Landscaping, terraces and balconies: description and dimensions, including arrangements for waterproofing and drainage

6.00 Sections

6.01 Grid/modular section lines

6.02 Datum levels

6.03 Ground levels: existing and new finished ground and floor levels

6.04 Floor levels and identification

6.05 Room/space names when applicable

6.06 External dimensions
(a) changes of direction, openings, etc.
(b) overall of building

6.07 Internal dimensions
(a) door/window openings, staircases, ducts, built-in furniture, guard-rails, etc.
(b) room heights and suspended floors or ceilings

6.08 Foundations: description and dimensions, including composition, size and stepping details

6.09 Foundation walls: materials, thickness, bonding, backfilling, damp-proof course, holes for drainage and ventilation

6.10 Fill and hardcore material, thickness, and layering

6.11 Solid ground floor construction: thickness, composition, additives, reinforcement, damp-proof membrane, insulation, screeds and finishes

6.12 Suspended ground floor construction: type, materials, dimensions, fixings, insulation, damp-proof membrane, ventilation and finishes

6.13 Intermediate floor construction: type, materials, dimensions, fixings, insulation, and finishes, including ceilings

6.14 Roof construction: type, materials, dimensions, fixing, insulation and finishes, including soffits and ceilings

6.15 External wall construction: type, material, dimensions, fixings, ties, insulation and finishes

6.16 Internal wall construction: type, material, dimensions, fixings, ties, insulation and finishes

6.17 Staircases: dimension and/or description of direction (up/down), treads (and number), rise/going, flight width, handrails, landings, balconies, etc.

6.18 Doors and door frames/linings: dimensions and/or description cross-referenced to schedules (including ironmongery)

6.19 Windows and window frames: dimensions and/or description cross-referenced to schedules (including ironmongery)

6.20 Lintel types, sizes and materials cross-referenced to schedule when appropriate

6.21 Location of movement joints

6.22 Air bricks and ventilation grilles

6.23 Layout of horizontal and vertical service ducting: dimensions and description of services carried (gas, water, electricity, telephone, TV and cable)

6.24 Fireplaces: dimensions and/or description of openings, hearth, projections, etc.

6.25 Flues: dimensions and/or description of type, position, size, lining, etc.

6.26 Mat wells and other changes in floor surface

Depending on the size of the project, the following items may be included either on the main drawings or on separate 'specialist' drawings *cross-referenced* with main drawings

6.27 Built-in furniture, cupboards, units, etc.: dimensions and/or description, cross-referenced to schedule/detail drawings

6.28 Heating units/radiators, etc.: dimensions and/or description cross-referenced to schedule

6.29 Sanitary fittings: dimensions and/or description cross-referenced to schedule

6.30 Specialist equipment, fire hoses, alarms, mechanical plant

6.31 Electrical installation: description and dimensioned position of intake, consumer unit (main switch/fuses), power socket outlets, lighting points and switches, emergency lighting, etc.

6.32 Gas installation: description and dimensioned position of intake, meter and supply points

6.33 Water installation: description and dimensioned position of intake, valve controls, rising main, supply points, and cold water storage cistern, draindown points, etc.

6.34 Heating installation: description and dimensioned position of boiler (equipment, model, size and rating), hot water cylinder, expansion tank, heating units, draindown points, etc.

6.35 Drainage installation: description and dimension of outfall, manholes, drainpipes, soil stacks, wastes, vent pipes, gullies, branch drains from fittings to include details of material size, gradient and direction of flow when appropriate

6.36 Rainwater drainage installation: description and dimensions of outfall, manholes, drainpipes, rainwater pipes and gullies, surface water gullies and direction of falls, etc.

6.37 External works and landscaping: steps, ramps, paving, grass planting, trees and built features, such as terraces, playgrounds (including equipment), amenities, etc.

6.38 Details, dimensions and/or description of external fuel storage, oil-tanks and sheds, garages, car ports, etc.

7.00 Elevations

7.01 Grid/modular section lines

7.02 Datum levels

7.03 Ground levels: existing and new finished ground and floor levels

7.04 Foundation lines

7.05 General description of facing materials including type, material, texture and colour

7.06 Door positions and cross-reference to schedules

7.07 Window positions and cross-reference to schedules

7.08 Direction of opening of door and windows

7.09 Air bricks and ventilation cowls

7.10 Description and position of movement joints

7.11 Soil stacks, gutters, rainwater pipes, and other exposed services, including alarms, lighting, power points, standpipes, etc.

7.12 Description and position of special features: signs, sculptures, decorative displays, etc.

7.13 Description and location of sheds, storage rooms, etc., detached from main building

7.14 Description and location of landscape features: trees, bushes, plants, earth ramps and mounds, etc.

8.00 Detail drawings, axonometrics and isometrics

8.01 Grid/modular planning/section lines

8.02 Location of doors, windows and other openings

8.03 Wall, floor, roof thicknesses

8.04 Location of furniture and other fixings

8.05 Details of construction and finishes

9.00 Schedules

9.01 Reference number (from drawings), location, type and size of items

9.02 Manufacturer's reference if applicable

9.03 Ancillary information concerning colour, iron-mongery, methods of fixing, applications and details of immediately surrounding elements or components, etc.

9.04 Cross-reference with other schedules unless fully described; colour to be painted, ironmongery involved, etc.

Appendix II

Examples of standard contract
administration forms

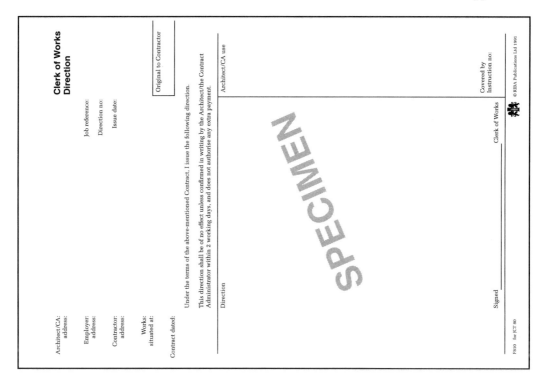

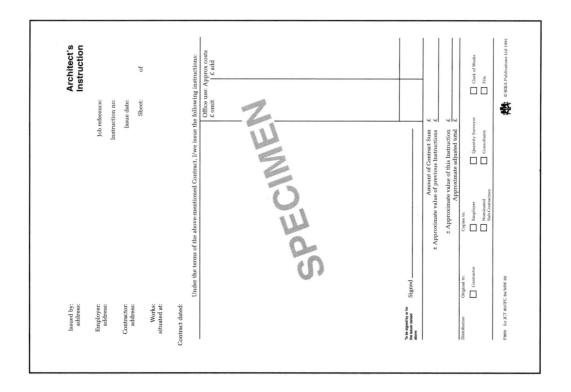

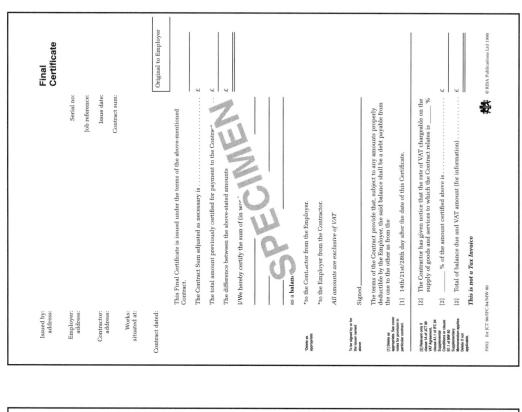

Certificate of Non-Completion

Issued by:
address:

Employer:
address:

Contractor:
address:

Works:
situated at:

Contract dated:

Job reference:

Certificate no:

Issue date:

Under the terms of the above-mentioned Contract,

I/we hereby certify that the contractor has failed to comp...

SPECIMEN

*1. the Works

*2. Section No.

*3. by the ...

*4. by the Date for Completion or within any extended time fixed under the contract conditions.

*Delete as appropriate

3. This item applies to JCT 80 only.

4. This item applies to IFC 84 and MW80 only.

Signed

To be signed by or for the issuer named above

Distribution	Original to:	Duplicate to:	Copies to:	
	☐ Employer	☐ Contractor	☐ Quantity Surveyor	☐ Clerk of Works
			☐ Consultants	☐ File

F854   for JCT 80/IFC 84/MW 80

© RIBA Publications Ltd 1991

Practical Completion of Nominated Sub-Contract Works

Issued by:
address:

Employer:
address:

Contractor:
address:

Works:
Situated at:

Contract dated:

Nominated Sub-Contract Works:

Serial no:

Job reference:

Issue date:

Under the terms of the above mentioned Con...

I/We certify that:

SPECIMEN

Practi... ...red to

above _____ 19____

Signed

To be signed by or for the issuer named above

Distribution	Original to:	Duplicate to:	Copies to:	
	☐ Employer	☐ Contractor	☐ Quantity Surveyor	☐ Services Engineer
		☐ Nominated Sub-Contractors	☐ Structural Engineer	☐ File

© 1985 RIBA Publications Ltd

Certificate of

Practical Completion

of the Works

Issued by:
address:

Employer:
address:

Contractor:
address:

Works:
Situated at:

Serial no:

Job reference:

Issue date:

Contract dated:

Under the terms of the above mentioned Contract,

I/We certify that Practical Completion of th‾

19

SPECIMEN

To be signed by or for
the issuer named
above

Signed

The Defects Liability Period will therefore end on:

19

Distribution	Original to:	Duplicate to:	Copies to:	
	☐ Employer	☐ Contractor	☐ Quantity Surveyor	☐ Services Engineer
		☐ Nominated Sub-Contractors	☐ Structural Engineer	☐ File

© 1985 RIBA Publications Ltd

Statement of

Partial Possession

by the Employer

Issued by:
address:

Employer:
address:

Contractor:
address:

Works:
situated at:

Job reference:

Statement no:

Issue date:

Contract dated:

Under the terms of the above-mentioned Contract,

I/we hereby state that a part of the Works?''

19

SPECIMEN

was taken into possession by the Employer on

19

To be signed by or for
the issuer named
above on behalf of the
Employer

Signed

Distribution	Original to:	Duplicate to:	Copies to:	
	☐ Contractor	☐ Employer	☐ Quantity Surveyor	☐ Clerk of Works
			☐ Consultants	☐ File

F812 for JCT 80/IFC 84

© RIBA Publications Ltd 1991

DIRECTION

THE CHARTERED SOCIETY OF DESIGNERS

SUPERVISING OFFICER'S NAME & ADDRESS

AGREEMENT DATE
DIRECTION NO
JOB NO
DATE

TO CONTRACTOR

We direct you under the terms of the above Agreement to make the following payments to the Nominated Sub-Contractors listed below, which amounts are included in the Certificate.

NO   DATE

NOMINATED SUB-CONTRACTOR

TOTAL TO DATE   CERTIFIED PREVIOUSLY   BALANCE INCLUDED IN CERTIFICATE

SIGNED (Supervising Officer)

The above amounts are exclusive of VAT and are GROSS amounts. No account has been taken of any retentions, discounts or other deductions to which the Contractor may be entitled. Copies to be sent to all Nominated Sub-Contractors listed.

© 1988 C.S.D. STANDARD APPROVED FORM

INSTRUCTION

THE CHARTERED SOCIETY OF DESIGNERS

SUPERVISING OFFICER'S NAME & ADDRESS

AGREEMENT TITLE
INSTRUCTION NO
JOB NO
DATE

TO CONTRACTOR

We issue the following instructions under the terms of the Agreement. Where applicable the Contract Sum will be adjusted in accordance with the relevant terms of the Agreement.

ITEM   INSTRUCTIONS

£ OMIT   £ ADD

SIGNED (Supervising Officer)

CONTRACT SUM £
VALUE OF PREVIOUS INSTRUCTIONS £
VALUE OF THIS INSTRUCTION £
ADJUSTED TOTAL £

DISTRIBUTED TO
Contractor
Employer
Quantity Surveyor
Structural Engineer
Architect

© 1988 C.S.D. STANDARD APPROVED FORM

FINAL CERTIFICATE

THE CHARTERED
SOCIETY OF DESIGNERS

SUPERVISING
OFFICER'S
NAME &
ADDRESS

AGREEMENT DATE
CERTIFICATE NO
JOB NO
DATE
CONTRACTOR'S
NAME
EMPLOYER'S
NAME

We certify that under the terms of the above Agreement the Defects Period has ended, all/any defects have been made good to our satisfaction and final payment is due as listed below:

Contract Sum   £
(Adjusted as necessary in accordance with Instructions issued)

Less the sum of the amounts paid to the Contractor under Interim   £
Certificates and the amount of Retention monies paid to the
Contractor at Practical Completion; and the amounts of any monies
paid to other contractors to rectify defects under Clause 2.5 of the
Agreement.

Balance Now Due   £
to be paid from/to the Employer to/from the Contractor within 14 days
of the date of issue of this certificate
(In words)

The above amounts are exclusive of VAT
This is not a Tax Invoice

SIGNED (Supervising Officer)

© 1988 C.S.D. STANDARD APPROVED FORM

INTERIM CERTIFICATE

THE CHARTERED
SOCIETY OF DESIGNERS

SUPERVISING
OFFICER'S
NAME &
ADDRESS

AGREEMENT DATE
CERTIFICATE NO
JOB NO
DATE
CONTRACTOR'S
NAME
EMPLOYER'S
NAME

We certify that under the terms of the above Agreement interim payment as detailed below is due from the Employer to the Contractor:

Total Value   £
(Including the value of works by nominated sub-contractors as
detailed on any accompanying direction form)

Less Retention   £
(As applicable)

Balance   £
(Cumulative total certified for payment)

Less cumulative total previously certified   £

Amount due for payment on this certificate   £
(In words)

The above amounts are exclusive of VAT
This is not a Tax Invoice

SIGNED (Supervising Officer)

© 1988 C.S.D. STANDARD APPROVED FORM

EMPLOYER/SUB-CONTRACTOR AGREEMENT

THE CHARTERED
SOCIETY OF DESIGNERS

SUPERVISING
OFFICER'S
NAME &
ADDRESS

THIS AGREEMENT
is made the _____ day
of _____ 19

BETWEEN "the Employer"
of

and "the Sub-Contractor"
of

WHEREAS
The Sub-Contractor has by Tender dated _____ tendered as a Nominated Sub-Contractor
for the following works ("the Sub-Contract Works") forming part of the works under a CSD Works Agreement dated
_____ between the Employer and
_____ as the Contractor

and known as _____

NOW IT IS AGREED
Where the Sub-Contractor is (either at the date of this Agreement or later) nominated as Sub-Contractor for the
Sub-Contract Works described in the said Tender
1.0   THE SUB-CONTRACTOR WARRANTS AS FOLLOWS (in this Agreement "he", "him" and "his" refer to the
Sub-Contractor):
1.1   He has exercised and will exercise reasonable and proper care and skill in any of his design work, in his
selection of materials and goods for the Sub-Contract, and in meeting any design specification or requirement
which is his responsibility in terms of the Sub-Contract.
1.2   He will, as soon as reasonably practicable, provide the Contractor with such information as will enable the
Contractor to give written notice to the Supervising Officer in terms of Clause 2.2 of the Works Agreement of any
circumstances which have delayed or will or are likely to delay completion of the Works under the Works
Agreement.
1.3   He will in the performance of the Sub-Contract Works do nothing that may give rise to an entitlement on the
part of the Contractor to an extension of the Date of Completion under Clause 2.2 of the Works Agreement.
1.4   Nothing in his Tender excludes or limits his liability for any breach of the warranties in this Agreement.

2.0   THE EMPLOYER UNDERTAKES AS FOLLOWS:
2.1   In the event that the Contractor fails or refuses to pay to the Sub-Contractor within 14 days of receiving the
appropriate Certificate from the Supervising Officer the amount certified as due to the Sub-Contractor (less any
deduction by the Contractor for claims the Contractor may have against the Sub-Contractor or the specified
retention or cash discount) the Employer will pay such an amount (less such deductions) to the Sub-Contractor
together with the amount of Value Added Tax properly chargeable by the Sub-Contractor.
2.2   Once the Supervising Officer has issued a Certificate of Practical Completion the obligation of the Employer
under Clause 2.1 above shall be limited to the amount due from the Employer to the Contractor after allowing for
such sums as are reasonably necessary to pay for remedying defects and omissions in the Works which have
appeared within the Defects Period, losses suffered by the Employer by reason of any breach by the Contractor,
and sums payable to other nominated sub-contractors.

AS WITNESS The Parties: Signed For and on behalf of _____ Employer
in the presence of _____ Witness
for and on behalf of _____ Sub-Contractor
in the presence of _____ Witness

© 1988 C.S.D. STANDARD APPROVED FORM

TENDER FOR NOMINATED SUB-CONTRACTORS

THE CHARTERED
SOCIETY OF DESIGNERS

SUPERVISING
OFFICER'S
NAME &
ADDRESS

PROJECT TITLE
JOB NO
DATE ISSUED

You are invited to tender for the proposed nominated sub-contract as
described below in accordance with the terms of the CSD WORKS
AGREEMENT 1989 ISSUE and the tenderer agrees by signature of
this form to observe, perform and comply with all the provisions of
that Agreement on the part of the Contractor so far as they relate to
the sub-contract works, to comply with all Conditions of Tender as
listed overleaf and to execute in favour of the Employer before this
Tender is accepted an Agreement in the attached form.

PARTICULARS OF MAIN AGREEMENT (if known at date of enquiry)
Date for Possession (Clause 2.1) _____
Date for Completion (Clause 2.1) _____
Bonus for early completion (Clause 2.4) _____
Liquidated Damages for delayed completion (Clause 2.3) _____
Defects Period (Clause 2.5) _____
Interim Payment Interval (Clause 4.1) _____
Retention (Clause 4.1) _____ %
Insurances
a   Persons (Clause 6.1) _____ pounds (£ _____)
b   Property and Works (Clause 6.2) _____ pounds (£ _____)
TO BE COMPLETED BY THE TENDERER
Subject to the execution of a sub-contract with the main Contractor we undertake as follows:
1   To carry out and complete the works described or referred to in the drawings, specifications and/or bills of
quantities supplied to us, in accordance with the Conditions of Tender listed overleaf, and the conditions of the main
Agreement so far as they relate to the sub-contract works for the sum given below which includes 2½% cash
discount to the main Contractor for payment within 14 days of the date of the issue of the Direction form.
2   To commence the works at any time reasonably stipulated
by the main Contractor but not less than _____ weeks
from the date of written acceptance of this tender or from receipt of
such further drawings of particulars as agreed with us as necessary
for a start to be made, whichever is later.
3   To complete the works within _____ weeks
of our unimpeded working time (whether worked continuously or
staged to suit the progress of other trades).
4   To make good as instructed at our own cost any defects, shrinkages or other faults which appear within the
Defects Period under the main Agreement, and which shall be due to materials or workmanship not in accordance
with the Conditions of Tender, or to frost occuring before Practical Completion of the works, and to pay any costs
incurred by the main Contractor and/or any other sub-contractor arising from the making good of such defects,
shrinkages or other faults.
5   For the sum of £ _____
6   This offer is open for acceptance until the expiration of _____ weeks
from the date of this tender and is conditional upon the terms of a
sub-contract being agreed with the main Contractor.

SIGNATURE _____ DATE _____ 19
ADDRESS
TELEPHONE NO.

© 1988 C.S.D. STANDARD APPROVED FORM

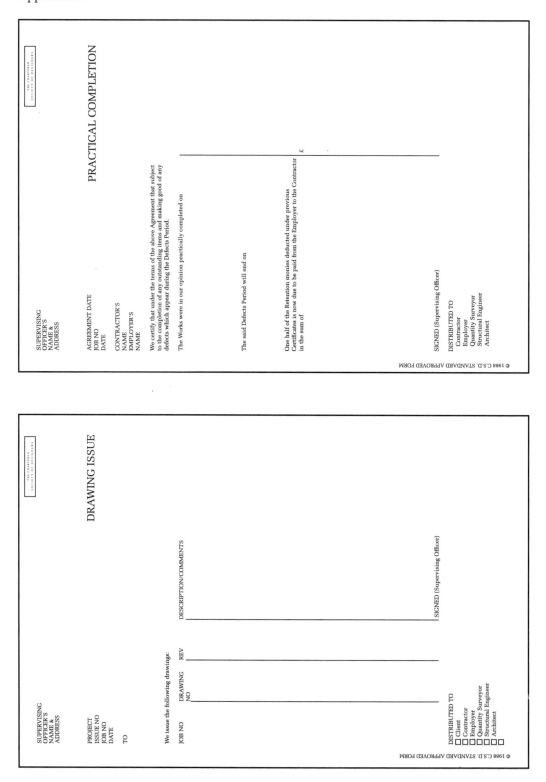

THE CHARTERED SOCIETY OF DESIGNERS

PRACTICAL COMPLETION

SUPERVISING
OFFICER'S
NAME &
ADDRESS

AGREEMENT DATE
JOB NO
DATE

CONTRACTOR'S
NAME
EMPLOYER'S
NAME

We certify that under the terms of the above Agreement that subject
to the completion of any outstanding items and making good of any
defects which appear during the Defects Period.

The Works were in our opinion practically completed on

The said Defects Period will end on

One half of the Retention monies deducted under previous
Certificates is now due to be paid from the Employer to the Contractor
in the sum of   £

SIGNED (Supervising Officer)

DISTRIBUTED TO
Contractor
Employer
Quantity Surveyor
Structural Engineer
Architect

© 1988 C.S.D. STANDARD APPROVED FORM

THE CHARTERED SOCIETY OF DESIGNERS

DRAWING ISSUE

SUPERVISING
OFFICER'S
NAME &
ADDRESS

PROJECT
ISSUE NO
JOB NO
DATE

TO

We issue the following drawings:

JOB NO　　DRAWING　REV　　DESCRIPTION/COMMENTS
　　　　　　NO

SIGNED (Supervising Officer)

DISTRIBUTED TO
☐ Client
☐ Contractor
☐ Employer
☐ Quantity Surveyor
☐ Structural Engineer
☐ Architect

© 1988 C.S.D. STANDARD APPROVED FORM

Appendix III

Examples of planning and
Building Regulations application
forms

PLANNING APPLICATION

OFFICE USE ONLY
APPLICATION NUMBER

DATE RECEIVED

DATE VALID

Use this form to apply for **Planning permission** for:
Outline Permission
Full Permission
Approval of Reserved Matters
Renewal of Temporary Permission
Change of Use

Please return:
*6 copies of the Form
*6 copies of the Plans
*a Certificate under
  Article 7
*the correct fee

1. NAME AND ADDRESS OF APPLICANT

Post Code _____

Day Tel. No. _____

2. NAME AND ADDRESS OF AGENT (if used)

Post Code _____

Tel. No. _____

3. ADDRESS OR LOCATION OF LAND TO WHICH APPLICATION RELATES

State Site Area _____ Hectares
This must be shown edged Red on the site plan

4. OWNERSHIP

Please indicate applicants interest in the property and complete the appropriate Certificate under Article 7.

Freeholder ☐   Other ☐

Leaseholder ☐   Purchaser ☐

Any adjoining land owned or controlled and not part of this application must be edged Blue on the site plan

5. WHAT ARE YOU APPLYING FOR?   Please tick one box and then answer relevant questions.

☐ **Outline Planning Permission**

Which of the following are to be considered?   ☐ Siting   ☐ Design   ☐ External Appearance   ☐ Means of Access
☐ Landscaping

☐ **Full Planning Permission/Change of use**

☐ **Approval of Reserved Matters following Outline Permission**

O/P No. _____ Date _____ No. of Condition this application refers to: _____

☐ **Continuance of Use without complying with a condition of previous permission**

P/P No. _____ Date _____ No. of Condition this application relates to: _____

☐ **Permission for Retention of works**

Date of Use of land or when buildings or works were constructed: _____ Length of temporary permission: _____

Is the use temporary or permanent? _____ No. of previous temporary permission if applicable: _____

6. BRIEF DESCRIPTION OF PROPOSED DEVELOPMENT

Please indicate the purpose for which the land or buildings are to be used. _____

What is the present use of the land/buildings? _____

What buildings are to be demolished? _____

_____ Gross Floor Area _____ m²

7. NEW RESIDENTIAL DEVELOPMENTS.   Please answer the following if appropriate.

Please state what type of building you are proposing i.e. house, flat: _____
Please give the following:

No. of dwellings _____ No. of surveys: _____ No. of Habitable rooms: _____

No. of Garages: _____ No. of Parking Spaces: _____ Total Gross Area of all buildings: _____

How will surface water be disposed of? _____

How will foul sewage be dealt with? _____

8. ACCESS

Does the proposed development involve any of the following? Please tick the appropriate boxes.

New access to a highway ☐ Pedestrian ☐ Vehicular

Alteration of an existing highway ☐ Pedestrian ☐ Vehicular

The felling of any trees ☐ Yes ☐ No

If you answer Yes to any of the above, they should be clearly indicated on all plans submitted.

9. BUILDING DETAILS

Please give details of all external materials to be used, if you are submitting them at this stage for approval.
_____
_____
_____

List any samples that are being submitted for consideration. _____
_____

10. LISTED BUILDINGS OR CONSERVATION AREA

Are any Listed buildings to be demolished or altered?   ☐ Yes ☐ No

If Yes, then Listed Building Consent will be required and a separate application should be submitted.

Are any non-listed buildings within a Conservation Area to be demolished?   ☐ Yes ☐ No

If Yes, then Conservation Area consent will be required to demolish. Again, a separate application should be submitted.

11. NOTES
Part 2 of this Planning Application Form should be completed for all applications involving Industrial, Warehousing, Storage, or Shopping development.
An appropriate Certificate must accompany this application unless you are seeking approval to Reserved Matters.
A separate application for Building Regulation approval is also required.
Separate applications may also be required if the proposals relate to a Listed Building or non-listed building within a Conservation Area.

12. PLEASE SIGN AND DATE THIS FORM BEFORE SUBMITTING
I/We hereby apply for Planning Permission for the development described above and shown on the accompanying plans.

Signed _____

On behalf of (if agent) _____ **Date** _____

HOUSEHOLDER PLANNING APPLICATION

Use this form to apply for **Planning Permission** for:
* an Extension
* a High Wall of Fence
* a Loft Conversion
* a Garage or Outbuilding
* a New or Altered Access
* a Satellite Dish

Please return:
* 5 copies of the Form
* 5 copies of the Plans
* a Certificate under Article 7
* the correct fee

1. NAME AND ADDRESS OF APPLICANT

Post Code _____

Day Tel. No. _____

2. NAME AND ADDRESS OF AGENT (if used)

Post Code _____

Tel. No. _____

3. ADDRESS OF PROPERTY TO BEALTERED OR EXTENDED

4. OWNERSHIP

Please indicate applicants interest in the property and complete the appropriate Certificate under Article 7.

Freeholder ☐   Other ☐

Leaseholder ☐

Purchaser ☐

5. BRIEF DESCRIPTION OF WORKS (including any demolition works) e.g. erection of two a storey rear extension.

6. DESCRIPTION OF MATERIALS

7. ACCESS AND PARKING

Will your proposal affect?   Please tick appropriate boxes

Vehicular Access   Yes ☐   No ☐

A Public Right of Way   Yes ☐   No ☐

Existing Parking   Yes ☐   No ☐

8. DRAINAGE

a. Please indicate method of Surface Water Disposal

_____

b. Please indicate method of Foul Water Disposal
Please tick one box

Mains Sewer ☐   Septic Tank ☐

Cesspit ☐   Other ☐

9. TREES

Does the proposal involve the felling of any trees?

Please tick one box   Yes ☐   No ☐

If Yes, please show details on plans

10. PLEASE SIGN AND DATE THIS FORM BEFORE SUBMITTING

I/We hereby apply for Full Planning Permission for the development described above and shown on the accompanying plans.

Signed _____ Date _____
Date
On behalf of (if agent) _____

BUILDING REGULATIONS
APPLICATION

Use this form to give notice of intention to erect, extend, or alter a building, install fittings or make a material change of use of the building.

Please return:
*3 copies of the Form
*3 copies of the Plans
*the correct fee

1. NAME AND ADDRESS OF APPLICANT
Applicant will be invoiced on commencement of work.

Post Code _____

Day Tel. No. _____ Fax No. _____
Email _____

2. NAME AND ADDRESS OF AGENT (If Used)

Post Code _____

Tel. No. _____ Fax No. _____
Email _____

3. ADDRESS OR LOCATION OF PROPOSED WORK.

4. DESCRIPTION OF PROPOSED WORKS

5. IF NEW BUILDING OR EXTENSION PLEASE STATE PROPOSED USE.

6. IF EXISTING BUILDING PLEASE STATE PRESENT USE.

7. DRAINAGE

Please state means of:

Water Supply _____

Foul Water Disposal _____

Storm Water Disposal _____

8. CONDITIONS

Do you consent to the Plans being passed subject to conditions where appropriate?   Yes ☐   No ☐

Do you agree to an extension of time if this is required by the Council?   Yes ☐   No ☐

9. COMPLETION CERTIFICATE

Do you wish the Council to issue a Completion Certificate upon satisfactory completion of the work?

Yes ☐   No ☐

10. FIRE PRECAUTIONS ACT 1971

Is the building put, or intended to be put to a use designated under the Fire Precautions Act 1971?

Yes ☐   No ☐

11. FEE

Please state estimated cost of the work (at current market value) £ .... Amount of Fee submitted £ ....

Has Planning Permission been sought?   Yes ☐   No ☐   If Yes, please give Application No _____

12. PLEASE SIGN AND DATE THIS FORM BEFORE SUBMITTING

I/We hereby give notice of intention to carry out the work set out above and deposit the attached drawings and documents in accordance with the requirements of Regulations 11.   Also enclosed is the appropriate Plan Fee and I understand that a further Fee will be payable when the first inspection of the work on site is made by the Local Authority.

Signed _____ Date _____ On behalf of (if agent) _____

Index